Recycled Aggregate Concrete

Recycled Aggregate Concrete (RAC) as a sustainable material is gaining increasing importance in the construction industry. This book discusses properties, specifications, and applications of RAC and offers readers insight into current research and advances in the development and utilization of RAC. It shares information gathered about concretes that use RCA (Recycled Concrete Aggregate, a component of RAC), as well as findings and conclusions.

This book:

- Presents principles of RAC, including theories and experiments.
- Describes advanced behavior and properties.
- Covers specifications and codes.
- Highlights best practices.
- Summarizes the use of RAC in sustainable concrete construction.
- Features scientific findings, citations of reliable sources, conclusions, and recommendations that ensure the book is accessible to various levels of expertise.

This book will be useful for researchers, concrete scientists, technologists, practicing engineers, and advanced students interested in reusing construction waste for sustainable construction practices; it will help them strive toward meeting the UN Sustainable Development Goals (SDGs).

IOM3 Publications

Series editor: The Institute of Materials, Minerals, and Mining (IOM3)

Materials and the associated fields of minerals and mining are indispensable to modern society. To optimize our consumption of materials, to minimize the impact of mining, and to ensure disposal of used materials in a responsible way it is more essential than ever to fully understand their production, properties, applications, and alternatives. The IOM3 Publications Series seeks works on a range of topics relevant to materials, minerals and mining science, and engineering in the 21st century as societies around the world evolve to meet the universal challenges of awareness, sustainability, adaptability, and responsible use. Experts from industry, research institutes, and academia are invited to submit proposals for books within the fields of materials science, minerals, and mining, especially those focused on providing commentary on or solutions to contemporary societal issues, such an energy, the environment, health care, and emerging topics in automation, AI, and machine learning.

Recycled Aggregate Concrete
Technology and Properties
Natt Makul

For more information about this series, please visit: https://www.routledge.com/IOM3-Publications/book-series/CRCIOM3

Recycled Aggregate Concrete
Technology and Properties

Natt Makul

CRC Press
Taylor & Francis Group
Boca Raton London New York

CRC Press is an imprint of the
Taylor & Francis Group, an **informa** business

First edition published 2023
by CRC Press
6000 Broken Sound Parkway NW, Suite 300, Boca Raton, FL 33487-2742

and by CRC Press
4 Park Square, Milton Park, Abingdon, Oxon, OX14 4RN

CRC Press is an imprint of Taylor & Francis Group, LLC

ISBN: 978-1-032-18941-3 (hbk)
ISBN: 978-1-032-18987-1 (pbk)
ISBN: 978-1-003-25709-7 (ebk)

DOI: 10.1201/9781003257097

Typeset in Times
by MPS Limited, Dehradun

Contents

Preface

There is no room for doubt regarding the fact that this book will inspire discussion. It is never simple to question an established system, regardless of the nature of the system or the context in which it functions. Because of this, we are going to make every effort to keep things the way they are. After working in the cement industry for the past 15 years, during which time I have also been involved in the research and development of alternative cement ingredients, I believe it is my duty to share the information that I have gathered about concretes that make use of recycled aggregates, as well as my findings and conclusions with regard to these concretes. For our benefit, the benefit of our children, the benefit of our loved ones, and the benefit of society as a whole, which will live in a safe and sustainable environment free of emissions, I hope that this report has the capacity to raise public awareness and begin a conversation that might lead to substantial changes and a rethinking of construction materials. I also hope to launch a debate that may lead to significant changes. In addition, it is my sincere wish that this book's publication drives heightened general awareness and sparks a debate that will lead to the introduction of significant policy changes. In spite of this, a scientific approach has to consider the various contaminants that may be found in recycled aggregates like glass, metal, and gypsum that come from the demolition of ancient buildings and structures. Even though I would want to see recycled concrete aggregates become a common building material that is employed in the construction of structures and foundations, I do not believe this will happen unless the presence of these pollutants is addressed first.

This book is divided up into a total of six chapters. In the first place, it is open to the general public, including those with a limited or nonexistent grasp of the process of producing Portland cement or recycled concrete aggregates. This accessibility is of the utmost importance. This book is aimed at a specific demographic of readers in particular.

Regarding recycled concrete aggregates, there is much information to process. This includes the characteristics and concepts of recycled concrete aggregates, their applications, best practices, and sustainability. Specific chapters may present qualities, ideas, and behaviors of recycled concrete aggregate (RCA) that are more difficult to comprehend than others; however, the scientific arguments presented, the citations from authentic sources, the conclusions formed, and the recommendations made will guarantee that it is relatively easy to understand. Anyone who possesses a fundamental grasp of physics will have no trouble comprehending recycled concrete aggregate's characteristics and behaviors. Because this book is not structured like a textbook, it is straightforward to read in the order in which the chapters are laid out.

Natt Makul
September, 2022
Bangkhen, Bangkok, Thailand

1 Introduction to Recycled Concrete Aggregate (RCA)

1.1 RCA FOR SUSTAINABLE CONSTRUCTION

Sustainability issues like climate change must be addressed in order to make progress. Societal alterations are becoming required on a big scale, at a systemic level, and throughout time. Sustainability transitions describe changes that lead to a more environmentally friendly future. There must be significant changes in the systems that provide for the needs of humans in order for them to be met in the future. Scholars who study sustainability transitions have done much research over the past 20 years and said that the significant, big changes that people hope for are not going to happen, a lot of significant changes have not happened yet, and that transitions research has to widen its reach to accommodate new viewpoints and re-energize its creative capacity. The way knowledge is made has been thought of as a possible reason for this, disseminated, and how they are applied in society has a significant effect on transition processes, and reconsidering knowledge creation systems may be a primary lever for facilitating revolutionary transformation (Jones, 2014). With the inclusion of a broader range of social actors, there is a growing emphasis on collaboratively creating information that is both policy-relevant and action-guiding in order to enable sustainability transitions (Friend et al., 2016). During the process of transition, it is thought to be essential to use co-creative ways to develop knowledge in order to: (a) recognize the multiplicity of values that influence the perceived attractiveness of competing transition approaches across society actors; (b) effectively disrupt and counteract entrenched power systems' conflicts of interest in established structures that stifle change, and (c) help people participate in the process of making changes.

There needs to be more creative ways to get people to help co-create knowledge about how to make the world a better place. When people work together on new knowledge, it is essential to point out that participation alone is not enough; it is also important to consider how the outputs of these processes are used. The vital importance of involving citizens in these processes is a necessary precondition for democratizing transitions and establishing the viability of co-creation of knowledge in the first place. Since the subject's introduction, different normative situations have been developed, and initiatives of all kinds and scopes involved in transitioning to a more sustainable future have adopted this standard technique.

There are scenarios regularly generated and discussed in the field by expert groups and a select set of interested parties; general people have to deal with the long-term

DOI: 10.1201/9781003257097-1

consequences of governmental decisions without any influence. According to the research, specialists are overrepresented systematically, and there is a general lack of representation of different people in collaborative knowledge-making processes in sustainable development (Musch and von Streit, 2020). Transitions to a more sustainable economy are difficult because of the high level of systemic uncertainty that characterizes them and the time it takes to complete those (Geels, 2010). To resolve problems like climate change and sustainability transitions, it has been suggested that different types of knowledge, like practice-based, tacit, and lay, should be incorporated into how the human resolves these problems.

Researchers in the still-new field of experiential futures are trying to create tangible memories of virtual events by combining future inquiry methods like scenarios with human-centered, immersive experiences with methods to create and design immersive, empathetic, and performative studies. The field of experiential futures is still in its infancy, both in terms of study and practice (Candy and Dunagan, 2017; Candy, 2010; Kelliher and Byrne, 2015; Kuzmanovic and Gaffney, 2017; Vervoort, Bendor, Kelliher, Strik, and Helfgott, 2015). Using human and future system relationships as its focal point, researchers in experimental futures study pay much attention to how humans experience things and how they make sense to them. It has been used to look at possible futures critically by using only one word at a time, installing things for a short time, and in projects dealing with social and technological aspects. It has not been employed consistently yet but can last for a very long time, and they have a lot of the more complicated and less certain things, similar to how experiential futures projects have been used in the past.

They were going back in time to several historical events that have shaped our current understanding of experience futures, then connecting the dots and drawing conclusions from the growth patterns of these parallel histories. In order to better understand experience futures, our primary purpose is to conduct a conceptual investigation that could be a valuable addition to the efforts for a more sustainable future. Then, applying what the humankind has learned to the study of how people create and construct their daily lives, as an example, the beauty of every day, cognition in the context of literary forms, to serve as a guide for the creation of new methods that make sense to people when they are going through changes, prior to the conclusion of the essay.

One-third of all new buildings are in the residential construction industry, where concrete is widely used. However, a large quantity of carbon dioxide (CO_2), a greenhouse gas, is released while manufacturing Portland cement, a key concrete component. On average, one ton of CO_2 and other GHGs are produced for every ton of Portland cement made during the cement manufacturing process. Table 1.1 shows that CO_2 emissions by developed countries could play a significant role in the long-term growth of the cement and concrete industries in the 21st century.

The World Commission on Environment and Development defines sustainable development as meeting present demands without jeopardizing future generations' ability to satisfy their own needs in the future. Sustainability is the concept of being concerned about our world's well-being while promoting continuing progress and human development.

TABLE 1.1

CO$_2$ Emissions by Developed Countries in 2002 (Naik and Moriconi, 2005)

Country/Union	CO$_2$ (%)
USA	25
EU	20
Russia	17
Japan	8
China	> 15
India	> 10

Designing for sustainability necessitates a departure from conventional methods that take these modifications from the designer's perspective into account in order to be effective. The new design method must consider each design decision's effects with reference to local environmental and cultural assets, as well as the regional and global surroundings.

For more than 2,000 years, concrete has been in use. Concrete's longevity and sturdiness are among its most lauded qualities. Additionally, concrete contributes to the advancement of social and economic well-being, as well as environmental conservation; on the other hand, it is frequently disregarded. Concrete structures outperform steel ones in terms of energy efficiency. They offer greater design freedom and cost than steel or aluminum structures; they are also better for the environment than construction made of steel or aluminum (Cement Association of Canada, 2004).

The concrete industry will be required to meet two critical human needs, namely, environmental preservation and the supply of infrastructure to sustain the world's rising industrialization and urbanization. In addition to its sheer magnitude, concreting waste products such as fly ash and slag is by far the most cost-effective and ecologically friendly way to recycle millions of tons of industrial waste, if required, which has high pozzolanic and cementitious qualities for the production of concrete. Cement substitution (60%–70%) by industrial waste materials in concrete is unquestionably beneficial in conserving money, efficiency in energy use, durability over the long run, and concrete's overall environmental character. Because of this, "The use of by-product supplementary cementing materials should become more important in the future." In Malhotra's 2004 book, he discusses the need for a strong sense of community.

Both old and new buildings coexist in various European historical towns: those made of stone, bricks, and wood that may have been around for a long time, and those constructed in the last 40 years, often made of reinforced concrete, are unlikely to last more than 150 years (Mehta and Burrows, 2001). In certain instances, ancient structures have been rehabilitated to meet current requirements, extend their useful lives, and help the community stay healthy due to their restoration. However, despite recent breakthroughs in the ability of reinforced concrete structures to increase their longevity, there is a lack of expectancy, and it is not likely that old buildings will last as long as they did before. A modern structure can be created within a short amount of time; reinforced concrete and steel are used to support the structure and have a life expectancy of less than a 100 years. For instance, the Alhambra in Spain contains towers dating back to the 11th century; although the brick used to construct these structures is of poor grade, none of the things that need to be fixed are massive.

Despite its best efforts, modern culture has been unable to discover a mechanism for creating attractive, functional structures, constructions that will last for current and future generations while reaping the benefits of the legacy structures constructed by earlier generations. The preservation of heritage but not the construction of heritage implies a shift in societal ideals and a shift in view on the function of society. Increased emphasis on long-term construction will aid in alleviating many existing issues; for example, incapacity to sustain the ecosystem, unemployment at an all-time high, social isolation, and spiritual void.

To develop new historical structures, it is necessary to solve problems in material technology, structural engineering, architecture and art, smart city and planning, philosophy, public policy, and management and finance. Additional efforts must also be made to revive historical trades and methods, which are becoming increasingly rare. In Figure 1.1, the critical synergies are depicted in a diagram and other things to consider when making new historic concrete structures.

When it comes to establishing future heritage through the construction of exceptionally durable structures, during this time, there are new instruments as well as situations that need to be taken into account:

- Philosophical—while these novel architectural principles do not provide a solution for all of society's woes, participants in the endeavor must believe they are contributing to producing an enduring masterpiece that will be appreciated for long periods.
- The structure and construction processes require updating; new building concepts are required.
- Financial—by investigating them, determine if new architectural ideas will favor our bottom line. New financial tools and government laws may be necessary.

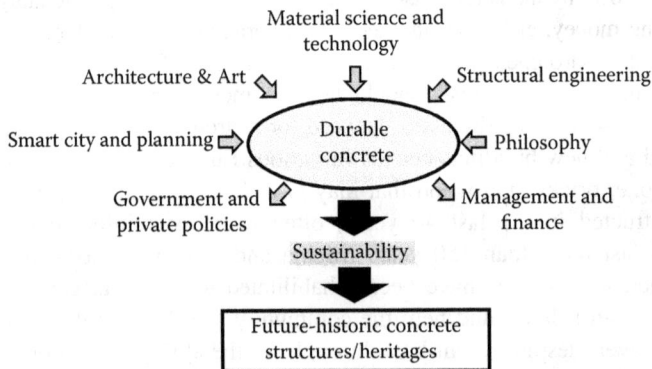

FIGURE 1.1 Aspects that influence the design of long-lasting concrete structures (Adapted from Gil-Martin et al., 2012).

Many interconnected fields are required to meet the challenges of creating dura-bility constructions that are practical in today's world. It entails the creation of financial instruments and inducements, to the extent that society's goals justify them, to see the worth and allow for lengthier payback periods that represent the most significant issue. The remaining difficulties are solvable by a community disillusioned with the current economic situation and a belief that an attitude shift is required. The remaining issues can be sensibly resolved by a society that recognizes using a mindset transformation. Maybe the 21st century will herald the start of a new historical circle, with social and artistic reforms that will enable society to move aside from consumption and the obsession with relatively brief advantages that characterized the second period of the 20th century.

Industrialization and urbanization had reached their zenith due to the increasing population's depletion of the earth's resources. In contrast, it contributes to a scarcity of natural resources and the development of massive amounts of solid waste. The construction industry, among others, is responsible for a significant portion of the acquisition of natural resources and the generation of wastes, which we refer to as construction and demolition waste (C&DW). Construction and demolition waste is generated when roadways and historic structures are demol-ished and disposed of in landfills. Aggregates are inert, granular parts that hold the concrete together and keep it from falling apart. Depending on their origin, such aggregates can be igneous, sedimentary, or metamorphic. The C&DW can be used as aggregates in concrete, making up for the shortage of aggregates in the en-vironment. Compared to rural areas, urban areas have a lot of raw materials because technology has made it possible to build things faster and better. In this case, re-cycling will be the most cost-effective option out of the three R options (reduce, reuse, and recycle). Instead, changes in technology have made it necessary to make close to tons of concrete per person. About a third of the materials are turned into C&DW. According to the World Business Council for Sustainable Development (WBCSD), the world's aggregate demand has increased to 3.8 tons per person, as demonstrated in some countries in Figure 1.2. This means that natural coarse ag-gregates (NCA) need to be replaced with a better material for the environment.

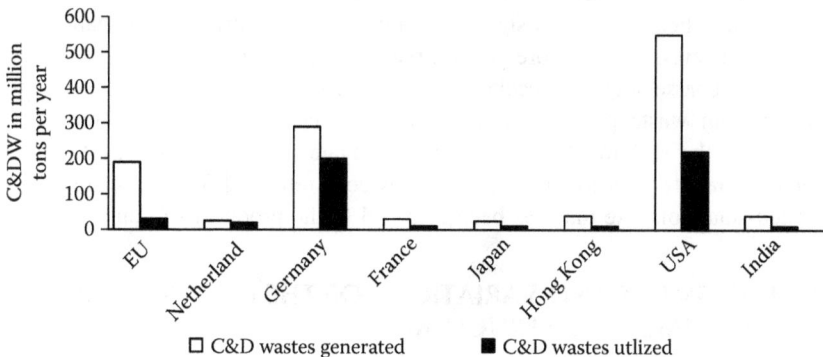

FIGURE 1.2   Generation of C&DW (Jagan et al., 2020).

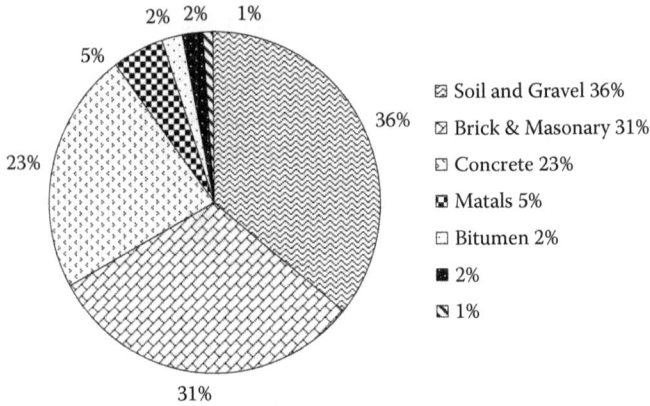

FIGURE 1.3   Compositions of C&DW in India (Jagan et al., 2020).

Infrastructure development would be improved by using the 3R strategy in the construction industry, road connectivity, and other related areas of concern. When building or tearing down the trash, you might find steel rods, broken bricks, wood, concrete, and other things, or a mix of these things, as shown in Figure 1.3). Concrete is the most common filler material, and it comes in both coarse and fine forms. Using recycled concrete aggregate (RCA) with a finer or coarser mesh size as a partial or total replacement for NCA started a new age of concrete called RCA. RCA is a non-living material with cement mortar smeared on it as if it were the original concrete. Much research has been done to learn more about RCA properties, mechanical properties, and how long it lasts when used in raw RAC. Many qualities of recycled concrete are required before it can be used in concrete, including adherent water absorption, grading, crushing and abrasion resistance, specific gravity, mortar content, density, recycling process, and specific gravity. Comparing RCA to NCA, many things contribute to RCA's poor physical properties in order to find out how much it is used in concrete.

Crushing demolition and construction trash into smaller pieces results in re-purposed aggregate. When it is mainly composed of broken concrete, it is classified as RCA, but when it contains significant amounts of materials other than crushed concrete, it is classified as more general recycled aggregate (RA). At this time, the only type of coarse aggregate acclaimed for usage in the creation of new concrete is obtained from waste products associated with construction or demolition. RCA differs from the reclaimed or recovered aggregate. Reclaimed aggregate and re-covered aggregate refer to aggregate that has been retrieved from a fresh-state or a hardened-state concrete that has been returned to the producer of concrete.

1.2   ASSESSMENT AND VARIATIONS OF THE EXISTING CONCRETE MATERIALS AND STRUCTURES

Among the most serious objections are the absences of a formal model development process and the absence of explicit definitions and mathematical analysis methodologies for essential aspects. There have been some novel models proposed. However, there has been no evidence of a link between sustainable development and industrial competitiveness. Porter's Diamond Model was used as a starting point to develop a new conceptual model incorporating theories and concepts from internationalization, sustainable development, and industry competitiveness.

Because of the fast growth across the globe and the ever-growing population, infrastructure and structural engineering are becoming more crucial. Most building systems depend on natural raw materials, which use much energy, are costly, and produce much waste throughout the material processing and construction process. Aside from that, the negative environmental consequences of particular building materials are growing increasingly severe in some instances. Furthermore, due to its low-cost requirements, cement concrete is the most often used material in homes and constructions, owing to its long-term quality, high mechanical strength, and simplicity of application (Nuaklong et al., 2020). In contrast, since the main cement element is limestone, it is costly, requires a lot of energy, and emits CO_2 during the process (Rattanachu et al., 2020). According to the World Bank, Ordinary Portland Cement (OPC), production accounts for around 5%–8% of worldwide CO_2 emissions (Khan et al., 2017; Sani et al., 2020). Exploring possible waste-based cement solutions as environmental defense awareness grows, such as reducing the cost impact of energy conservation measures, leads to environmentally friendly and sustainable building.

The anticipated deterioration mechanisms for reinforced concrete structures in service are various, and each one has the potential to affect the structure's level of durability and anticipated operational life. Reinforcement corrosion is widely regarded as the most crucial factor contributing to the deterioration of reinforced concrete structures. As a result of the carbonation process, CO_2 is one of the most prominent sources of environmental degradation.

Typical goal service civil engineering constructions, such as bridges, maritime structures, and even residential buildings, have life spans ranging from 50 to 100 years, with the potential for much longer service lives in the case of monumental projects. Maintaining such structures in the field may be difficult at times. Damage to reinforced concrete structures involves the creation of models for estimating how much of their service life remains. In the literature, there has been a significant advancement in the creation of service life models for RC constructions affected by corrosion. For example, a semi-probabilistic life model based on probability and statistics has been constructed and presented for use by the general public, which is well known and can be used to figure out how long reinforced concrete structures that have been corroded by carbonation last. The model is likely to be effective in predicting the remaining service life of reinforced concrete structures, and it will do so by including well-known software as well as readily available and easy-to-measure elements as inputs, among other things, as shown in Table 1.2 (Ebrahim and Abdel-jawad, 2020).

To develop a modified semi-probabilistic model for evaluating the remaining service life of carbonated reinforced concrete buildings using reliability theory, the proposed model can be utilized to account for circumstances where the probability distributions of carbonation depth and concrete cover do not always match. It is an

TABLE 1.2

Parameters for Analysis of the Semi-Probabilistic Model of Carbonation Corrosion Provided (Ebrahim and Abdel-Jaward, 2020)

Parameter	Unit	Distribution Type (Parameters)
Resistance parameter (concrete cover)	mm	Normal distribution (μ,σ)
Load parameter (carbonation depth)	mm	Normal distribution (μ,σ)
Age of structure	years	Discrete

excellent method for confirming and estimating a product's service life. It is demonstrated that it may be utilized for both objectives. It was discovered that by adjusting the process for determining the coefficient of variation of carbonation depth data acquired, the coefficient of variation could be calculated more accurately, which was carried out on the prototype, and the results for the anticipated residual service life were directly influenced. This alteration indicated that the method used to calculate the coefficient of variation of the acquired carbonation depth data directly impacted the conclusions drawn about the remaining service life. The efficiency of the concrete used in the structure under discussion significantly affects the rate at which the structure's dependability index decreases with age. It is possible to use an inspection to evaluate structural deterioration in reinforced concrete structures. It is relevant to a broad class of buildings that reflect a structure or component of a structure that poses a threat to human life.

As a result of its 100% recycling of end-of-life (EoL) concrete, the Netherlands outperforms countries such as China (Zhang et al., 2018), Australia (Tam et al., 2010), Canada (Yeheyis et al., 2012), and other European Union member states in terms of concrete recycling and C&DW management systems (Eurostat, 2018). Simply crushing concrete and using it as a base for road construction is the most typical practice in the Netherlands regarding concrete recycling as a low-grade or low-added-value method of concrete recycling. At the moment, the wet procedure is the most extensively utilized approach for recovering high-grade concrete to create concrete-cleaning fines by washing coarses and leaving the tiny proportion for road base filling (sieved sands); while doing so, sludge is generated, which must be treated before being reused. The disadvantage of the wet procedure is that it necessitates the construction of a large washing plant, which is expensive. The result is that in the Netherlands, more than 90% of all waste concrete continues to be treated as road base materials of a low-quality level.

Improved evaluation can be achieved by integrating the most recent information about the structure, such as including information regarding pre-existing fissures in the assessment, which may be achievable because of digital twin (DT) models, which are projected to play a vital role in the future optimization of critical infrastructure management. It is a virtual representation of the structure built by a DT. Moreover, it has the potential to serve as a decision-making assistance tool. Testing techniques have recently been developed due to recent improvements in inspection procedures. New and improved methods for obtaining information regarding cracks will be implemented. Therefore, the finite element simulations used in structural evaluations must integrate data on pre-existing cracks into their calculations.

Despite this, not using either the discrete or spread cracking methods approaches has had sufficient development to be used to characterize pre-existing cracks in concrete buildings. Discrete cracks are widely used while examining dams to include crucial pre-existing fissures. A fundamental frictional rule in the shear transfer model that does not account for aggregate interlock may be sufficient for large-scale structures with high self-weight. However, this is not the case for other types of structures. For beams with a low shear reinforcement ratio, aggregate interlock, for example, is recognized as an effective shear transfer mechanism. On the other hand, the treatment of aggregate interlock was not universally applicable but was specifically designed to mimic the fatigue response of the slab.

The variation in damage produced by individual fracture breadth and shear retention (aggregate interlock) was incorporated, addressing weaknesses high-lighted. The feasibility of representing the influence in the assessment of anchorage capacity and cracking fractures caused by corrosion while confined under a concrete cover, and the development was based on that study. Unlike other methodologies that require knowledge of the load history, the suggested methodology does not require this information in practice. For one reason or another, no one has attempted to chart it. These measurements (crack patterns and widths) may be viewed at a glance, unlike this. Furthermore, when compared to modeling the entire load history, direct integration of pre-existing cracks is more efficient in computing. Capacity estimation of structures with pre-existing fractures is possible using the proposed methodology, which is appropriate for use with DT models.

In addition to weakening or inserting unique crack elements at the locations of cracks, other options may be considered. Several studies have used controlled contraction and tensile loading to pre-crack reinforced concrete beams, which were subsequently examined in four-point bending to validate the methodology. Adding pre-existing cracks in weaker elements resulted in better estimations than those obtained using traditional finite element analysis (FEA), notably in terms of the ductility and failure properties of the elements. It was possible to get a more accurate idea of the material's ultimate strength and flexibility. However, they provided more precise predictions of failures restricting deformation capability. In comparison to studies with discrete crack analyses with feeble parts required less implementation and computation time. Moreover, a combination of decreased strength and toughness of reinforcement bonds (to represent damage from previous tensile loading) and a change in the weaker components' compressive behavior (to show that cracks have closed in the compressive zone) was demonstrated to affect the bending stiffness (Samir, 2000).

For a variety of reasons, in many aspects, assessing existing structures and in-frastructures differs from designing new systems. The remaining useful life (e.g., less than 50–100 years) is a good representation of the reference service life connected with an existing building in which the building should perform its intended function, which may be considerably different from the structure's planned service life (e.g., 50–100 years). Furthermore, the expenses of retrofitting existing structures with safety measures are more significant than the cost of designing new ones with the same precautions, which is a significant savings. As a result, the goal levels of dependability associated with economical optimization and human safety criteria should be distinguished while evaluating current structures and designing new systems. A further advantage of existing structures is that they had frequently demonstrated "acceptable prior performances" when they were subjected to pertinent loading scenarios during their operational life. This understanding should be incorporated into the evaluation process, even though it is difficult to account for this component.

Over the last several decades, a substantial study has been conducted on chloride-induced corrosion. In the field of structural deterioration, Tuutti's paradigm is primarily acknowledged as the appropriate conceptual model. When plotted against time, this model distinguishes between two phases: the beginning and the propagation phases. On the other hand, the propagation phase is relatively brief compared to the starting phase. As a result, determining the initiation time appears to be particularly critical.

Probabilistic techniques are commonly used to anticipate the time at which corrosion will begin to occur in concrete structures. In a probabilistic approach, Monte Carlo Simulation (MCS) is one of the most straightforward and straightforward techniques. This approach treats the underlying variables as if they were random. The first-order and second-order moment approaches, respectively, and the response surface method are all probabilistic methodologies, and so on.

Pellizzer and Leonel (2020) conducted a study that demonstrated the usefulness of probabilistic techniques in predicting corrosion onset time. Enright and Frangopol looked into determining the time it took for corrosion to start and discovered that the coefficient of variation of random factors had an impact. Song and Pack looked into the corrosion start time of a seabed-installed concrete tunnel-box structure; they considered the temporal dependence of chloride transport while taking the diffusion decay index as a constant number, as opposed to previous studies. Saassouh and Lounis investigated the efficiency of MCS, first- and second-order dependability approaches for predicting corrosion start time, as well as the effect of random variables on time required for corrosion to initiate. They also looked into the effect of random variables on corrosion start time. On the other hand, the diffusion coefficient's time-dependent property was not considered, which is still necessary to have a better knowledge of the corrosion appraisal influenced by random variables' onset times, mainly when the diffusion decay index's uncertainty is included. A better understanding of how random variables affect how long corrosion takes to start is also needed.

Using a probabilistic approach for evaluating future service life, the probabilistic modeling of fatigue resistance is based on large amounts of fatigue tests and probabilistic modeling of the service duration. Identifying and measuring the uncertainty related to vehicle weight, the probabilistic modeling of activities incorporates variables such as vehicle location and temperature-induced strain changes. In this case, the modeling can only verify the fatigue limit state using data from a year of monitoring at the Crêt de l'Anneau Viaduct and weigh-in-motion

(WIM) measurements provided by Swiss authorities. For fatigue resistance, a stochastic fatigue resistance model is developed using data from an extensive fatigue test database accessible in the literature. A general technique for modeling the interaction of design parameters for extant structures to recalibrate partial safety factors is based on a recent model developed for stochastic fatigue resistance. A case study of the Crêt de l'Anneau Viaduct and the structure's assessment of resistivity part safety parameters are found that the validation of partial safety factors for resistant materials is beneficial for existing structures. Code-based designs can be modified to meet performance requirements by including measured safety considerations rather than extensive reliability calculations (Reto et al., 2021).

With the help of clean, recycled aggregates, ultrafine particles (cement-rich powder and glass powder), and mineral fibers found in construction waste, these things make concrete work better and affect its strength. Unambiguously, the study will use clean recycled aggregates. While determining how much EoL material can be utilized to build a new kind of concrete product that is strong, robust, and long-lasting is also a significant aspect; concrete may be manufactured using recycled materials derived from construction and demolition waste projects. In order to reuse the majority of the C&DW produced by the construction business, a new concrete mix incorporates recycled products such as aggregates, cement paste, glass powder, and mineral fibers that are prepared with the intention of recycling as much of the waste as possible. In place of natural aggregates, the components' RCA, fine recycled aggregates, and recycled cement-rich ultrafine (RCU) are being employed as the primary building blocks for C2CA technology. Using glass powder and mineral wool in concrete could be better because they could be pozzolanic and make concrete more resistant to cracking and mechanical stress, which has not been looked at in much other research. C&DW can be used to make the most environmentally friendly and long-lasting concrete design. In addition to reducing the consumption of natural resources, such a product also reduces the CO_2 emissions related to concrete manufacture.

1.3   ENDING CEMENTITIOUS MATERIALS (DEMOLITION AND DECOMMISSION)

Concrete aggregate, also known as reclaimed concrete material, is recycled from construction waste left over after Portland cement concrete (PCC) roads and other building concrete structures have been demolished, as shown in Figures 1.4 and 1.5. After demolition and excavation, a stockpile facility (such as an aggregate provider) or landfill is the most common destination for RCA. The RCA is crushed to a specific gradation for use in the final product as a high-quality base or subgrade material after the reinforcing steel is removed at the stockpile facility or construction site.

Inert materials such as concrete, plaster, metal, glass, wood, and plastic compose most of the trash generated during construction and deconstruction. These wastes are typically disposed of in illegal landfills, which have negative consequences for

FIGURE 1.4   Concrete wastes from building structure.

FIGURE 1.5   Demolition of concrete waste from existing foundation concrete structure.

the landfills themselves and the environment. The Central Pollution Control Board estimates solid waste creation in India. Annual waste creation is around 43.5 million tons (48 million tons), and waste produced by the construction industry accounts for 25% of the total waste output. Between 181 and 272 million tons of solid waste are generated annually in the United States (200 to 300 million tons). Along with the World Bank, Shanghai, China, creates 19.1 million tons (21.1 million tons) of such waste annually, accounting for 45% of the city's yearly solid waste production. Building garbage accounted for 26% of Hong Kong's daily solid waste production, or 3,251 metric tons per day (3,584 metric tons per day). Building and demolition debris can be turned into recycled concrete aggregates with various environmental advantages. It has been reported that characteristics such as water absorption, specific gravity, and density can affect recycled aggregate concrete mix quality. Grabiec and colleagues (Grabiec et al., 2012) found that recycled aggregate quality has been enhanced using the bio-deposition of calcium carbonate. As a result, RCA's water absorption was lowered. Using Tam et al. acid presoaking approach (Xiao, 2018), the water absorption was only reduced by 7.27% to 12.17% when the acid concentration was 0.1 Molar (M). According to Akbarnezhad et al. (2011), the acid concentration of 2M indicated in their approach resulted in less than 1% water absorption of the RCAs.

Much research has been conducted to describe the essential qualities of RCA. Given that coarse aggregate composition is 65% of concrete mixes and that a large amount of building and demolition waste is commonly available at different construction sites, the use of RCA as a coarse aggregate has a great deal of potential. However, there are still some gaps and weaknesses in the present research and experimental work that must be addressed to incorporate all of the tests that must be performed to evaluate RCA qualities. Because of this, several studies have been chosen to assess and determine RCA and basalt's viability as coarse aggregates in concrete mixes; researchers looked at their physical and mechanical qualities.

When RCA is compared to basalt, it shows a smaller amount of flaky and distorted particles. It outperforms other concrete combinations. As a result, concrete mixtures with a specific proportion of RCA can be more workable in terms of deformation resistance and compaction. Due to fractures and the adhering mortar and cement paste used during construction, the RCA has substantially more excellent absorption and fluctuation in wet/dry strength than standard aggregates. These must be taken into account before mixing. In addition, other testing results have revealed that RCA continues to encounter the specifications for aggregates used in concrete mixtures.

For highway engineers, using RCA is an enticing alternative because of the potential cost and time savings that may be obtained by recycling C&DW. In order to make use of significant volumes of building and demolition debris, it is necessary to obey to the minimum standards reputable by the American Association of State Highway and Transportation Officials (AASHTO), as well as local criteria. Because these recycled materials may be manufactured on the construction site, the quality control of recycled materials may decrease throughout the project. As a result, highway engineers may opt to combine recycled materials with thick, graded aggregate purchased from quarries, as has been done in the past for base and sub-base applications. More than a few researchers have looked into the benefits of RC in applications. However, only a small number of investigations into the physical characteristics and the mechanical behavior of mixed materials that have been used in the construction of roadways, they must be treated with cement when used as sub-bases or bases to (a) enhance the usability of road materials, (b) increase the mixture's strength, (c) add to the strength, and (d) enhance the capability for load distribution.

As defined by the American Concrete Institute (ACI), this type of crushed stone contains only a trace amount of cement as a coarse aggregate binder, along with just enough water to ensure adequate compaction and cement hydration. In general, cement-treated aggregate as a road base material is prepared by cement by combining with coarse natural aggregate, or it is made of crushed aggregates and cement and is designed to be used as a busy traffic base or heavy traffic course for vehicles hauling heavy loads. It has recently been proposed that recycled aggregate be used on road bases to conserve natural resources and prevent environmental pollution caused by solid waste (Pourkhorshidi et al., 2020).

1.4 CHARACTERIZATIONS AND CLASSIFICATIONS OF RCA

A number of properties of RCA are affected by the presence of cement paste (adhered mortar). Crushed recycled concrete aggregates may be distinguished from natural aggregates by the presence of cement paste on their surfaces. Because of its chemical composition, attached mortar has higher reactivity than the natural aggregate (NA), allowing it to bond and interact with other binding agents. Because of this, the reliability of RCA should have been examined in light of the technical standards used in its final application. Aggregates must meet minimum physical and mechanical specifications to be suitable for hot mix asphalt (HMA) manufacture, which is laid forth in Colombia in the National Standards for Highway Construction-INVIAS (INV), as well as undergo appropriate quality control. For example, porosity and absorbance as crucial characteristics of RCA physical qualities; the bulk specific gravity, in reality, of RCA, is typically lower than that of NA. Furthermore, the dispersion of coarse and fine aggregate particles considerably affects the design and efficiency of the concrete mixture. In contrast to RCAs, which are granular materials available in various sizes, when the gradation requirements for concrete mixtures are not reached in an as-received condition, the RCA material must be crushed again. When RCA materials are compared to natural aggregates, it has been demonstrated that RCA materials are more absorbing than the NA. Moreover, the water absorption of RCA materials increases when the substance's particle size is lowered. In the coarse fraction, the amount of mortar that sticks to the RCAs aggregate tends to make it less dense, and the amount of water it can hold depends on how porous the cement matrix holds the recycled concrete together. The water absorption of RCA materials increases when a substance's particle size is lowered. In the coarse fraction, the amount of mortar that sticks to the aggregate in RCA tends to make it less dense, and the amount of water absorbed is proportional to how porous the cementitious matrix sticks to the recycled concrete.

In contrast to RCAs, which are granular materials available in a variety of sizes, when the gradation requirements for concrete mixtures are not met in their raw state, the RCA material has a different size distribution; it needs to be crushed again. It has been shown that RCA materials have a greater capacity for absorption than natural aggregates when compared to one.

1.5   UTILIZATION OF RCA IN CONCRETE

Furthermore, RCA from C&DW makes up a significant portion of the solid waste generated during rehabilitation and construction processes. According to the Asphalt Institute, it is one of the recyclable aggregates that can be utilized in concrete mixtures. According to a literature survey (e.g., RCA), it has been utilized successfully as material for base and sub-base courses in several contexts. However, RCA has been documented in a few sources (for example) when hot mix concrete is utilized. As a result, the purpose of this lab investigation is to examine the RCA features in detail, consisting of physical and mechanical characteristics. RCA and a mixture of RCA and basalt were subjected to several tests, to enable the future use of RCA as dense graded concrete associated with coarse particles.

The use of RCA as a valuable component in concrete mixtures has been demonstrated, and it can enhance certain qualities of concrete. Adding RCA to a natural aggregate that does not meet the gradation guidelines is a rapid and informal method to improve its quality, making it possible to use it to make concrete. RCA dry from the air can be used for natural aggregate up to 50% of the natural aggregate without needing technological methods or changes in the concrete mixture composition. It is predicted that some minor consistency adjustments will occur. Adding RCA improves the gradation of the aggregate, which allows for more compression strength, as shown in Figure 1.6. It has been shown that there is a strong link between how much water the used aggregate absorbs and how much water the concrete absorbs. Although RCA results in enhanced water absorption values in concrete, as shown in Figure 1.6, this concrete property can be predicted with reasonable accuracy. When subjected to

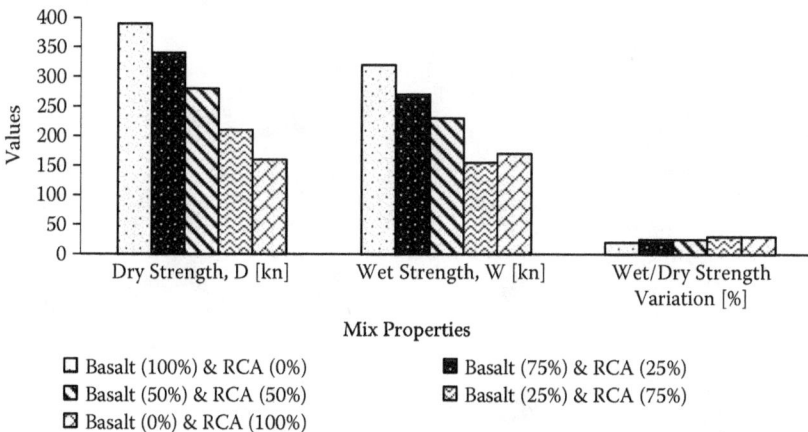

FIGURE 1.6 Wet strength and dry strength of different mixes of basalt and RCA (Tahmoorian and Samali, 2018).

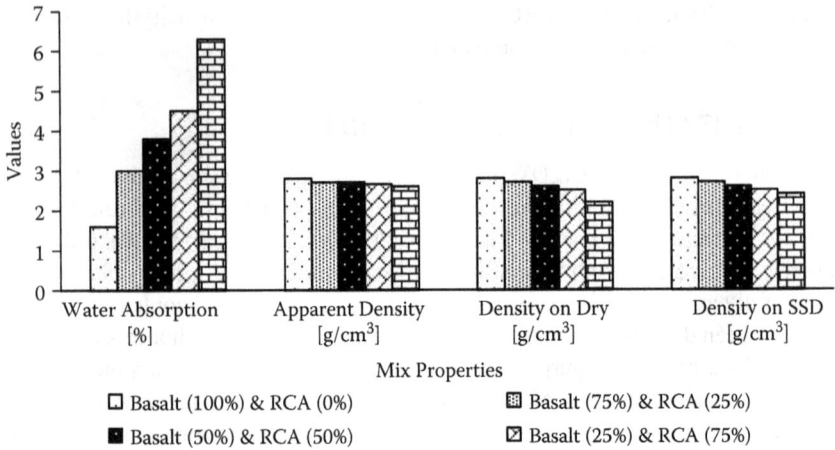

FIGURE 1.7 Water absorption and particle density of basalt and RCA mixtures (Tahmoorian and Samali, 2018).

pressure, concretes containing low-quality RCA have a very high-water penetration depth and hence cannot be used when waterproof concrete is required. Nevertheless, using high-quality RCA can help to ensure that the concrete has the required water resistance (Figure 1.7 (Tahmoorian and Samali, 2018)).

2 Principles of Recycled Aggregate Concretes (RACs): Theories and Experiments

2.1 RHEOLOGY OF FRESH RAC

Understanding the rheological qualities and behavior of freshly mixed concrete helps enhance the quality control of the concrete. Recently, one of the unique concretes, such as self-compacting concrete (SCC), has stricter regulations, and the need to measure new concrete's flowability is growing. For this reason, fresh concrete is crucial to the placement process, which comprises hauling, pumping, casting, and solidifying.

To determine the rheological characteristics of fresh concrete, for example, yield stress and viscosity of plastics, at the moment, no conventional test procedures are available. On the other hand, various types of concrete rheometers and associated models to determine the rheological qualities of freshly poured concrete (Ferraris et al., 2001). Upon examination of these testing techniques, it was discovered that the proposed approaches provide valid ties that are statistically equivalent. However, their values for the traits differ significantly (Ferraris and Brower, 2003). Once created, concrete qualities can be managed using standard test procedures, but they are not beneficial for developing the concrete mixture. Ferraris and deLarrard (1998), Roshavelov (2005), and Mahmoodzadeh and Chidiac (1998) have developed these models (2011). The models proposed by Ferraris and deLarrard, as well as Chidiac and Mahmoodzadeh (2009) accurately explain concrete rheological parameters, such as yield stress and viscosity of plasticity, according to a comprehensive evaluation.

To predict the flow behavior of special concretes, several plasticities and viscoplasticity models have been presented in the literature (Chidiac and Habibbeigi, 2005; Lu et al., 2008) that no mathematical models in the literature describe how fresh mortar behaviors regarding the fundamental principles of significant field limitations. In the constitutive equations for describing the flow of fresh mortar and concrete, mathematical analysis is utilized. Brief results indicate that the first stage presents a brief overview of rheology in the concrete and mortar building context. Afterward, based on prior work, the constitutive models for mortar and new concrete are explained in detail by Lu et al. (2008) and Mahmoodzadeh and Chidiac (2011). Furthermore, a new cement and mortar model should have three parts: collisions, static, and dynamic interactions. The first and second parts have been

DOI: 10.1201/9781003257097-2

derivatived from the effort of Mahmoodzadeh and Chidiac (2013) based on the yield stress and plastic viscosity. A variant of the Lu et al. (2008) model is included in the final component, which incorporates the concept of the cell method, indicates that shear stress is a combination of static, dynamic, and particle collisions, and fresh concrete flows according to the Bingham model for typical slump concrete.

Many tests, most of which are based on actual evidence, have been devised to assess the concrete's ability to be worked. The workability of concrete can be predicted on-site during the mix design of concrete in the laboratory and during concrete testing. It is used for quality control throughout the construction of the structure, making it possible to take the required safeguards before laying the concrete in case of an accident during production. As a result, the test helps to avoid or minimize economic damage.

Based on a two-phase substance (cement paste and aggregates), it is important to note that the interaction between these two phases does not consist uniquely of lubrication between paste and aggregates but of a complex mix of the paste's viscous effect and the aggregates' mass. As a result, it is beneficial to understand the rheology of cement paste, which shares many characteristics with the rheology of concrete. Conversely, the complexity of the concrete combination makes it challenging to investigate the rheology of this material. Several variables might affect the rheology of freshly poured concrete, including the nature of the ingredients and the water absorption capacity, which is more significant for recycled aggregates than natural aggregates. Other aspects of the concrete's composition are taken into consideration as well. As a result of these elements affecting its rheology, concrete formulation (w/c, gravel-sand ratio, paste volume, and additives) must be considered. Considering the time factor in evaluating concrete rheology is necessary because it is not an intrinsic component of the concrete's composition. Because of the hydration process, the amount of time fresh concrete may be used before it becomes also hard to work is limited. Whenever this limit is exceeded, the workability of the concrete is reduced, and consequent to improve the rheological behavior of the concrete. Therefore, the production of fresh concrete could benefit from using RCA from C&DW. Besides, it has many environmental benefits, like cutting down on the need for primary resources and producing embodied energy and emissions. It could also cut down on the amount of pressure on landfill sites.

A substance's physical, mechanical, and chemical qualities must be understood appropriately to use the RCA. Following the findings of several researchers, some unique things about RCA have the potential to impact the performance of concrete, either favorably or unfavorably. It is essential to keep in mind that RCA is distinct from natural aggregate (NA) in the sense that the cement paste that is attached on the surface of RCAs after they have been crushed remains on the recycled concrete aggregates after they have been crushed. This distinction segregates RCA from NA. The presence of this cement paste, also known as glued mortar, is typically regarded as the quality indicator for the quality that affects many of the RCA's traits. The attached mortar is more receptive than a NA regarding chemical reactions, which makes it more likely to bond and interact with other binding agents than a NA. It is vital to ensure that the aggregates used to create concrete comply with specific physical and mechanical requirements and quality.

Natural aggregate concrete (NAC) and RCA can utilize an imposed continuous flowability at the fresh state and a desired compressive strength of approximately 35 MPa after 28 days, respectively. Several different percentage replacements include the NAC and RAC at 30, 65, and 100 wt% (Xiao et al., 2018). While using recycled aggregates up to 30% of the total mix does not increase the water demand of concrete, it does result in a drop of 14% in compressive strength. When the replacement ratio is increased, the cement content is increased to retain the same w/c that increases compressive strength, which improves to mitigate the detrimental impact of recycled aggregates. There is a substantial difference in the splitting tensile strength and elastic modulus between recycled aggregate concrete and conventionally sourced aggregates, as was the flexural tensile strength.

2.2 PHYSICO-MECHANICAL CHARACTERIZATION OF RAC

The physical and chemical characteristics of recovered RCA compared to the physical and chemical characteristics of the original RCA were performed by Chen et al. (2012, 2013). RCA materials have a higher absorbency than natural aggregates. RCA materials' water absorption rises as their particle size decreases. There is a tendency for the quantity of mortar adhering to the aggregate on the RCA to lower density, and the amount of water absorbed to be subject to the porosity of the cement matrix linked to the recycled concrete. Some coarse aggregate is added to the crushing process to make the fine particles, but most of the material is dissolved mortar or cement paste. Because of this, the RCA coarse fraction has excellent quality since the aggregate's absorption capacity is lower when there is less mortar.

In general, the RCA properties of small coarse fraction (particle size > 4.0 mm) and a fraction (particle size < 4.0 mm) are different from one another. Because the coarse and fine fractions come from different quarries, this property is significant for NA. RCA is composed of natural coarse aggregate plus a linked mortar layer of cement paste, whereas fine RCA has more crushed mortar than the natural coarse aggregate component. The differences in composition between the fine fraction and the coarse fraction of the RCA determine the unique qualities of each fraction.

It is well-known that the porosity of the cement pastes by applying the findings of Powers and Brownyard's work and the porosity of the coarse aggregate is determined by experimentation. The mixing law is applied to compute the concretes' total water porosity. It has been demonstrated in several studies that the incorporation of RCA into concrete produces a more remarkable improvement in water permeability than merely increasing the volume of the paste. Moreover, Archie's law is tested by creating a connection between electrical resistivity and water porosity independent of the examined resistance class. The electrical resistivity studies, as well as the estimations of the formation factors for the cement matrices and concretes, indicate that the C25/30 cement has a more linked pore structure than the C35/45 cement. Additionally, this demonstrates that RCA significantly impacts this concrete transfer property as the formation factor of both sets of concretes improves as the replacement ratio of natural aggregates with recycled aggregates increases.

When using RCA associated with natural aggregate and cement, calcium silicate hydrate (C–S–H), ettringite, monosulfate, and calcium hydroxide ($Ca(OH)_2$) are produced conventionally. These are in the amorphous phase and have sufficient activity to generate modest physical and chemical reactions with the cement from which they originated. Due to the chemical composition and microstructural surface characteristics of RCA, this case complicates the bonding law between the two materials and can be affected by many different things. As a result, CSRCA exhibits mechanical characteristics, stiffness, and fatigue qualities that differ from typical cement-stabilized materials in which if the CSRCA is designed; traditional cement-based materials are made this way results in unsuitable strength criteria will result in either too high or deficient strength. Some the traditional design indicates that the concrete pavement structure is based on the splitting strength and modulus of gravel stabilized with cement.

Physical characteristics of RCA, such as high porosity and absorption, have been discovered to be important in the literature. These are just two of the many factors. Although GSB is typically lower than the NA, absorption values can be moderate to extremely high compared to the NA. There is a layer of mortar that is bonded to the concrete, and this layer reduces the GSB while also increasing the absorption of the concrete. This is because each particle has a significant amount of adhering mortar, and the strength of the initial concrete also plays a role in determining how much adhering mortar is present in each particle.

In concrete that contained RCA, Gomez-Soberon saw an increase in the pore size distribution, which he attributed to the chemical. Mercury intrusion porosimetry (MIP) was utilized to ascertain each specimen's porosity level. After 91 days of wet curing, it was discovered that the pores in concrete containing RCA at a replacement percent of around 3.8% were approximately larger than those found in regular concrete (Gomez-Soberon, 2002) due to the angularity of the RCA particles. RCA combinations are less compacted and contain more voids than rounded aggregate mixtures when comparing packing density. There is more water in the mix because there is less dense packing, which makes the paste more porous. Filling voids in concrete requires that the matrix, water, air, and powder particles that is smaller than 125 μm in size, work together. This is important when pouring concrete into a mold. The proportion of water to cement, the reactivity of the mixture, the form of the particles, and the size morphology of the nanoparticles all affect the microscopic qualities (strength or durability) of concrete. Moosberg-Bustnes et al. said that superplasticizer is vital for keeping the low w/c and ensuring the cement and filler are mixed well (Gomez-Soberon, 2002; Moosberg-Bustnes et al., 2004). According to the authors, a dense particle system has a better superplasticizer efficiency than a porous low-density system, which is not what most people think. Due to the strong demand for superplasticizers by 100% RCA mixtures, there is still an essential factor to improve the reinforced RCA quality. It is possible to produce high-quality RCA concrete, if the properties of RCA are well known and the proper mix design method is used.

It has been discovered that recycled fine aggregate (RFA) can be used in combination with RCA to develop recycled aggregate RCA proportioned using a novel approach that can be accomplished by using the same replacement ratio derived from the equivalent mortar volume (EMV) design approach. This approach can be obviously used in the case of the production of pervious concrete.

To develop RCA based on pervious concrete with the lowest possible manufacturing costs while simultaneously providing the best possible environmental performance within the boundaries of the technical parameters is identified in this review while adhering to all applicable standards and regulations. It is developing concrete that complies with all applicable standards and regulations. Pervious concrete manufacturing has the potential to have adverse environmental consequences, and these consequences must be analyzed and compared to the negative environmental consequences of using RCAs or natural aggregates. Depending on the material used, its density ranges from 1,600 to 2,000 kg/m^3, and flexural strength ranges from 0.1 to 3.8 MPa. On the other hand, pervious concrete has a compressive strength lower than regular concrete and ranges from 2.8 to 28 MPa because pervious concrete has a high void index. Pervious concrete is used in a variety of applications. As a result, pervious concrete is only used when minimal mechanical performance is required (Gebremariam et al., 2021). One of the compositions of cement paste and its physicochemical properties is the recycled aggregates be composed of the original aggregate and the cement paste when removed from old concrete in a standard reprocessing procedure. It is possible to think of these aggregates as a binary system in which natural aggregates and cement paste are the two ends of a mixing sequence. Calculating bulk and actual densities across the mixing series can generate a simplified phase diagram for the necessary features using the values for pure natural aggregates and pure cement paste.

2.3 MICROSTRUCTURES OF RAC: INTERFACIAL TRANSITION ZONE (ITZ)

Conventionally, the grain-scale microstructure was responsible for controlling the micromechanical behavior and driving the complex macroscopic reaction. Heterogeneity (such as mortar, interfacial transition zone (ITZ) and aggregate) is well known to determine a significant influence in the creation of tensile stress at a specific area, which ultimately leads to initiation of cracks. The ITZ, an area of concrete between the mortar and aggregate, is particularly susceptible to mesoscopic fluctuations in strength, according to this study, i.e., between microscopic and macroscale dimensions size (100 μm–100 mm) between mortar and aggregate.

2.4 MECHANICAL PROPERTIES OF RAC

To suitably utilize waste concrete, it is required to incorporate it into fresh concrete as recycled aggregate, known as RAC, as shown by yhe typical properties in Table 2.1. As a structural concrete, RAC has gained much attention, mainly from construction departments. Some investigation into the material's physical properties has been conducted. Among other things, Zhu et al. conducted several experiments to find out how concrete blocks made from low-grade recycled aggregates react to heat. Belen et al. (2020) found that the concrete with 100 wt% RCA content for the structural RAC and Xiao et al. conducted a series of tests on shear transfer across a

TABLE 2.1

Material Properties of Aggregates for the Concrete Mixture (Tamana et al., 2020)

Aggregates	Bulk Density, kg/m^3	Specific Gravity			Absorption Capacity, %	Moisture Content, %
		Bulk Dry	Bulk SSD	Apparent		
NCA	1,636	2.60	2.65	2.72	1.71	1.64
RCA	1,335	2.16	2.33	2.61	8.02	2.82
NFA	–	–	2.60	–	1.52	–

crack in RAC. Apart from that, some researchers, such as Levy and Helene, 2004, Abbas et al. (2020), and Patrcia et al. (2020), have undertaken studies into the durability features of RAC. As a result, to widen the scope of research, it is vital to report about RACs on the mechanical properties and the different ways they can be used, particularly in building constructions. It is worth noting that the RAC strength indexes listed above were all evaluated after a regular curing age (28 days). It can be generally concluding that the peak strain value of RAC to go up as the amount of RCA in the sample increased. As RCA replacement ratios of 80, 90, and 100 wt% are used, the peak strain values rise by approximately 15% compared to the NCA. The compressive strengths of RAC concrete at two years are more significant than the compressive strength of the specimens at one year. Moreover Poisson's ratio of RAC is in the range of 0.17 to 0.24.

Construction waste from demolished concrete structures was used to create RCA, which were then crushed into smaller particles to achieve the desired sizes. It was common practice to incorporate recycled aggregates into concrete mixtures to save natural aggregates while reducing the amount of concrete trash in landfills. Recycled aggregate concretes have a higher modulus of elasticity and compressive strength than that of concretes containing natural aggregate. Additionally, they have a lower water impermeability, abrasion resistance, durability, and other features that make them less engaging than natural aggregate concrete (Choi et al., 2016; Ying et al., 2016; Lei et al., 2020). Nevertheless, as previously mentioned, numerous pozzolanic components have been included in concrete mixtures and ordinary Portland cement (OPC) to enhance the qualities of recycled aggregate concretes.

Laboratory studies (Somna et al., 2012a, 2012b; Tangchirapat et al., 2012) show that adding pozzolanic materials like bagasse ash, palm oil fuel ash, or fly ash to recycled aggregate concrete can improve its compressive strength, water permeability, and resistance to sulfate and chloride over time. Also, adding ground bagasse ash to recycled coarse aggregate concrete (Somna et al., 2012a, 2012b) can reduce water permeability and chloride penetration by up to 20%. This can be done mechanically or in terms of durability. Also, Tangchirapat et al. (2012) found that crushed palm oil fuel ash at a concentration of up to 20% wt% was used in place of OPC in recycled coarse aggregate concrete. This decreased the compressive strength by 7% compared to control concrete but decreased chloride penetration by a significant amount. You can also use up to 35% ground fly ash as a binder with a w/c ratio of 0.45. This gives recycled coarse aggregate concrete with any w/c ratio a strong compressive strength. In their study on the stress and strength of concretes, Lei et al. (2021) found that concrete made from recycled aggregate has a wide range of life spans.

In civil infrastructure and building structures, aggregate is a naturally occurring mineral that is mined. The aggregate content of concrete paving mixtures is typically between 90 and 95 wt%, or 75 and 85% by volume, with about 12,500 tons of aggregate being utilized for every kilometer of flexible pavement created under conventional industry practices. Aggregates in this amount are mostly made from natural resources. Environmental and financial considerations have driven many civil engineers to advocate for new construction using recycled materials. To date, RCAs have been successfully used for structural and non-structural reasons in new concrete constructions, according to current investigations. There have also been investigations on using RCA in bases and sub-bases as unbound materials, bitumen-treated granular materials, or cement-treated granular materials. Road sub-bases can be improved by including recycled concrete, bricks, and RCA. Poon and Chan (2006) reported CBR values of over 30% during the testing, indicating that these materials might be used as sub-base. Recycled aggregates have differing physical, chemical, and mechanical properties from natural aggregates because of the cement paste applied to their surface. It is partly because the cement paste's composition differs from that of natural aggregate. RCAs are lighter, absorb more water, and have lower abrasion resistance than other concrete forms because of the cement paste. Wong et al. investigated whether recycled concrete could be used as a partial aggregate substitute in hot mix asphalt (HMA) compositions (Paranavithana et al., 2006). There were 6% natural granite aggregates, 45% untreated, and 45% heated RCA in the Marshall mix design. In all possibilities, all combinations would meet the minimum requirements of the Singapore Land Transport Authority (Paranavithana et al., 2006). Compared to control mixes, the mixture with a higher concentration of RCA had a greater resilience modulus and creep resistance than the control mixture. If RCA could be used on a low-traffic road, it is reasonable for total aggregates replacement of 25, 35, 50, and 75 wt% of natural aggregates. Moreover a lower amount of energy was required to compact HMA mixtures with RCA compared to the control mixtures. In contrast, the voids in mineral aggregate (VMA) and voids filled with asphalt (VFA) values of the combinations fell when the amount of RCA was increased. All HMA combinations with RCA can meet the minimum rutting standards regarding the rut depth (Paranavithana et al., 2006). The tensile strength ratio of 80% was achieved by all the mixes, except for a 75% RCA mixture, when subjected to a moisture-induced damage test. Furthermore, RCA-containing mixtures were less stiff than control mixtures in dynamic modulus tests.

RCA as a coarse aggregate in HMA mixtures by Paranavithana and Mohajerani (2020) had an impact on HMA mixture performance. Coarse RCA contributed to 50% of the total dry weight of the HMA mixtures, which were composed of fine RCA. According to the research, HMA mixes containing RCA have lower resilience modulus and creep resistance than the control mixture. RCA can be used as a partial or entire replacement for coarse aggregate (CA), fine aggregate (FA), and filler in hot mix concrete mixtures, as indicated by Arabani et al. in their research. RCA-FA and CA deception were found to be the most effective combinations in the MS, fatigue, persistent deformation (rutting), and robust modulus tests demonstrated that this mixture outperformed all other mixtures. It was possible to compute the mechanical parameters of concrete mixtures with steel slag and concrete aggregate waste. Tests such as Marshal stability, dynamic creep, and indirect tensile fatigue indicated that the most efficient method utilized steel slag as the coarse material and wasted concrete as the fine aggregate.

Wu et al. (2017) experimented on the impact of C&DW on concrete mixture performance by dividing into two distinct groups based on the aggregate size. Coarse recycled aggregates bigger than 4.75 mm in particle size and fine recycled aggregates smaller than 4.75 mm in particle size are acceptable. One mixture consisted solely of limestone aggregates; two combinations contained fine and coarse C&DW limestone; and a third mixture contained only C&DW limestone aggregates, both of which were coarse and fine (100 wt% limestone aggregates). The effectiveness of the concrete mixtures was evaluated using scanning electron microscopy, immersion Marshall testing, freeze-thaw split testing, low-temperature bending tests, and high-temperature rut testing. The product met or exceeded expectations in performance testing and was found to meet or exceed China's technical specifications (Kim et al., 2022). Another study examined whether precoated RCA might be used in hot mix concrete at the thicknesses of 0.25, 0.45, and 0.65 mm before being coated with slag cement paste. A coating paste volume of 0.25 mm was found ideal for HMA at a thickness of 0.25 mm in the testing. HMA specimens were subjected to moisture-induced damage and rutting tests as well as indirect tensile strength testing to see how long they would last with various substitution ratios of precoated RCA in the percentage replacements of 25, 50, 75, and 100 wt%. By increasing the RCA replacements of stone mastic concrete (SMA) mixtures resulted in increased bitumen absorption. Owing to the more significant porosity of the concrete and the presence of connected mortars on top of the crushed concrete surfaces. However, by immersing the RCA in water and cleaning them before using them in the concrete mixes, we were able to considerably improve the RCA performance in SMA combinations by lowering binder absorption and enhancing stickiness between the RCA. Moreover, adding RCA to concrete mixes increases the voids in the mixtures' total mix (VTM) simultaneously, decreasing their density values. RCA performs better than virgin aggregates. The main reasons for this phenomenon are the decreased density, lower specific gravitational, and increased porosity. In addition to showing acceptable trends, SMA mixtures comprising 20 and 40 wt% RCA are sufficient to meet the Marshall properties' standard criteria for medium traffic loads. Nonetheless, SMA mixtures containing more than 40 wt% RCA replacement are preferable for use on low-traffic pavements such as parking lots. However, greater attention should be given to the qualities of concrete mixtures due to raising the RCA. Thus, the mechanical strength and fatigue properties of CSRCAs depends on how the pavement responded to the load, the mechanical and fatigue properties of the CSRCA, and the theory of fatigue accumulation that CSRCA material's fatigue qualities leads to developing a fatigue equation specific to the material based on pavement load response. Addtionally, when the curing age is less than 60 days, a drop in UCS and resilient modulus of CSRCA is realized when the curing age is higher than 90 days. In addition, the splitting of CSRCAs and flexural-tensile strengths rise slightly as the RCA dose increases as well as the pavement's material qualities and service life.

RCA may be used as a partial replacement for natural aggregates, and the resultant RAC has comparable mechanical qualities to NAC. The presence of connected mortar from the source concrete, which adheres to the surface of recycled particles, significantly reduces the compressive strength of RAC. However, RAC has comparable or superior performance than NAC's compressive and flexural strengths. Recycled concrete's rougher surface roughness is attributed with enhancing the adhesion and interlocking of recycled concrete and mortar (as opposed to NCA). It has higher compressive strength than the NCA. On the other hand, the cause of RCA's negligible influence on the flexural strength of concrete has not been completely examined.

In recent years, RCA concrete's mechanical characteristics can be improved by using ternary concrete mixed with fly ash, silica fume, or fiber, as shown in Figures 2.1–2.3. Such ternary concrete may incorporate the advantages of both components, depending on the country, accelerate the pozzolanic ash response time, and improve the concrete's working performance and durability.

The existence of two interfacial transition regions, which are usually found between the aggregate and the new cement mortar, but in the case of RAC, the old mortar that, is still attached, and the new cement mortar contributes to the degradation of mechanical qualities. The highest coarse NA to RCA replacement ratios in the concrete mixture that can be reached without sacrificing compressive strength

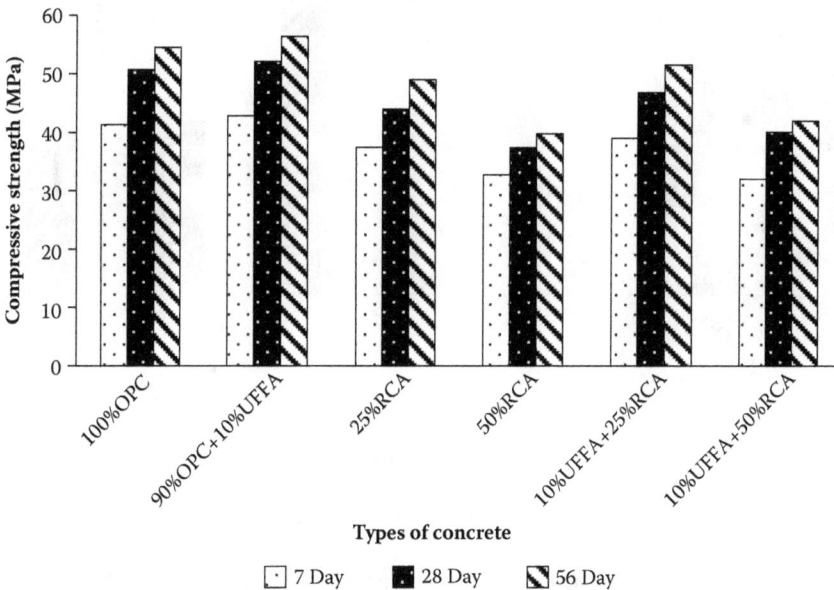

FIGURE 2.1 Compressive strength of concretes containing RCA (Shaikh, 2016).

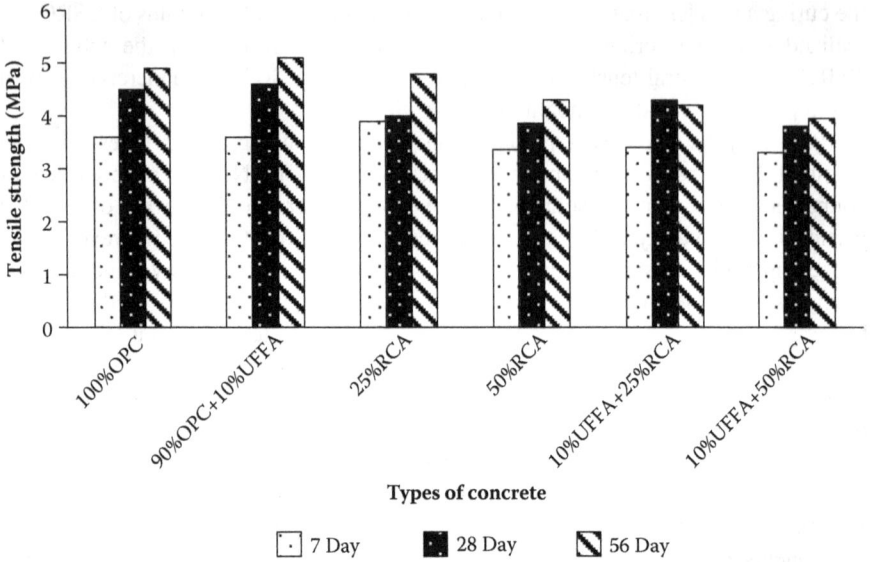

FIGURE 2.2   Splitting tensile strengths of concretes containing RCA (Shaikh, 2016).

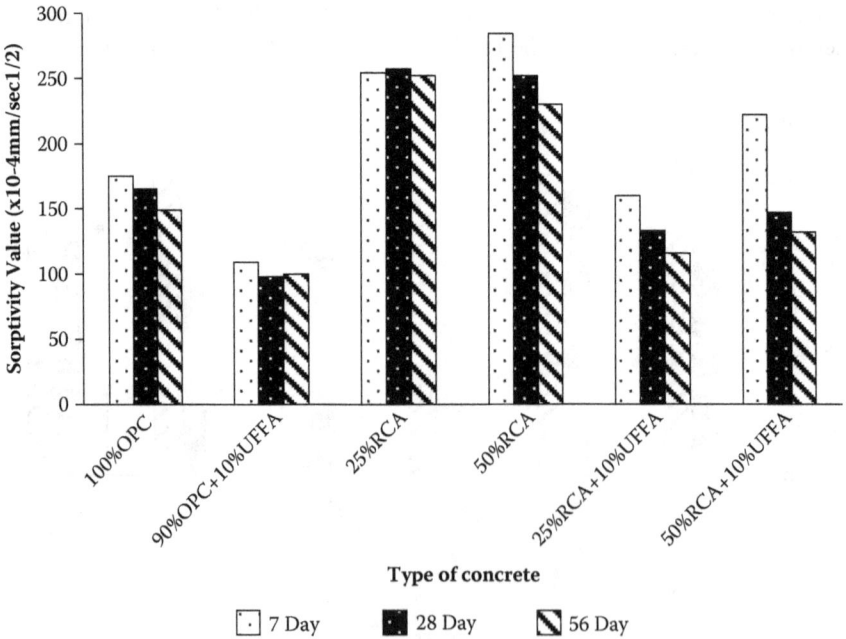

FIGURE 2.3   Chloride ion permeability of concretes containing RCA (Shaikh, 2016).

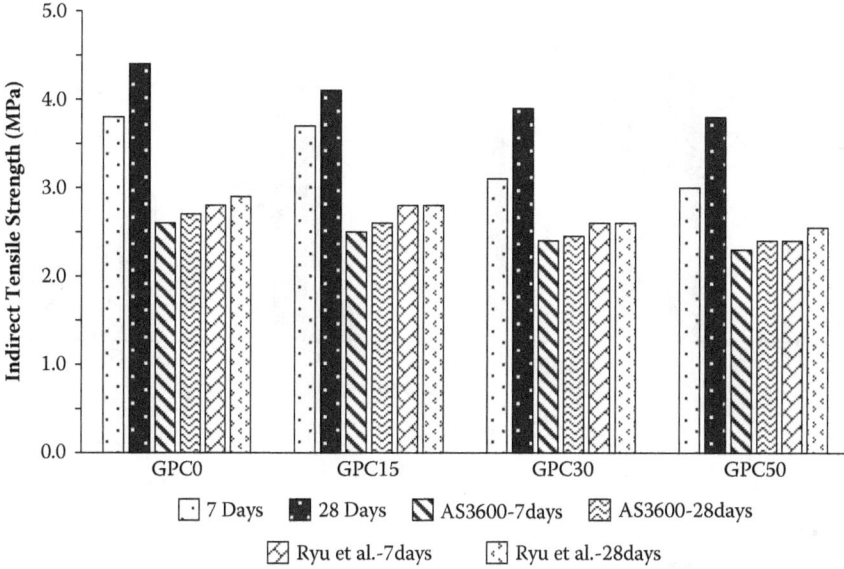

FIGURE 2.4 Indirect tensile strength of geopolymer concrete containing RCA and NCA (Shaikh, 2017).

are examined. The maximum replacement rate of NA by RCA was determined to be 30 or 50 wt%, which is consistent with the European Standard definition of the maximum replacement rate for structural segments (Gebremariam et al., 2021). Using only additives, the maximum allowable percentage of replacement is 50 wt%; however, using additives (cementitious materials, etc.) improves the qualities of concrete products. Another crucial feature of concrete investigation is its long-term durability in various procedures and other things that can make concrete more durable and have long-term ability.

It is possible to claim that ultrafine fly ash (UFFA) is an effective technique to increase concrete's mechanical and durability qualities, hence its performance. In addition, utilizing 10% UFFA as a partial cement substitute in concrete created from C&DW can produce high-performance concrete (Shaikh, 2016).

For geopolymer concrete, recycled coarse particles can significantly impact geopolymer concrete's mechanical and durability qualities. RCA affects significantly fly ash geopolymer concrete's mechanical and long-term durability. The compressive and indirect tensile strengths (Figure 2.4) and the elastic modulus are reduced when RCA partially replaces NCA at all ages. Unexpectedly, the elastic modulus of geopolymer concrete with RCA is slightly less after 28 days than after 7 days (Figure 2.5). In addition, indirect tensile strength and elastic modulus of geopolymer concrete containing RCA exhibit excellent correlations with compressive strength throughout both aging stages.

Because of the presence of RCA in the mix, the durability characteristics of geopolymer concrete, including sorptivity and immersion absorption, chloride ion penetrability, and the volume of permeable voids (VPV) are drastically impacted.

(a)

(b)

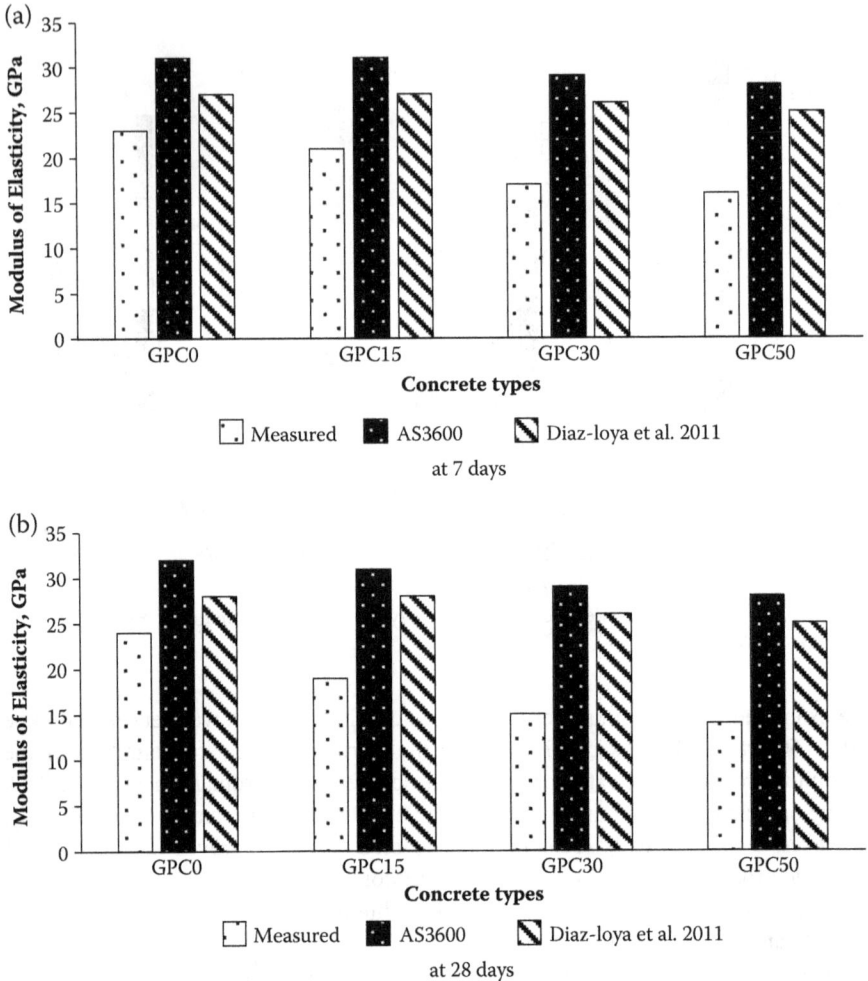

FIGURE 2.5 a) Elastic modulus of geopolymer concrete containing RCA and NCA at 7 and 28 days with exiting models (Shaikh, 2017). b) Elastic modulus of geopolymer concrete containing RCA and NCA at 7 and 28 days with exiting models (Shaikh, 2017).

The fact that geopolymer concrete containing RCA is more durable than geopolymer concrete without RCA due to the microstructure of geopolymer paste is more refined than OPC paste. There is a strong association between compressive strength, VPV volume, and water absorption.

There are some inconsistencies in the mechanical properties of recycled concrete at high temperatures (600 and 800°C) on compressive strength and modulus of elasticity of concrete that contains 50 and 100 wt% recycled material. To evaluate whether Eurocode 2 (Mostert et al., 2021) is relevant to RAC, the recycled aggregate concretes lost more compressive strength than their control counterparts that were made with natural coarse aggregates (NCAs). In addition, recycled aggregate concretes are lower residual compressive strengths than conventional concrete at high and low temperatures. However, when recycled aggregate concretes and control concrete are subjected to elevated temperatures, the remaining compressive strength is larger than the compressive strength recorded at elevated temperatures. Furthermore, when the compressive strength and elastic modulus of recycled aggregate concrete are studied at high temperatures, it has been found that there is a strong relationship between the two properties (Shaikh, 2017).

A number of studies on the mechanical behavior of rubberized recycled aggregate concrete (RRAC) have demonstrated a reduction in compressive and flexural strengths. The inadequate adhesion of the rubber to the cement paste decreased the compressive strength of RRAC, with rubber particles reckoned the principal contributing cause. The decreased flexural strength results from the rubber particles' weaker rigidity, which results in the lower flexural strength. When an external load is applied to the concrete, the load carried by the rubber particles is substantially lower than that carried by the surrounding aggregates due to the stress concentrations resulting in the flexural strength is reduced. The deformation parameters of RRAC were superior to those of RAC under static and flexural fatigue loads with a rubber replacement rate of up to 20%. They were demonstrating superior deformation properties. As a result of their ability to delay the formation of new fractures and operate as dispersed micro-spring units, rubber particles significantly improved the flexural deformation behavior of RRAC (Mostert et al., 2021). By replacing aggregates in concrete with crumb rubber (CR) and RCA led to unusual mechanical behavior, as measured by the compressive and flexural strengths of the resulting concrete mixture. The greater normalized strain capacity under compression and load-deflection behavior under flexure of RRAC, in conjunction with its lower stiffness, improves the likelihood that it will be utilized in future structural applications.

For the lowest mechanical resistance, the abrasion resistance coefficient and aggregate crushing values indicate that recycled concrete aggregate of a building is the best material. The properties of waste concrete, which may be similar to those used in road construction, may be one possible explanation for recycled concrete aggregate from a pavement superior mechanical performance to the recycled concrete aggregate of a building (obtained by crushing same strength class concrete and same thickness), but for building demolition waste, which may not be similar (depending on the structural elements), can result in significantly different mechanical properties.

In contrast, the mechanical qualities of RAC are typically lower than those of conventional concrete (CC). Because RCA covered in old mortar may contain flaws that degrade the joint contact within RAC, this is the case, making it less durable. It was discovered in early studies by Nixon (1978) that the strength loss caused by RAC might be up to 20% greater than the strength loss caused by conventional chemistry. RACs mechanical qualities are adversely affected by RFA, which is primarily owing to the significant water absorption of the RFA composite. When compared to RAC that was produced without RFA, RAC that was manufactured with 25 and 100 wt% RFA had compressive strength losses of 15% and 30%, respectively. Khatib (2017) corroborated these findings. Another significant drawback of RFA is that it has a higher degree of drying shrinkage and a lower degree of durability when compared to conventional concrete made with small natural particles. As a consequence of these restrictions, the capacity to practically apply RAC to structures that exist in the actual world is severely restricted.

The recycled masonry aggregate, also known as RMA, can have an impact on the mechanical properties of recycled asphalt concrete (RAC). Several properties of RMA set it apart from NA, indicate that coarse RMA can absorb up to 19% of the water it is exposed to, while fine RMA can absorb 12% to 15% of the water it is exposed to, which is above 10 times as much as natural sand. Natural gravel and sand typically have a dry density of between 1,800 and 2,700 kg/m^3, while coarse RMA has a density between 2,000 and 2,500 kg/m^3, which is less than the density of coarse RMA (Shaikh, 2016).

These aspects of aggregates, such as their physical, mechanical, and thermal properties and endurance, are influenced by them. RMA affects the workability of freshly cast concrete due to its enhanced water absorption. Workability and effectiveness w/c affect the compressive strength of concrete. The sum of the cement and water interacting determines the w/c level. RCA concrete's water absorption capacity is determined by the amount of recycled aggregate included in the mix. As NA is substituted more frequently, the compressive strength of RAC decreases. RAC prepared by 15 wt% coarse RMA has no noticeable decrease in compressive strength. Concrete compressive strength can be lowered by up to 35% if coarse NA is completely replaced with RMA, according to the study. Using fine RMA to replace natural sand in RAC partially has not been shown to impact compressive strength. It is possible that the high silica and alumina content in broken bricks could lead to pozzolanic reactions, which would cause them to decay (Shaikh, 2016).

The mechanical and durability properties of RCA must be evaluated before they can be used effectively in concrete construction projects. RAC has recently been studied for its mechanical and durability features, as stated that "To get a good product out of recycled aggregates, you must first get a good product from the concrete from which they were taken and vice versa." The quality and quantity of mortar used on RCA determine their physical and mechanical qualities. Concrete strength is influenced by various factors, including w/c, amount of adhering mortar used, and particle size and crushing technique employed. Because of an impact crusher, RCA can be produced free of cement mortar adhesion. Depending on the cement mortar adhered to RCA, it can change its water absorption and density. RCA's water absorption increases in lockstep with the strength of the concrete from which they are made. As stated by the manufacturer, recycled aggregates should not be used in recycled aggregate concrete when their water absorption exceeds 7% for coarse and 13% for fine by 5% more free water to get the same slump as conventional concrete created from natural particles. In order to get the same slump with RAC that has been prepared with both coarse and fine recycled aggregates, approximately 15% more free water is need with conventional concrete manufactured with natural aggregates. It is possible for the compressive strength of RAC produced with 100 wt% RCA to be lower than that of regular concrete when the w/c is the identical. It is possible that a weaker relationship between the recycled aggregate and the old cement paste that's attached to it is to blame for this circumstance. According to the research that has been conducted, the compressive strength of RAC concrete that contains 100 wt% RCA is approximately 60%, 75%, and 95% of the compressive strength of concrete made with natural aggregates, respectively.

Although not unusual, the modulus of elasticity of concrete manufactured from 100% recycled concrete aggregate is 20%–25% lower than that of ordinary concrete. This decrease may reach 30–50% when recycled fine and coarse RAC is used in the RAC. When 100 wt% RCA was utilized, the split tensile strength of reinforced concrete was reduced by 26%; when RCA was manufactured from 30 MPa strength concrete or an unknown source, the loss ranged from 25% to 30%. There is no discernible difference between the flexural strength of RAC and that of concrete containing natural particles when natural aggregates are used (Mikhailenko et al., 2020).

The development of concrete mixing methods with RCA and the investigations of their physical and mechanical qualities of durability have been the subject of numerous researches in recent years. Ravindrarajah and Tam's research (Mikhailenko et al., 2020) shows that using recycled gravel and natural aggregate in concrete results in 5% more water consumption than making concrete with natural material. Natural aggregates have a lower density and higher water absorption than RCA, mainly attributable to an ancient hanging mortar used in its place. Using recycled aggregates in their dry state may necessitate extra mixing water and early workability.

Several aspects, including cement quantity, w/c, recycled aggregate quality, replacement ratio, and amount of old paste adhered, influence mechanical qualities in concrete that contains RCA. According to a recent study by Wardeh and colleagues (Mikhailenko et al., 2020) the cement content must be altered while the w/c remains constant, maintaining the same compressive strength class. Depending on the mixing procedure employed, the composite's strength and elastic modulus will be altered to a greater or lesser degree. According to the available research in the literature, the compressive strength is unaffected by adding up to 30 wt% RCA to a mix with equal mix parameters. Exteberria et al. (2020) discovered that when the w/c is detained constant, concrete containing 25 wt% coarse RCA performs similar to the conventional concrete. The w/c was lowered from 4% to 10% in designs containing 50 and 100 wt% gravel RCA, respectively.

There is a substantial difference between RCA and NC in terms of the amount of old mortar linked to the RCA. This is usually recognized as one of the primary reasons for the inferior qualities of RCA compared to NC. Old mortar re-interfaces with new mortars, resulting in a more significant number of interfaces in RCA when the mix design is the same as in NC. Because the interface is the weaker zone in concrete, it allows the RCA to crack, reducing its overall durability efficiently. The first damage to the RCA, such as internal damage accumulated during the crushing process, impacts the properties of RCA; thus, it is crucial to understand this. However, the unhydrated cement still presents in the prior mortar can continue to hydrate and benefit the internal curing process. Hydration products, in particular, can improve the aggregate's and old mortar's overall RCA qualities by increasing not just the bonding strength between the aggregate and old mortar but also by covering and filling microcracks. In the past, researchers have investigated the impact of old mortar on reinforced concrete (RC). In addition, mixed recycled aggregate (MRA) and modeled recycled concrete (MRC) were developed to simplify the genuine RC in order to get ready for the experiment that was carried out. Xiao et al. (2005), for instance, conducted an empirical and analytical investigation to evaluate the stress-strain characteristics of MRC when it was subjected to uniaxial tension. These factors included the density of the adhering old mortar's coverage, the first MRA damage that occurred during the original concrete's service operation, and the crushing procedure used in recycling unwanted concrete. More importantly, prior research has focused chiefly on MRC's mechanical properties, with only a few studies reporting on MRC's endurance in a complicated or hostile environmental condition.

In 2007, Rahal (2007) found that the cubical and cylindrical compressive strengths of RAC and its indirect shear strength were 10% lower than those of NCA. In light of these circumstances, the moduli of elasticity have decreased by no more than 3%. Both RAC and NCA have similar trends in compressive and shear strength development on top of the strain at peak stress.

The strength and toughness of RAC are about the same as those of NAC for ensuring that RAC is of the right quality for concrete and any weaknesses can be easily fixed by making simple changes to the way the mixture is produced or the way the structure is built. As a result, more RAC should be used in concrete, reducing the social and environmental problems caused by the substantial part of C&DW.

Studies have also revealed that the inclusion of RCA has no negative impact on flexural strength; nevertheless, a complete replacement with RCA resulted in a 10% drop in flexural strength, according to the findings. According to Wardeh and colleagues (Reiner et al., 2010), when the Eurocode 2 was used to design the RAC mix, the RAC made with 100% NCA had a 20% drop in flexural and tensile strengths. Based on the generated model came up with results that were more like those that came up when they used the equations in Eurocode 2 than when they used the equations in Eurocode 2. Other than using the equivalent mortar volume (EMV) approach, they claimed that no mix design strategy has been able to mitigate RA negative impact on RAC and elastic modulus (EM).

For steel fiber-reinforced recycled coarse aggregate concrete (SFRCAC), where the desired compressive strength and flexural strength and slump are taken into account, steel fibers can be used to make recycled coarse aggregate concrete (RCAC) more flexible as well as the steel fiber's typical coefficient by the amount of water it contains. The absolute volume method should be used to figure out the material component of the SFRCAC. Because RCA has a lower apparent density than the NCA and the mass of RCAC is highly dependent on the apparent density and replacement ratio, it is recommended that this method be used.

The early-age compressive strength of concrete is influenced by the aggregates used in its construction. The aggregate size depends on what the concrete will be used for and how the materials are mixed: Less cement and water are needed for aggregates with larger diameters than smaller ones. Their diameter, less than 4.0 mm, is usually used to classify them as coarse or fine aggregates. The quality and kind of aggregates used in concrete impact its durability and workability; extra care is taken to ensure that no chemicals that could cause damage are used.

Recycled concrete, including coarse RCA, has gotten minimal attention for post-fire mechanical qualities (heated, cooled, and then loaded at air temperature), the residual compression strength of concrete made with natural coarse granite aggregates and concrete with 75% by volume fine RCA. One and four hours of immersion in warm water at 500°C followed by gradual cooling in an oven at room temperature completed the testing. Residual strength decreased by 16% and 10%, respectively, in the conventional and RAC after one hour of heating. About 26% of the remaining strength of both concretes had been lost after four hours of heating. An alternative explanation suggested by some authors suggested that more excellent RCA concrete performance was due to the contact between old cement on a RCA and a zone that transitioned from old to new cement, ITZ. The same thermal expansion coefficients make micro and macro cracks in mortarless prone to ascend, it was thought at the time. Mortar is used more in RCA-type concrete than other concretes.

In concrete, RCA are used frequently, connected with a reduction in the physical and mechanical qualities of the finished product. As a result of their reduced density, more significant porosity, and excellent water absorption, RCAs are more water-resistant.

Generally, when fly ash (FA) and RCA were used to replace more OPC and NA, the mechanical characteristics of the resulting concrete decreased. RCA concrete with FA and RCA concrete without FA had less of a mechanical difference over time. As the pozzolanic interaction between FA's SiO_2 and recycled concrete aggregates (RCAs) and $Ca(OH)_2$ progresses, the concrete becomes more durable as a result. In contrast to OPC concrete, which takes a long time to build strength, RCA concrete is more vital since it has a shorter curing period.

As stated before, the capacity of the superplasticizer (SP) to disperse FA particles in the mixture further increases the rate. Therefore, it is simpler for the FA particles to reach the RCAs and $Ca(OH)_2$. The mechanical performance of MPR mixes including SP was the best, followed by mixes with low quantities of FA and coarse RCA. Mixed concrete with a high proportion of FA or RCA aggregates performs the worst. SP improves the mechanical qualities of concrete, and then FA and RCA, when used in excess of a certain inclusion ratio, may be detrimental to the mechanical properties of concrete. In addition, substantial levels of coarse RCA in concrete are less damaging than substantial amounts of fine RCA, even if it is less supporting than using FA in a concrete construction. Although mechanical behavior is vital, rating concrete mixtures only on their strength may not be "business as usual." Cost is an essential factor in determining the concrete strength class, and it must be weighed against the intended application since an economic life cycle evaluation was completed on all concrete combinations.

To better understand the mechanical and economic features of mixtures, including high concentrations of FA and RCA, mechanical proporties such as compressive strength, splitting tensile strength, modulus of elasticity, and modulus of rupture are all negatively affected by the addition of FA and RCA to the concrete mix. Integrating FA results in a considerable reduction in compressive and splitting

tensile strengths, but incorporating RCA significantly increases the modulus of
elasticity. When comparing blends containing FA and RCA to blends containing
only OPC, the high growth of all the listed characteristics continued after a certain
age. The addition of SP further increased the strength development of mixtures
containing FA and RCA. It is therefore strongly recommended to combine FA and
RCA in the same concrete mix, including SP.

The cost of concrete mixtures decreased marginally due to the inclusion of RCA
in the mix. Although SP was used sparingly, the overall cost was significantly
increased. A 60% FA addition to NA or 30% FA addition to RCA concrete can
reduce the high cost of SP-containing cement. However, if ordinary Portland
cement is used instead of RCA, the cost of FA concrete is lower than it would be
with conventional concrete. Accordingly, these conclusions should not be applied
universally because various approaches or predicted situations in life cycle inventory modeling, as well as nations or locations other than Portugal, may produce
differing outcomes.

In higher efficacy, some combinations are more cost-effective than others,
although it is not always so. It is crucial to consider the mechanical qualities of
modern concrete mixes while picking the optimal concrete mix. The cost of concrete mixtures increases significantly when SP is used. Only when 100% fine RCA
is used alone or in combination with 100% coarse RCA can any mixtures incorporating SP (regardless of the binder and aggregate type) be considered suitable
solutions (located in the cost-efficient zone, i.e., less expensive than the reference
mix). As far as mechanical and financial considerations are concerned, a small
amount of FA is strongly advised. For this reason, even if there is a significant loss
in mechanical characteristics due to a high amount of FA, SP-based mixtures are
still considered suitable (and hence decrease within the cost-effective range).
Coarse aggregate content in RCA concrete mixtures leads some to conclude that
they are more cost-effective. The cost and mechanical qualities of fine RCA concrete are determined by the ratio of fine RCA concrete to coarse RCA concrete
utilized in the project. Large replacement of fine RCA without FA in concrete is not
guided, primarily because mechanical performance is significantly reduced. A mix
that incorporates all three signals together is more effective mechanically than one
that does separately.

Research on the impact of FA and RCA taken together on the tensile strength
and abrasion resistance of concrete has been looked into, but the results have shown
that the stronger the concrete is when these two chemicals are added, the less
durable it is when they are not, but FA makes the concrete mix design more durable
over time.

The geopolymer concrete incorporating recycled coarse aggregate possesses
greater compressive strength and elastic modulus than its OPC counterpart containing RCA. Moreover, the aforementioned mechanical properties decrease when
the mixture's proportion of RCA increases. In addition, they discovered that the ITZ
of the geopolymer concrete was superior to that of the OPC concrete. Posi et al.
(2013) investigated the mechanical properties of geopolymer concrete with recycled
lightweight particles. They found that the compressive strength of geopolymer
concrete falls as the percentage of recycled lightweight aggregate increases,

comparable to that of the RCA of normal weight. However, their inquiry about the modulus of elasticity results is contradictory. The compressive and indirect tensile strengths of geopolymer pervious concrete, which includes crushed concrete, have been determined, and crushed bricks as coarse aggregates are weaker than natural coarse aggregates. In addition, the compressive and indirect tensile strengths of all three types of geopolymer concrete increase when the sodium hydroxide solution concentration rises. Also published in 2016 was a study of geopolymer concrete containing coarse particles of crushed concrete, which revealed a loss in compressive strength due to the addition of RCA (Nukalong et al., 2016). In addition, a study on geopolymer concrete with recycled concrete fragments as coarse aggregates was published; using recycled concrete fragments as coarse aggregate resulted in reduced compressive strength. The effects of varying RCA contents produced from a mixture of construction and demolition debris on compressive strength, stiffness, and stiffness, as well as geopolymer concrete's indirect tensile strength and elastic modulus, and a comparison of these results with those obtained from 100% NCA. In addition to water absorption, the volume of permeable spaces, water sorptivity, and chloride ion penetration, geopolymer concrete, including RCA, is evaluated and compared to geopolymer concrete containing only NCA in order to determine the effects of different RCA on durability. RCA is a form of polymer utilized in geopolymer concrete production. Currently, just a few studies demonstrate that geopolymer concrete containing RCA has exceptional durability. There are statistics on the compressive strength and elastic modulus of geopolymer concrete, including RCA, but they are inconsistent since they are reliant on various source materials, alkali activators, and RCA source, as shown in Figure 2.6, and other variables.

In addition to aggregate type and thermal conductivity, aggregate density and porosity influence the thermal conductivity of concrete, as shown in Table 2.3. The w/c and the amount of cement in the mixture significantly affect the thermal conductivity of concrete. According to an earlier study, aggregates with a lower thermal conductivity produce concrete with lower thermal conductivity. This factor could have a favorable effect on masonry aggregate (RMA)-reinforced concrete due to the reduced thermal conductivity of the concrete due to the weak thermal conductivity of masonry aggregate (RMA), which ranges between 0.60 and 0.78 W/(mK). NAs' thermal conductivity is determined by their mineralogical properties, chemical composition, and degree of crystallinity. Crystalline NA transmits heat more effectively than its amorphous and vitreous counterparts of the same composition (Table 2.2).

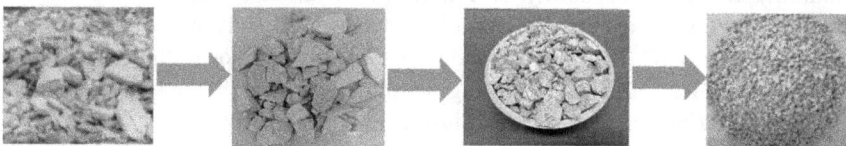

FIGURE 2.6   Resource and manufacturing of RCA.

TABLE 2.2

Thermal Properties of Various Materials in the Dry State (Pavlu et al., 2019)

Type of aggregate Aggregate	Thermal Conductivity–λ (W/(m·K))	Type of Concrete	Thermal Conductivity–λ (W/(m·K))	Compressive Strength (MPa)	Density (kg/m³)
NA-Basalt	4.03	Basalt concrete limestone	2.26	N/A	N/A
NA-Limestone	3.15	Limestone concrete	2.03	N/A	N/A
NA-Silestone	3.52	Siltstone concrete	2.21	N/A	N/A
NA-Quartzite	8.58	Quartzite concrete	2.77	N/A	N/A
RA-Concrete	2.22	RCA 50%	0.90	34.7	2050
		RCA 70%	0.80	39.1	2040
RA-Masonry	0.8	RMAC-various replacement	0.60–0.78	4.0	1480
RA-EPS	0.04	Concrete EPS content 55%	0.56	11.85	1140
		Concrete EPS content 65%	0.50	7.74	1070

RCAC: Recycled concrete aggregate concrete; RMAC: Recycled masonry aggregate concrete

TABLE 2.3

Effect of Recycled Aggregate on Concrete Properties

Compressive Strength	Tensile Strength	Modulus of Elasticity	Drying Shrinkage	Creep	Permeability	Freeze-Thaw Resistance	Depth of Carbonation
⇔	⇔	⇓	⇑[1]	⇑[2]	⇔	⇔	⇔

Notes
[1] In comparison to the qualities of a reference concrete containing natural aggregate with the same w/c
[2] Effect more noticeable as the proportion of fine recycled aggregate increases

2.5 DURABILITY OF CONCRETE CONTAINING RCA

Durability refers to the ability of concrete to withstand various environmental conditions for its service life. Because of its increased porosity and water absorption than NAC and RAC, RMAC's durability is typically inferior to that of these two materials. The feasibility of increasing the durability of a product has been demonstrated in prior research at first, just a few tested and proven methods for increasing the freeze-thaw resistance of RAC. However, the porous RAC can be improved by immersion in water before being mixed with the concrete. This allows the water in the RA to slowly be released for more cement hydration, which makes the concrete mix more durable and robust. This process is called compensating for the water absorption of the RCA, which means that more water is used because of the RCAs ability to absorb water. Another reason to add fly ash or metakaolin to the concrete mix is that mineral admixtures can react with calcium hydroxide $(Ca(OH)_2)$ to make an additional C–S–H gel that makes the concrete more dense and robust. Second, it was found that SP assisted the carbonation depth of RAC.

Concrete containing RCA is a composite material of various constituents with various properties, it has more complicated thermal properties than many other materials. However, moisture and porosity levels in the concrete mix impact its qualities. Concrete's mechanical and physical qualities are affected by prolonged exposure to high temperatures. When concrete is exposed to high temperatures (for example, as a result of an accidental fire), it deteriorates rapidly and goes through a series of changes and reactions; as a result, the cement gel structure gradually breaks down, decreasing durability, increasing structural cracking, drying shrinkage, and changing aggregate color (Yang & Lim, 2018). Some of the most significant consequences of increased temperature on concreteare increased porosity, due to dehydration of the cement paste due to moisture content fluctuations; expansion due to heat; changes in pore pressure; loss of strength; incompatibility; thermal creep and spalling as a result of high pore pressure; and thermal cracking as a result of incompatibility. Regardless of whether it is a gas or a liquid, the distribution and movement of water have been demonstrated to be critical in causing localized damage to concrete structures. Noumowe and Debicki (2002) discovered that the endothermal nature of vaporization results in huge temperature gradients and high vapor pressure locally during heating, which can cause tensile strains to exceed the concrete's strength. Since the early 20th century, when the first studies on heat-exposed concrete were done, reinforced concrete structures have exhibited excellent fire resistance due to cement-based materials' physical properties, including inflammability and limited heat conductivity. As a result of this latter ability, even in a lengthy fire, the outwardly heat-damaged layers can prevent the inner core from reaching an unacceptably high temperature.

Thermomechanical, mechanical, and deformational characteristics of RCA concrete are used to build a structural part impact how the member behaves when exposed to fire in a controlled environment. When subjected to the extreme temperatures associated with building fires, concrete's thermophysical, mechanical, and deformation properties alter dramatically, as with other construction materials. These qualities alter with temperature and are influenced by the concrete mix design's composition and attributes. High strength concrete (HSC) fluctuates differently as a function of temperature than those of normal strength concrete (NSC). This variation is more evident in mechanical properties because they are influenced by factors such as tensile strength, moisture content, density, heating rate, silica fume concentration, and porosity (Kodur, 2014). According to the type of fire, the loading system, and the structure, different kinds of RCA concrete failure exist when exposed to fire. More importantly, the failure could be caused by a variety of factors, a weakening in bending or tensile strength, a weakening in shear or torsional strength, and other weaknesses, and a decrease in compressive strength, among others.

Adding FA to natural aggregate concretes and exposing them to the sea for an extended period boosted the concrete mix's durability dramatically (Cheewaket et al.,

2014; Githachuri et al., 2012). At 30 to 40 wt% of binder and a w/c ratio as low as 0.40, the concretes were less likely to be degraded and that reinforced concrete structures exposed to maritime conditions lasted longer when fly ash was employed as a binder. FA replacement and w/b significantly affects compressive strength and corrosion resistance in concretes made using RCA that had been submerged in salt water for seven years.

To examine how temperature affects the mechanical properties of conventional and high-strength concrete containing RCA in practical. Concrete's mechanical qualities, precisely its compressive and tensile strengths, are two aspects that should be considered while designing for fire resistance and its modulus of elasticity and stress-strain response when compressed based on the Eurocode (EN 1991-1-2–2004) (Hou et al., 2020) and the ASCE (American Society of Civil Engineers, 1992) (Hou et al., 2020). As stated, the standards of the presence of attached mortar were the most significant factor contributing to the fact that the RCA is good at its work because of what happened after it was heated. It took away about 70% to 80% of the adhered mortar on the RCA. In addition, RCA has a high bulk density and low specific gravity following treatment. It increased by approximately 9% and 6%, respectively. Following heating and rubbing, RCA's ability to absorb water was reduced by roughly 66%. RCA was exposed due to the removal of weakly connected mortar. Compared to traditional concrete, the mechanical parameters (modulus of elasticity, compression strength, split tensile strength) utilizing FA and RCA to produce high-performance concrete/silica fume significantly improved. When FA or silica fume (sf) were added to treated recycled concrete aggregate (TRCA)-produced concrete, the barrier to water absorption increased, along with the concrete's permeability to chloride ions, and the water absorption rate.

Several environmental processes have the potential to take place, including a variety of different things, such as salt solution, salt-solution freeze-thaw cycles, mechanical stress, and the combined impacts of all of these factors. In addition, there was a modeled recycled concrete (MRC) that had varying degrees of damage to the RCA, as well as variations in the thickness of the old mortar and the amount of coverage provided by the old mortar. This was done to understand better how the long-term durability of RC is affected by an adhered mortar attached to RCA. The durability of RC reduced more rapidly as the number of times it was subjected to a combined mechanical load and salt-solution freeze-thaw cycles increased. There were first cracks when the old mortar was mixed with new mortar, which was the weakest part of the RCA at that time. The RAC endurance decreased as the thicknesses of the previous mortar increased, the covering of the old mortar and the first damage went up, and so did the durability of RC. When the old mortar was damaged in the first place, it was also more critical than how thick the old mortar was in terms of causing damage to the old mortar. However, adding 1.5% nano-SiO_2 to the concrete and 10% fly ash to the cement might make the RC more resistant to mechanical stress and salt-solution freeze-thaw cycles, which can harm the concrete, could make it more durable. The concrete's compressive strength loss is reduced from 19.8% to 9% when the treatment methods are applied; 10.3 and 7.5%, respectively. Cracks and enormous recycled concrete fragments can be seen at both the old and new ITZ after mechanical stress and salt-solution freeze-thaw cycles, resulting in the rough and untidy micromorphology seen at the old ITZ (Tamanna et al., 2020).

In order to improve the durability and mechanical qualities of RCA, SP can accelerate the crystal growth, making the concrete's structure denser. On the other hand, this treatment becomes less effective over time. Another way to produce RAC more resistant to aggressive environment is to use mineral admixtures, such as large amounts of FA, which can fill the pores and improve the microstructure of RAC. When minerals are used as a cement replacement, for example, the pH of the concrete decreases, and its resistance to carbonation increased. In addition, it takes more binding to make the coarse RCA hold together, but the RCA sand and filler fractions are not significantly higher than the control sand and filler. This means that when coarse RCAs are replaced by mass, the energy needed for compaction increases significantly. This is true even when the coarse aggregates are replaced by RCA sand and RCA filler, but not when they are replaced by mass. Even though replacing RCA with mass can lead to an increase in the maximum indirect tensile strength, this can also make the new mixture less ductile and resistant to fractures because of the RCA used in the replacement. However, this attribute has no significant difference if the RCA filler is changed. Higher levels of RCA replacement result in a reduction in the water sensitivity of the concrete mixtures, while lower levels of replacement result in little or no reduction in this property. Addtionally, rutting resistance increases as sand and coarse RCA are substituted for conventional sand and aggregates, with rutting with 32% RCA being at half that of the control.

For the case of sorptivity, submerged absorption, chloride ion penetration, and volume of permeable void (VPV) are negatively affected by the incorporation of RCA in geopolymer concrete; however, these values are superior to those of OPC concrete, containing the exact amount and kind of RCA. The geopolymer concrete with RCA has more extraordinary durability qualities than geopolymer concrete that does not contain RCA and shows that the microstructure of geopolymer paste is more refined than that of OPC paste and that there is a strong association among compressive strength and VPV and water absorption. Using the current sustainable concrete, which has some OPC replaced with other materials and some RCA in place of NCA, it can be made even more sustainable by using 50% less NCA (Gunasekara et al., 2020).

Zhu et al. (2018) reported that the durability of RCA with varying replacement percentage by putting it through freeze-thaw cycles and exposing it to chlorine indicating that frost damage has a significant role in chloride penetration. The mechanical properties steadily deteriorate as the number of cycles used and the rate at which they have been replaced increases. On the other hand, in an atmosphere that features freezing and thawing cycles and chloride assault, recycled aggregate-included concrete has a duration of 50 years. Furthermore, incorporating RCA in the building sector can help safeguard natural aggregate resources and the disposal costs of RCA and generate considerable economic benefits.

The concrete's mechanical characteristics and long-term durability are seen when using rubber powder as a cement alternative. It is also one of the critical goals of the research project to examine the influence on concrete quality when the cement and coarse aggregate are substituted with rubber powder by using genetic programming to derive equations for anticipating concrete with WRP and RCA's compressive/tensile/flexural strengths reduction, durability parameters, and the concrete's durability factors.

Because the widespread uses of RCA aggregates are of low quality, there have been little opposition to their employment in the past. On the other hand, in many studies, repurposed aggregates have shown their adaptability in all concrete-related endeavors (Gunasekara et al., 2020). There are a few issues as long as RCA is used in layered mortar on the surface of the RCA. Consequently, low specific gravity, increased water absorption, accelerated shrinkage, and lower bulk dry densities are attained (Gunasekara et al., 2020). Natural aggregates have a potential for water absorption that is two to three times that of RCA. These problems are primarily due to the manufacture of RCA from RAC.

The incorporation of RCA into high-strength concrete is limited due to the small number of studies conducted. When coarse RCA replacements of 5, 10, and 12.5 wt% were used in high-strength concrete at a specific w/c, the strength was determined to be comparable to that of natural aggregates with the same w/c. Several other studies have found similar results to ours. This research show that RCA can be used in high-strength concrete, but they mention durability or long-term performance, which are critical in practice.

With its NCA and cement mortar, RCA appears to have a lower density than other materials; however, this is actually due to its high water absorption and reduced apparent density. A greater crushing index and poorer cohesiveness in the recycled concrete aggregate result from the vast number of micro-cracks created during the waste concrete crushing process. In the face of diminishing supplies of NCA, using RCA offers a long-term alternative that also reduces the need for landfill trash disposal.

The permeability tests performed with a dropped head and permeation tests with a constant head are used to figure out the permeation of OPC concrete prepared by sustainable aggregates; it is possible to maintain the constant head permeability testing while considering the time required for preparation. Porosity and permeability have a direct connection. The dropping head and constant head permeable tests are 10% of the estimated value obtained from the dropping head and constant head permeable tests, respectively. This indicates that it cannot calculate the true hydraulic potential of the materials being used. On the other hand, the continual head permeation test demonstrates how well the head can be kept open and can estimate these parameters in real time.

To find out how coarse crushed concrete aggregate (CCA) from known structural elements affects the fast chloride migration coefficients, as well as how coarse CCA accelerates the time before corrosion begins and causes cracks in structural concrete. There was some potential for performance in the water absorption, chemical analysis, and petrographic investigation, but the course CCA from the source was the least effective aggregate. Preliminary testing of coarse CCA sources prior to use in structural concrete are recommended for identifying potential concerns.

Numerous studies have demonstrated that RAC mixed with freshwater has superior fresh, mechanical, and long-term performance characteristics to ordinary, pre-cast concrete. Using RCA instead of NCA reduces the CS by 10%–20%, according to some research (Gebremariam et al., 2021). Although the w/c and cement volume remained constant, the compressive strength of RAC was reduced by 20%–25% when compared to conventional concrete. To provide the best possible quality control, RCA from a single source is essential. Since aggregate attributes fluctuate depending on which resource the RCA is drawn from, the difference in compressive strength (CS) may become more noticeable when multiple resources are used to create the RCA. The strength qualities of RAC are also influenced by the mortar used to glue the coarse aggregates together. To put it another way, the compressive strength of the RAC decreased by as much as 10%, which equaled a maximum of 34% of adhering mortar when utilizing smaller sizes.

Recycled wash water exhibited no adverse effects on the formation of alkali-activated RCA concrete. The strength of mortar made using waste treatment plant water, and those made with freshwater was virtually the same. In-depth research into the CS of concrete made from treated sewage revealed that increasing the volume of treated effluent increased the compressive strength by as much as 28 days. A further 1.5% increase in compressive strength content was found when treated effluent was used for curing. With tap water instead of freshwater, there is a 9% increase in the concrete strength; with treated household sewage, there is no change in the setting time.

Recycled aggregate geopolymer concrete created with a wide range of effluents was the focus of this investigation and analysis (sugar mill effluent, manure mill effluent, and textile mill effluent). Several RCA mixtures were compared using a one-way analysis of variability (ANOVA) test to see what effect differences in their qualities had on the results. The results shown that the recycled aggregate geopolymer concrete (RAGC) mixes formed with fracture energy (FE) had 91% compressive strength and that the RAGC mix was 90% stronger than the RAGC mix of freshwater. Moreover, concrete created using textile mill effluent exhibited greater tensile strength (approximately 17% greater) than freshwater concrete. A total of 97% of the split tensile strength (STS) was achieved by RAGC mixes made using fertilizer mill effluent, 92% by mixes produced using sugar mill effluent, and 95% by mixes produced using SE effluent, as opposed to freshwater-derived concrete.

In terms of CIM, sugar mill effluent ranks highest among the RAGC mixtures examined at 120 days compared to freshwater concrete, which has a CIM of 105.9%. For RAGC mixes made with SE, the maximum CIM is 102%, since chloride ions could not get through the concrete matrix because of the high geopolymer concrete (GPC) density in the RAGC microstructure took place.

In addition, the analysis of variability test revealed significant differences in the impact of different RAGC mixtures on the condition of compressive strength. Similarly, there were no significant differences across RAGC blends regarding the split tensile strength, CIM, or acid attack resistance. As a result, the waste materials used and the environmental impact of the effluents investigated can be used to make sustainable concrete.

The RAC's waste effluent can be used to make green concrete, which has significant mechanical and durability properties and can be used to build large structures out of concrete in different parts. When natural aggregates are completely replaced using aggregates made from RCA, the developed composite had high draining potential; the coefficients of all the plates examined exceeded the legal limit, which corresponds to 0.1 cm/s. Having a high hydraulic conductivity is a desirable characteristic linked to having many voids in the pervious concrete, which is linked to having a lot of hardened concrete density. Other things make it less vital, but these things also make it less intense.

The RCA based on pervious concrete that has been made could be beneficial. The mechanical tests carried out in the laboratory agreed with the published research and ensured that the required level of functionality was achieved. On the other hand, the type of aggregate that goes into concrete plays a significant role in determining its strength; therefore, RCA should be used. The need to develop a way to make pervious concrete made with RCA stronger is therefore significant. In this case, if you want to make the pervious concrete mix more durable while still maintaining its hydraulic capacity and environmental performance, it might be a good idea to include a small proportion of fine aggregates. Using the physical characteristics like density, water absorption, or abrasion resistance to categorize recycled aggregates is essential to keep performance features from being spread out. This is because recycled aggregates are so diverse. The stronger the concrete used to make recycled aggregate, the more resistant the final product would be a general rule.

When RCA is contrasted with traditional concrete and pervious concrete made with natural aggregates, previous experiments show that RCA is superior. It has been discovered to contribute significantly to pervious concrete's environmental performance. Before making a selection, however, it is vital to consider the distance the recycled aggregate must be transported from the concrete plant to the waste processing plant, where it will be accepted. It is crucial to note the potential for the long-term sustainability of SF incorporation. The RAC mixtures made of this substance outperformed the others in terms of functional and environmental performance.

Both RCA and natural aggregate were prewetted before being batched into fresh concrete to determine the aggregates' influence on the workability of the concrete. Even though the aggregate was prewetted, the RCA had a lower slump than the reference concrete when using the same ratio of water to cement. For the concrete to keep its appropriate consistency as the amount of recycled aggregate was raised, it required an increased amount of SP. This phenomenon was ascribed to the angular shape of the RCA and the possibility that it continued to absorb water.

By evaluating the properties of the RCA, some use recycled aggregate for reinforcing. RCA has historically been used mostly as fill or sub-base materials in the United States. The modulus of elasticity of the concrete prepared with 100% RCA was roughly 35% lower than that of ordinary concrete. The addition of RCA had no appreciable effect on the concrete's compressive strength or splitting tensile strength. When indicated in Table 2.3, the modulus of elasticity is influenced negatively as the quantity of RCA increases. Using recycled materials, especially fine aggregate significantly increased drying shrinkage and creep in the concrete. After two years, the drying shrinkage of concrete made with recycled aggregate was between 60% and 100% more than that of conventional concrete, while the creep rate was between 30% and 50%. In terms of its resistance to freezing and thawing, carbonation trend, and oxygen permeability, no apparent changes were detected between the concrete created with 100% recycled aggregate and the concrete used as a standard.

By evaluating the properties of the RCA, some use recycled aggregate for reinforcing. In the United States, RCA has historically been used mostly as fill or sub-base materials and relatively rarely as aggregates in freshly built concrete roads, as shown in Table 2.4. The use of recycled aggregate concrete in markets other than

TABLE 2.4

Compositions and Use of Recycled Aggregate (DIN 4226-100, Aggregates for Concrete and Mortar, DAiStb-Guideline: Concrete with Recycled Aggregates)

Type of Aggregate	Composition, %						Max. Amount of Soluble Chlorides, % of the Mass	Min. Particle Density, kg/m³ (lb/ft³)	Max. Water Absorption, % by Mass	The Maximum Amount of Coarse Recycled Aggregate, % of Total Aggregate		
	Concrete, Natural Aggregate	Brick	Calcerous Sandstone	Other mineral constituents[1]	Asphalt	Contaminants[2]				Reinforced Concrete		Fill or Sub-Base Material
										Interior Element	Exterior Element	
Type 1: Concrete aggregate	≥90	≤10		≤2	≤1	≤0.2	0.4	2000 (125)	10	50	40	100
Type 2: Buildings aggregate	≥70	≤30		≤3	≤1	≤0.5		2000 (125)	15	40	–	100
Type 3: Masonry aggregate	≥20	≥80		≤5	≤5	≤1	≤0.5	1800 (112)	20	40	–	100
Type 4: Mixed aggregate		≥80		≤20	≤20	≤1	0.15	1500 (93)	NR	–	–	100

Notes

[1] Lightweight brick, concrete, cellular concrete, plaster, mortar, porous slag, and pumice are all examples of materials.

[2] Glass, ceramics, paper, gypsum, plastics, rubber, wood, and metal are all examples of materials.

[3] Includes a fine RCA

the pavement is in its infancy. The German Committee of Reinforced Concrete study findings may give the U.S. concrete industry important information for taking this next step in the green building movement. However, the most visible impacts are shrinkage and creep because recycled aggregate includes mortar, a previous concrete component, and is far more porous and absorbent than most natural aggregates. The water absorption rate of RCA was between 5% and 6%, whereas that of fine RCA was between 9% and 10%, while the absorption rates for the natural aggregate range between 1% and 2%.

3 Advanced Behaviors and Properties of RAC

3.1 RCA IN FIBER-REINFORCED CONCRETE

Concrete is the most widely used building material in the world due to its low cost, high durability, ease of obtaining its parts, and ability to be molded into virtually any shape. Population growth and urbanization will inevitably lead to a rise in the use of concrete in construction in the foreseeable future. As a result of this concern, new sustainable ordinary Portland cement (OPC)-free cement binder and supplementary cementitious materials (SCMs), a partial replacement for OPC in concrete, have been developed. OPC is a crucial concrete component primarily responsible for holding the particles together. On the other hand, the production of OPC requires a great deal of energy and is responsible for more than 5% of global CO_2 emissions. According to an alternative computation, 1 ton of OPC releases 1 ton of CO_2 into the atmosphere (Naik and Moriconi, 2005). This is the most significant source of global climate change (Naik and Moriconi, 2005). In addition, the removal of natural aggregates has a severe effect on the natural environment, becoming more widespread as the use of concrete increases yearly. In response to increasing concern about the environmental effect of C&DW disposal, various academics across the globe are studying novel methods to recycle it in order to alleviate the existing burden on landfill space and minimize the industry's dependency on natural aggregates and minerals (Kou et al., 2012; Tabsh and Abdelfatah, 2009; Zaharieva et al., 2003). To minimize greenhouse gas emissions, RCAs from construction and demolition waste may drastically reduce the amount of concrete, which is 75%–80% coarse and fine particles (Corinaldesi and Moriconi, 2009). This is not a new theory, but many researchers throughout the globe have examined what it would mean that RCA is inferior to natural aggregates in terms of their physical qualities, and the general public mostly agrees (Etxeberria et al., 2007; Shaikh, 2013; Shaikh and Nguyen, 2013).

Because of cement's significant negative influence on the environment, it is cleared that its production should be increased, not decreased. There are three strategies to reduce the environmental impact of concrete buildings, as detailed below by Kromoser et al. (2019). It is helpful to consider how different materials interact while designing structures to gain a better sense of aesthetics and functional interoperability. Concrete structures, for example, can reduce cement consumption by using topology optimization because concrete structures are constructed with less material in mind. In some instances, hydraulic cement can be replaced with fly ash or slag sand. Another option to incorporate high-performance materials into concrete structures is to do so. To save on concrete, they have a high level of tensile and elongational strength and stiffness. As an alternative to steel reinforcement, fiber-reinforced polymer (also known as fiber-reinforced plastic (FRP)) has received

DOI: 10.1201/9781003257097-3

much attention in the last few years. Schladitz et al. (2012) used an overview of textile-reinforced concrete buildings that might be appropriate here. Reinforcement with fiberglass is a way to create lightweight, long-term concrete buildings (Kromoser et al., 2019). Carbon fiber-reinforced polymer has a comparable Young's modulus to steel compared to other fiber types, including basalt, glass, and aramid. This is a pretty powerful statement. When a building is nearing its serviceability limit, it is essential to reduce the deflections of its structures to a minimum, and carbon fiber-reinforced plastic (CFRP) as reinforcement is the most excellent option for this. In addition to its exceptional non-corrosive properties and high tensile strength of up to 3,000 MPa, CFRP reinforcement is an excellent choice for structural applications. This means that the maximum thickness of a concrete covering which can be used is 10-mm thick, and therefore projects made of concrete require a minimal quantity of the material.

Despite the fact that some of the mechanical properties of RAC are not as good as those of conventional concrete, these properties are sufficient for some concrete building applications in the construction industry. By adding the appropriate mineral additions to the mix, RAC can improve its mechanical properties and durability. Alternatively, RAC has the same pliability and fracture resistance as ordinary concrete. In order to increase the durability and toughness of the matrix, several kinds of fibers are commonly used in both conventional and reinforced concrete. Synthetic fibers and polypropylene (PP) are two typically used fibers in RAC that may improve the material's mechanical qualities, durability, and tensile strength and partially compensate for the material's deficiencies.

Adding fiber reinforcement to concrete, commonly known as fiber-reinforced concrete (FRC), to ordinary concrete is a very effective method for improving the concrete's mechanical qualities. Fibers may improve the anti-cracking performance of conventionally reinforced concrete (CC) composites by increasing the interface bonding between the two layers of the composite. FRC may often exceed CC in terms of compressive and tensile strength, suggesting it may be a good choice for recycled aggregate concrete (RAC) augmentation. For fiber-reinforced RAC, it is vital to evaluate the production costs and environmental effects from a practical aspect. Basalt fibers (BFs) produced from basalt outperforms commonly used fibers such as steel fibers (SFs), carbon fibers (CFs), and glass fibers (GFs). Basalt fibers are manufactured from volcanic rock basalt. BFs are more accessible to disseminate in concrete than SFs, but they are much cheaper than CFs over the long run. BFs exhibit improved thermostability, heat resistance, and alkali resistance compared to other fibers due to their more considerable molecular weight. The fact that basalt belongs to the class of inorganic silicates is the most important consideration, given that its consistency with cement is acceptable.

To fully understand the characteristics of fiber-reinforced recycled aggregate concrete (FRRAC), the splitting tensile strength and flexural strength of RAC reinforced with steel fiber and PP fibers containing mineral additives must be determined. Investigating the relationship between RAC's splitting, flexural, and elastic modulus is essential. Additionally, there is a need for research on the relationship between the compressive and tensile strengths of RAC. No research has been conducted to date on the influence of steel and PP fiber content on the slump, durability, and microstructure performance of RAC, as well as how the presence of SF or PP fiber affected the elastic modulus, stress-strain curves, ion permeability, freeze-thaw resistance, carbonization resistance, shrinkage performance, and microstructure of RAC. There has been substantial research into the manufacturing process of recycled aggregate (RA), employing construction demolition excavation waste (CDEW) as well as the performance of FRRAC. Due to the diversification of FRRAC preparation technology, the prepared FRRAC demonstrates exemplary performance and has been successfully applied in various applications.

Regarding fiber-reinforced concretes, a variety of research has been conducted to investigate the material's behavior at high temperatures. Several studies have examined what happens when different short fibers are added to RAC to strengthen the concrete. However, nowadays, the effects of temperature on fiber-reinforced recycled aggregate concrete have been limited to a few studies. This is the first study to examine steel fiber-reinforced recycled aggregate concrete's compressive and cracking properties at elevated temperatures ranging from 300 to 600°C. At 400 and 600°C, the compressive strength of recycled aggregate concrete with and without steel fibers decreased by 15%; the reduction was 59% and 31%, respectively (Zhang et al., 2020).

The preparation of RAC required the use of natural coarse aggregate (NCA's typical properties are shown in Table 3.1), whereas the use of RCA served as a suitable replacement for NCA. Because the emphasis on a single component (replacing coarse particles) may prevent other factors from interfering when being evaluated, Recycled Concrete Aggregates (RCAs) were not considered concurrently. In the context of comparative studies on fiber-reinforced RAC, five different fibers were evaluated: SFs, CFs, PF, BFs, and polypropylene-basalt hybrid fibers (HFs). To evaluate if CC was preferable, it was required to test fundamental mechanical characteristics such as compressive strengths, splitting tensile strengths, and elastic modulus.

Experimentally manufactured RACs, or fiber-reinforced composites, exhibited comparable conversion and strength for BF-reinforced RACs despite using a volume percent of basalt fiber that varied by type of fiber. The results of these formulae were then validated using back-propagation neural networks (BPNNs), which were used to identify the optimal volumetric percentage of BF for each of the RAC's mechanical properties. When compared to CC, RAC has the potential to possess better compressive and splitting tensile strengths than CC does in certain circumstances, provided that the combination ratios remain the same. This is because recycled coarse aggregates are better at absorbing water than fresh coarse aggregates, which makes recycled coarse aggregates better at absorbing water. In

TABLE 3.1

Physical Properties of Coarse Aggregates (Fang et al., 2018)

Aggregates	Grading (mm)	Bulk Density (kg/m³)	Apparent Density (kg/m³)	Water Absorption (%)	Crush Index (%)
NCA	5–20	1372	2790	0.4	5.1
RCA	5–20	1215	2491	3.4	10.0

addition, the inclusion of fibers may either increase or decrease the mechanical characteristics of RAC, depending on the fibers used. This latter issue may result from the fibers inside the concrete not being distributed regularly.

On the other hand, certain fibers each contribute in their unique way to the many failure mechanisms of the RAC. The ideal volume fractions of BF were 0.1, 0.2, 0.2, and 0.2 for cube compression, axial compression, splitting tensile strength, and elastic modulus of RAC, respectively. Because BPNNs were used, it was feasible to refine the ideal fractions to 0%, 1%, 5%, 10%, 15%, and 20%, respectively. These values were arrived at as a consequence of the use of BPNNs. In addition, to convert the cube compressive strength to other mechanical qualities whichqualities, it could be employed in developing design codes. In addition, one efficient strategy for enhancing the mechanical qualities of CC is to use FRC, also known as reinforced concrete. Fibers may enhance the anti-cracking performance of CC by strengthening the interface bonding between the two layers of the composite material.

FRC may generally surpass CC in compressive and tensile strength. This suggests that FRC may be feasible for RAC augmentation, provided that the appropriate fibers are used. In the case of fiber-reinforced recycled aggregate concrete (also known as RAC), on the other hand, it is essential, from a purely pragmatic viewpoint, to take into consideration both the costs of manufacturing and the influence on the surrounding environment. In testing, yarn manufactured from basalt performed better than used yarns such as steel, carbon, and glass. This demonstrates that basalt-made yarn is superior to the materials utilized in the past. However, BFs are more accessible to scatter in concrete than SFs, and over the long run, they cost far less than CFs. Steel fibers are more difficult to work with. Because of the increased molecular weight of these fibers, they are more resistant to heat and alkaline than other types of fibers. It is essential to bear in mind that basalt may be used in a manner similar to that of cement due to the fact that it is an inorganic silicate. When it comes time to produce RAC, RCA will stand in for NCA as the component of choice. Because concentrating on one aspect (the replacement of coarse aggregates) would impede the investigation of the effect of other aspects, RFA were not brought into simultaneous consideration either. A series of comparative studies on the impacts of five different fibers was conducted on fiber-reinforced RAC. These fibers were SFs, CFs, PF, BF, and polypropylene-basalt HFs, as indicated in Table 3.2. It was determined whether or not the elastic modulus

TABLE 3.2

Physical and Mechanical Properties of Fibers (Fang et al., 2018)

Fiber	Equivalent Diameter (mm)	Length (mm)	Density (g/cm³)	Tensile Strength (MPa)	Elastic Modulus (GPs)
BF	0.015	18	2.65	2630	88.9
CF	0.085	9	1.76	3700	230
PF	0.0182	19	0.91	556	8.9
SF	0.33	37	7.8	658	220

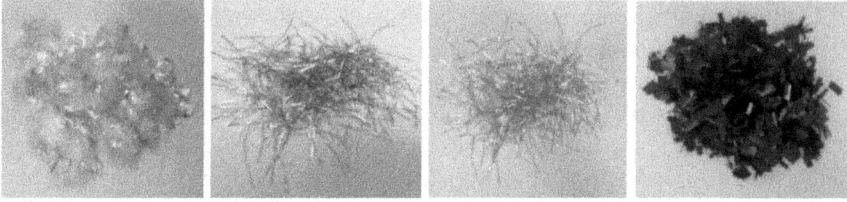

FIGURE 3.1   Fibers for RAC.

of CC was superior by comparing the compressive and splitting tensile strengths of CC to the elastic modulus of CC (Fang et al., 2018) (Figure 3.1).

The risk of fire or exposure to high temperatures is one of the most common dangers buildings face during their useful lives. In the case of a fire, researchers are looking at the high-temperature capabilities of concrete made using recycled aggregates. This is done to prevent the loss of life and property due to structural damage. Thermal stress brought on by a rapid rise in temperature and the presence of water vapor can cause reactive powder concrete (RPC) to have a high pore vapor pressure. This phenomenon has been the subject of several studies investigating how RPC behaves when subjected to extreme temperatures and the possibility of explosive spalling. It has been shown that using SFs, PP, or both types of fibers in conjunction with one another may prevent composite materials from detonating. Micro-channels are formed due to the melting of PP fibers at a temperature of around 167°C. These micro-channels facilitate the escape of vapor that has been confined and have a low vapor tension. The widespread dispersion of polypropylene fibers throughout the concrete mixture reduces drying shrinkage and fracture width in the cement matrix. This, in turn, increases the durability of the finished product. In addition, using steel fibers in the concrete mixture increases its tensile strength, making it more resistant to spalling brought on by the fire. Consequently, the use of hybrid steel and polypropylene fibers in RPCs results in an improvement of the material's performance in high-temperature conditions, while simultaneously decreasing the possibility of an explosive failure.

CFRP-reinforcing and CFRP-reinforced concrete structures, as well as future studies and applications of CFRP, will be built upon this foundation. The environmental statistics of the elements they include will not be fully documented. FRP reinforcement's environmental performance still necessitates reexamining existing environmental data (Zhang, 2014). An example of CFRP reinforcement illustrates the material's environmental performance.

Abiotic depletion of fossil fuel supplies, acidification of soils and water, and climate change are all possible outcomes of our continued reliance on fossil fuels. Life cycle assessment (LCA) of CFRP reinforcement, which includes a comparison to traditional steel reinforcement, is the topic of this essay. Next, we evaluate the environmental impact of the CFRP reinforcement in question by comparing it to built samples with similar design attributes. CFRP-reinforced concrete bridges are compared against more traditional concrete and mild steel structures. This framework is developed within the theoretical context, and computations are made under it. Results from three separate impact categories are shown here. The study's last paragraphs discuss the findings and ideas for further research.

CFRP reinforcement has a far more significant environmental impact than standard steel reinforcement regarding yield strength (more precisely, tensile strength). To put it another way, CFRP is made out of recycled products. These findings suggest that steel reinforcing should be employed in constructing ecologically friendly concrete structures. It was shown that the CFRP-reinforced bridge had the lowest global warming potential (GWP) and abiotic depletion of fossil resources (ADPf) of the three types of bridges that were eventually constructed, showing that it was the most environmentally beneficial alternative. While the typical concrete bridge's acidification potential (AP) is most significant, the carbon concrete bridge's AP is much higher since carbon fibers have a lot of AP, making them extremely strong.

The environmental effect of carbon fiber reinforcement in concrete projects, constructing a bridge using CFRP reinforcement instead of steel reinforcement might be a tremendous environmental decision if its full strength and stiffness potential are used. CFRP-reinforced structures might be made more stable. Research by Reichenbach et al. (2021) analyzed the advantages of pre-stressing CFRP rebars for the serviceability of CFRP-reinforced beams. They were satisfied with the outcomes. As a cautionary tale, it is essential to keep in mind that the application example discussed in this research cannot be used to generalize the environmental performance of CFRP-reinforced concrete buildings. Updated data on CFRP reinforcement in concrete buildings is essential if future environmental evaluations are accurate and transparent.

Reinforcement made of CFRP has a lot of benefits, but one of the significant downsides is that it requires a considerable quantity of crude oil and energy to produce. This is one of the most significant negatives. Reinforcement made of carbon fiber-reinforced plastic, often known as CFRP, is not the same as reinforcement made of steel. The performance of CFRP-reinforced structures in this environment is challenging to assess due to the fact that few design variables reflect how well they operate. Among the many studies that have been done on the topic, Inman et al. (2017) conducted research on the application of basalt fiber-reinforced polymer (BFRP) reinforcement in concrete beams. Even though this research demonstrated that the stiffness of BFRP reinforcement was lower than that of steel reinforcement, it is clear that there is a positive impact on the environment. In their study from 2015, Maxine and her colleagues investigated the environmental friendliness of CFRP post-strengthening reinforcement in contrast to the demolition and restoration of reinforced concrete beams. Structures made of standard concrete were shown to be more harmful to the environment, individuals, and the ozone layer than those made of carbon fiber-reinforced plastic, also known as CFRP-reinforced concrete.

3.2 RCA IN REACTIVE POWDER CONCRETE

Currently, there are many reports presenting how three different salt levels affected the mechanical qualities of ecologically friendly concrete. The mechanical parameters of an environment-sensible concrete contain 50 wt% less OPC and 50 wt% less natural coarse particles compared to 5% and 15% silica fume. Adding 10 wt% silica fume considerably increased the compressive strength, indirect tensile strength, and elastic modulus of 50 wt% RCA and 40 wt% slag-containing concrete at 3 and 7 days by 10 wt% SF is the best amount of SF for enhancing the mechanical properties of concrete with 50 wt% RCA and 40% slag. Additionally, the long-term mechanical characteristics of concrete with 50 wt% RCA and 40 wt% slag containing 10 wt% SF have improved significantly. The foundation of environmentally friendly concrete requires less than half the amount of OPC and natural coarse aggregates; after 28 days, it has a compressive strength close to 50 MPa, indicating that it is suitable for structural concrete in the majority of circumstances (Cwirzen et al., 2008).

Additionally, the mechanical qualities of RAC are comparable to those of natural aggregate concrete. In most instances, the adherence of mortar from the parent concrete to the surface of the RAC decreases the compressive strength of RAC, as measured by compressive strength 26–29 and flexural strength. Nevertheless, some investigations have demonstrated that RAC performs similarly or better than NAC. The rougher surface roughness of RCA has been linked to improved bonding and interlocking between RCA and mortar (as opposed to that of natural coarse aggregate (NCA)), which has been demonstrated to have greater compressive strength than the NCA. To determine the RCA effect on high-performance concrete (HPC) characteristics, several studies have concluded that reusing and repurposing concrete may be the most cost-effective solution and significantly reduce landfilling.

Furthermore, repurposing RCA to produce HPC recycling and reuse would reduce world CO_2 emissions. RCA can be used in fine and coarse aggregate applications, and its high replacement ratios make it ideal for both. It is also feasible to gain HPC by replacing the entire system. HPCs usability with RCA is severely hampered by the fact that RCA necessitates a significant volume of water to get the correct droop (Landa-Sánchez et al., 2020). In addition, HPC with RCA makes the concrete stronger because of strength improvement during early hydration, the high surface roughness of RCA, the hydraulic conductivity of concrete, and the amount of recycled aggregate (RA) in the mix. The amount of RCA substantially decreases tensile and flexural strength regardless of the time it has been in place. As opposed to natural aggregates, the use of RCA increases splitting and flexural strength as compared to the use of natural aggregates. Thus the presence of RCA in HPC composites increases the endurance of the composites; additionally, the ITZ between the aggregate and the cement paste. However, to use RCA in reactive concrete, how RCA particles of varying forms interact with the bonding mortar between them needs to be considered. If this were the case, the correct top-to-bottom relationship between molecular pushing and hard excellence would be one step closer to being understood. As it turned out, water absorption occurs more slowly than predicted during the mixing phase of RAC, resulting in free water being readily available to aid in the utility of the operation. If the w/c is extended, resulting in reduced compressive strength and addressing the abrasion resistance of HPC produced using RCA. Finally, it is also required to investigate the application of RAC in geopolymer and prefabricated concrete.

A new kind of HPC, called reactive powder concrete (RPC), is being developed at the moment. RPC has a compressive strength of 200 to 800 MPa. RPCs with a compressive strength of 200 MPa is good with fine quartz sand and ground-up quartz powder before being subjected to 90°C thermal treatment. Using steel microfibers improves the ductility and tensile strength of the material. It is also necessary to produce RPC with a strength greater than 600 MPa by applying compression loads of about 50 MPa while the material is still fresh, as well as heating it in an autoclave to a temperature of about 250 to 400°C, to make it. Using metallic aggregates instead of quartz aggregates can help make the concrete even more durable and strong. In addition, the impact resistance and shear strength of pillars are improved when RPC is utilized in the structure's construction. On the other hand, RPC and fibers were used to produce plates that could withstand much damage from bullets. Managing a high-range water-reducing admixture means that the amount of cement in this material is usually more than 800 kg/m^3, and the w/c is usually less than 0.20. That is not the only reason RPC is therefore durable. It comprises tiny pieces, such as crushed quartzite and silica fume. Optimizing the granular packing of these powders can make it possible to make a very dense matrix out of them. Another way to lessen the brittleness is to use steel fibers, which can also be used. Because RPC paste has such a high strength, the importance of the strength of small aggregates goes down a lot because of the strength of the paste. RPC also contains a high proportion of fine-grained cement and a low w/c, both desirable properties. This makes it challenging to produce high-performance concrete for structural applications using locally produced waste materials, which are more expensive. It has a lot of it with particle sizes between 150 and 600 μm of SF. In terms of particle size uniformity, porosity, and microstructure quality, RPC is superior to typical concrete. Furthermore, conventional or even HPC is not as strong or long-durability as RPC, which is formed by eliminating all coarse aggregates, using very little water, adding pozzolanic ingredients, very fine sand, and steel fibers to the mix, and then stressing and heating it.

The fact that RPC has a significant amount of SF is one of the things that set it apart from other companies. In addition, silica fume is used in place of less costly elements in traditional concrete to produce more modern concretes, such as RPC. Because it is the most important ingredient in a typical RPC, silica fume significantly impacts the rheological and mechanical characteristics of the concrete, which ultimately increases the material's durability. When the reactivity of the pozzolana is assessed by measuring the quantity of calcium hydroxide ($Ca(OH)_2$) in the cement paste at different intervals, SF is found to have a higher level of reactivity than SF as any other natural pozzolan.

With the addition of steel filaments, modified RPC surpassed original RPC in terms of strength and reduced drying shrinkage or creep strain (150 to 400 μm) when compared to the original RPC, which was made with a high-range water-reducing cement mixture containing silica fume and fine-ground quartzite. For UHPC, Ma et al. used crushed basalt and found that RPC with particle sizes ranging from 2 to 5 mm aggregate worked as well as RPC with a particle size range from 1 to 2 mm (Struble and Tebaldi, 2017).

In the past few years, researchers have looked into several new types of high-strength concrete, such as RPC, which is known for having excellent mechanical properties, lasting a long time, and very low permeability. Getting rid of coarse aggregates to get rid of the matrix's weak point, the addition of pozzolanic materials to create a dense microstructure, use of significantly lower water to binder ratio, such as 0.13, particle gradation to reduce porosity, improved curing processes, and the use of micro and/or macro steel fibers are the fundamental causes for the substantial improvement in RPC performance. It has compressive strengths between 150 and 800 MPa, rupture energies between 1,200 and 40,000 J/m^2, and ultimate tensile stresses between 0.1% and 1%. According to researchers, incorporating hybrid steel fibers, including straight and hooked steel fibers, can increase the flexural and tensile strength of RPC compared to micro steel fibers alone. It has also been proved that using discarded steel fibers in producing RPC is both practical and eco-friendly. While the cost of RPC has increased as a result of an increase in cement content and the elimination of lower-cost coarse aggregates, it has also provided benefits such as a reduction in the cross-section of members, which results in less dead load, and the partial or complete elimination of passive reinforcement due to the use of fibers, among others.

RPC is a precast concrete manufactured in a laboratory or factory setting with exceptionally high compressive strength. A short period at a controlled high temperature can be used to cure the material in this manner, as shown in Figure 3.2. As a result, the hydration process will be able to proceed and the compressive strength of the concrete should be obtained.

In contrast to RPC, which is placed on the pavement of the road, this material may either be cast directly on the floor or in situ. This suggests that special consideration should be given to the process of curing RPC after the casting has been completed in this case. It is feasible to execute an in-situ steam cure by covering the RPC and allowing steam to flow for three hours daily. This method requires wet sacks to be put over the concrete and allowed to cure for a few days in the field.

One of the essential components of RPC is river sand, which is used as a filler and is obtained by river mining. As a result of excessive and unplanned mining, the building sector's rapid development has threatened the sand reserves in the region. Although river sand takes a long time to make, it is worth the wait; the rate at which

FIGURE 3.2 Steam and water curing of RPC.

it is consumed now far exceeds the rate at which it is replenished. In order to protect navigation, flood management, and river ecology from the negative effects of river mining, it has become essential to look for other materials that can be used instead of natural sand in river projects. These methods can help solve the waste management problem while reducing the costs of RPC. One way to solve this problem is to use fine RAC instead of natural river sand to make RPC. Because fine RAC have destructive physical properties, they can only be used in very few construction projects because it does not have the same hardness, density, water-absorbing capacity, or strength as natural sand. RCA has been used in concrete for a long time and is used in many construction projects.

Furthermore, a new generation of concrete, an ultra-dense mixture called RPC, has been created with silica fume and other materials that make it very high strength. It is being used in the field to see how well it works. The qualities of reactive powder concrete combinations are superior to those of conventional and high-performance concretes because they are better at the micro-scale level than normal and high-performance concretes, which are better at the macro level (Ahmad, Zubair, and Maslehuddin, 2015). On the other hand, biological and construction wastes do not go away quickly and have been used in concrete production to improve environmental sustainability and quality (Zimbili, Salim, & Ndambuki, 2014). Second, it is not easy to produce RPC for structural applications using local materials. Modern concretes, such as reactive powder concrete, are produced by substituting the least expensive components with more expensive ones, such as silica fume, quartz sand, and quartz powder, a new type of RPC. For local construction projects, manufacturing concrete from these ingredients will result in increased raw material costs and the amount of time it takes to import them. RPC is more expensive than normal concrete because it has many more minerals.

Furthermore, it seemed that the heat curing process and the milling of quartz sand were the main reasons why RPC was not used more often (Gu, Ye, & Sun, 2015). Ordinary RPC costs more than other materials due to the use of particular technologies in preparation procedures and the high concentration of expensive components in the preparation methods. For example, in the construction of this structure, coarse materials were substituted with graded quartz sands (Tang, Xie, & Long, 2016). Furthermore, silica fume is one of the primary ingredients of RPC, with its dose typically ranging between 25% and 30% of the cementitious material in the mix. A large proportion of silica fume distinguishes RPC (Cheyrezy, Maret, & Frouin, 1995). Aside from the issue of availability, the use of silica fume impacts the costs of concrete manufacturing, even though it produces a substantial matrix. In addition to the cost issue, the shrinkage rate of silica fume concrete is high; its workability is poor and therefore it is possible to develop temperature cracks, which impacts the smoothness of the concrete's quality. It appears that substituting finely dispersed local wastes for the primary constituents of reactive powder concrete is a possible approach for solving economic and environmental issues while producing economically viable recycled RPC.

RPC is a fiber-reinforced, superplasticizer, silica fume. Several researchers used RPC to refer to a fiber-reinforced, superplasticizer silica fume-cement mixture with a very low w/c and containing very fine quartz sand (0.15–0.40 mm) in place of standard aggregate. Indeed, this is not concrete since the cement mixture contains no coarse material, indicating that it is not concrete. For the inventors, the lack of coarse aggregate was significant to the microstructure and performance of the RPC because it allowed them to keep the cement matrix and aggregate as stable as possible. When the RPC is made with very fine sand instead of normal aggregate, it has more cement with a range of 9,00–1,000 kg/m^3 (Al-Jubory, 2013).

When comparing the effects of various kinds of Portland cement and silica fume on the performance of RPC mixes, be sure to use cement that does not include C_3A and a grade of silica fume that is white. It took three days for the weight-to-compactness ratio to decrease as low as 0.18, while it needed 200 MPa for the compressive strength to reach its maximum level. It is not statistically significant that there was a considerable loss in compressive strength in the early phases, despite the fact that there was such a drop. Other varieties of silica fume, such as gray or dark SF, were used instead of white SF, and other types of Portland cement were used instead of the C_3A-free Portland cement (110–160 MPa), and the w/c increased (Al-Jubory, 2013). On the other hand, when these mixes were put through a stress test, there was no change in their early compressive strength. RPC is reinforced with corrugated steel and fibers to improve its compressive and flexural strengths. The fibers employed were corrugated steel fibers with a diameter of 0.4 mm and a length of 13 mm. Similarly, the concrete included recon 3s fibers with a triangular shape and a length of 12 mm. Both types of fiber were mixed. Ultra-high-strength concrete (UHSC) is a new construction material that uses RPC. For example, the success of RPC can be attributed to the absence of coarse aggregates, the low water-binder ratio, and pozzolanic fine components such as ordinary portland cement and silica fluoride, as well as micro-steel fibers. Additionally, the incorporation of pozzolanic and silica-rich fine ingredients, such as OPC, has contributed to the success of reactive powder concrete. Several researchers have extensively investigated silica RPC and employed earlier methods to investigate its different strengths and microstructural features. Because RPC performs better at high temperatures, it is employed in specific constructions subjected to high heat and temperatures. Additionally, research into the possibilities of using different materials for the binder and fine particles in concrete is progressing every day in order to create a more attractive and sustainable building material is gaining popularity. On the other hand, the materials' adoption for real-world applications in concrete structures depends on their economy and efficiency.

A low water-binder ratio resulted in ultra-high RPC strength by decreasing pores, the weakest connection between particles. Although it was discovered that adding a superplasticizer to the RPC mix improved workability, it was also found to be effective at increasing the mix % age, which is necessary to generate the stiff and dense matrix of RPC. It is possible to make high RPC by combining the principles of RPC with the theories of packing density. Regarding compressive strength at 28 days and rheology, the ultra-high-strength (UHS) mortars and concretes appeared almost identical (Rohden et al., 2020). Individually, when UHS mortars were used did the addition of steel fibers significantly increase flexural strength. Additionally, due to the coarse pebbles' higher water absorption capacity, it was necessary to employ a somewhat more excellent water-to-binder ratio than was initially intended.

When developing RPCs, it is crucial to consider several factors in their formulation. These include eliminating coarse aggregates, optimizing granular composition, pressing during casting, heat treatment, and metallic microfibre inclusion after hardening. Granular optimization techniques might be applied to improve recycled concrete powder's mechanical performance and consistency. Quartz and metallic aggregates, which have high mechanical strength, are the most commonly utilized fine aggregates in RCP. Other aggregates of basaltic, granitic, and calcareous origin, in addition to the ones specified, may be employed. Using low-strength elements such as limestone into RCP, compressive strengths above 200 MPa can be obtained since the aggregate does not serve as a limiting element in the process (83 MPa) (Rohden et al., 2020).

Because it is made of fine powders, RPC requires much energy to combine properly. The mixing procedure has an impact on the RPC characteristics as well. When it comes to fresh reactive powder concrete (RPC) properties, Hiremath and Yaragal evaluated the effects of mixing speed (ranging from 25 to 50 rpm to 100 to 125 rpm to 150 rpm) and mixing duration (10, 15, 20, 25, and 30 min) (Rohden et al., 2020). According to scientists, high speeds are required for optimal performance, while low speeds are detrimental to performance. On the other hand, high speeds resulted in high amounts of air being entrained. The 15-minute mixing time resulted in greater fluency and strength characteristics than the other.

Applying thermal cure to conventional strength concrete improves strength in the early ages, but tends to lower strength at 28 days and beyond. This state is not found in the RPC, and the thermal treatment produces significantly superior results to the RPC, even after 28 days.

3.3   RCA IN SELF-CONSOLIDATING CONCRETE

Regular concrete substitutes several parts, including its essential components, with self-consolidating concrete (SCC). This is done to improve the material's strength and durability. Gravel, crushed stone, and river or beach sand are all used in the construction of the SCC. Between 55% and 60% of the entire volume of the concrete is composed of the SCC aggregates. They significantly impact how simple it is to work with, how powerful it is, how reliable it is, and how long it will continue to function as intended. The aggregates have a substantial impact, as well, on the total cost of the SCC.

As a consequence of this, it is desirable to choose aggregates that are less costly in applications using SCC. Because of the construction boom in developing nations and the reconstruction in affluent countries like the United States, there is a shortage of natural aggregate in many areas of the globe. One of these rich countries is the United States. Consequently, more people are interested in constructing their projects utilizing various aggregates. In the production of fresh concrete, RCA, which is created from discarded concrete, has become more common in recent decades. The use of RCA in concrete buildings has become more common due to a lack of available NCA and increased expenses associated with disposal. In addition, contractors are looking towards RCA as an alternative to NCA (Figure 3.3) to compensate for the increased distance between natural aggregates and building sites

FIGURE 3.3   NCA.

compared to the past. As buildings are demolished, much concrete waste ends up in waste disposal, testing concrete, and using surplus or returned concrete. Environmental concerns are caused by the removal and disposal of these pollutants, which are highly harmful. A significant contribution to this goal is the reuse and recycling of RCA, which reduces pollution while reducing the massive consumption of natural aggregates in the building. For non-structural applications, many studies have shown that RCA can be a suitable replacement for natural NCA. Lately, researchers Tu et al., Kou and Poon, and Grdic et al. came up with SCC by replacing RCA with NCA in both partial and complete ways, and they called it SCC. In contrast, only a tiny amount of research has been undertaken on using RCA in SCC. These studies mainly looked at the hardened properties of SCC, but no systematic study was done to find out how RCA affects the important fresh qualities of SCC. Furthermore, no tests were done to see if RCA from tested (James, 2011), extra, or returned concrete could be used to make SCC.

SCC was produced by substituting 0–100 wt% NCA by weight with the RCA, as shown in Table 3.3. SCC by replacing RCA for NCA up to 50 wt% of the time, high filling and passing capabilities with appropriate segregated resistance can be generated. On the other hand, the effects of RCA on the essential fresh properties of SCC include filling capacity, passing ability, and segregation resistance.

RCAs surface roughness, angularity, surface porosity, and other physical properties were not conducive to improving SCCs fresh attributes. When SCC has a high concentration of RCA, these physical properties can significantly impact its filling

TABLE 3.3

Physical Properties of Fine and Coarse Aggregates (Safiuddin et al., 2011)

Physical Properties	RCA	NCA	FA
Saturated surface-dry based specific gravity	2.51	2.62	2.69
Oven-dry-based specific gravity	2.46	2.53	–
Absorption (wt%)	1.91	0.60	1.32
Moisture content (wt%)	1.32	0.17	0.31
Compacted bulk density (kg/m^3)	1366.2	1513	1618.5
Angularity number	9.47	7.45	–
Fines (< 4.75 mm) from aggregates* (wt%)	7.75	–	–

and passing capacity. Furthermore, these properties could have a role in the non-uniform dispersion of coarse particles, resulting in concrete diversity, which may result in segregation, particularly at greater RCA contents. J-ring flow, V-funnel flow time, T50 slump flow time, the separation index, and slump flow were all strongly correlated with each other because they changed with the RCA content. Using RCA in SCC instead of NCA can keep important fresh qualities of the concrete the same, even if it is up to 50 wt%, such as filling capacity, passage ability, and segregation resistance (Arezoumandi et al., 2015).

Lowered coarse aggregate content and high paste volume per unit aggregate content were why SCC filled better at 30 and 50 wt% RCA, respectively. A considerable drop in the ability to fill was seen with 100% RCA, primarily due to the higher amount of fine aggregate added after the first mixing process. On the other hand, the filling ability was diminished when the RCA content was more significant than 50 wt%. SCC mixtures with acceptable passing ability were present. Because of the lower coarse aggregate concentration in the 30 and 50 wt% RCA, a modest increase in passing ability was observed. The ability to pass at 70 and 100 wt% RCA was within the maximum level, despite the limited ability to fill because there was less coarse aggregate in the building, which led to more fine aggregate being used. The T50 slump flow time and V-funnel flow time results revealed that SCC mixtures containing RCA above 50 wt% were viscous, as indicated by the higher RCA content. The high viscosity of the SCC mixture, including 100 wt% RCA, produced is difficult to achieve in a continuous concrete flow during the V-funnel flow test. Due to the intermittent nature of the water flow, it took a long time for the concrete mixture to exit the V-funnel.

When the segregation resistance of SCC blends was evaluated using a Japanese sieve stability test (Belaidi et al., 2012), it was observed that all blends functioned admirably. When the concrete samples from this test were analyzed, the segregation indices were significantly lower than the stated maximum limit for this material (18%). While this discovery was consistent with the column segregation test results, it did not match all the segregation ratios. As a result, even though these mixes had the lowest degree of segregation index, their segregation ratios were much greater than the permitted maximum limit (15%) for SCC blends containing 70 and 100 wt% RCA. The substantial segregation ratio values of 70 and 100 wt% RCA were attained partially because of the substantial non-uniform dispersion of coarse materials during the concrete installation process, primarily as the result of aggregate collisions at low fluidity.

Monitoring the effects of RCA in structural applications is vital to keep an eye on how freshly cast concrete, especially SCC, puts pressure on vertical formwork when it is still wet in order to encourage their use. While NCA are the primary constituent, adhering mortar (which accounts for roughly 30 to 40 wt% of the aggregate skeleton) (Xu et al., 2020) provides specific properties to the aggregate skeleton, like rough surface texture, more angles, and better water absorption. As a result, the friction inside the granular phase increases, which makes it more difficult for the SCC to move and pass through things.

Over the last few years, it has been looked into how formwork pressure is created by virgin-aggregate SCC. While including RCA (as shown in Table 3.4) with specific surface roughness and water absorption characteristics may be beneficial, it has the potential to modify current knowledge and predictive models significantly. Generally, it has been demonstrated that when shear strength features, like internal friction and cohesion, start to form, they significantly affect the pressure and rate at which the SCC pressure decrease over time. It is the first property that stays the same over time and temperature; it has the most considerable effect on the maximum pressure that can be measured right after the casting process. It is possible to make the inside of a concrete block more slippery by increasing the coarse aggregate, lowering the water-binder ratio (w/b), or replacing portland cement with other cementitious ingredients like silica fume and blast furnace slag. Cohesion increases after the plastic SCC is cast. This allows the pressure to drop more quickly because the gel structure can hold more vertical force as time passes. Some chemical admixtures can also have physical effects on concrete, such as setting accelerator and specific mixes of viscosity-modifying admixture (VMA) and high-range water reducer (HWR), which can make it easier to work with, are associated with the development of cohesion in concrete.

TABLE 3.4
Physical Properties of NCA and RCA Materials (Assaad and Harb, 2017)

	Specific Gravity	Oven-Dry Rodded Bulk Density, kg/m³	Absorption Rate, %	Material Finer than 75 μm, %	Fineness Modulus	Adhered Mortar Content, %	ACV, %
NCA	2.72	1763	0.61	0.42	6.71	NA	17.8
RCA	2.43	1505	7.04	0.9	6.77	41.2	23.1

Note: NA is not available.

Stability and formwork pressure are controlled by the rheological behavior of SCC, which includes the thixotropic element of thixotropy and how it varies. A comprehensive study was conducted to determine the impact of RCA on these two properties. RCA substituted the virgin aggregates at various replacement ratios of 25, 50, 75, and 100 wt%; RCA substituted the virgin aggregates at various replacement ratios of 25, 50, 75, and 100 wt% (Assaad and Harb, 2017). It was close to hydrostatic pressure when the pressure gauge read following the casting process for virgin aggregate SCC. Due to the increased surface roughness and internal friction that resulted from the inclusion of increased RCA levels, initial pressure was lowered independent of mixture composition. When 100 wt% RCA materials were utilized in a 0.38-w/b combination, the initial pressure reduced dramatically to 88% (Assaad and Harb, 2017). For virgin-aggregate SCC, the rate at which pressure dropped over time depended on how much binder there was and how much weight there was. This is because more RCA water was absorbed by the concrete, making it more stable and minimizing the transformation of vertical forces into lateral stresses.

In the field of concrete technology, SCC has become one of the most important new things to come out. SCC has unique qualities that make it easier for a construction project to be done, like flowability, passability, and stability. Due to its various benefits, SCC has grown in popularity over the past few years, which include less labor needed to place and compact concrete, shorter construction times because of faster placement, less waste, reduced equipment and formwork maintenance costs, and higher fleet utilization due to faster round-trip times for transmitting mix. Reduced environmental impact SCC is more expensive than conventional concrete, and fine and coarse aggregates, as well as the binder and mineral additives, are rapidly accounting for a more significant portion of the total cost. Because natural aggregate supplies are becoming increasingly scarce, researchers worldwide are looking into alternatives, focusing on using RCA derived from construction and demolition waste (C&DW). By using these aggregates in SCC mixes, we can assist to raise the environmental value of concrete while simultaneously addressing the issue of dumping vast amounts of C&DW generated in cities across the world, which is a significant cause of air and water pollution.

Prior to incorporating RCA into concrete mixtures, it is necessary to remedy the material's poor characteristics. Adding mineral admixtures to concrete mixtures is one approach to addressing the poor characteristics of RCA. By adding mineral admixtures to concrete mixtures, you can increase the density while simultaneously decreasing the perviousness. Apart from that, adding mineral admixtures can aid in making up for the loss in mechanical and durability caused by RCA in concrete mixtures. Normal vibrating concrete (NVC) with coarse recycled concrete aggregate (CRCA) and metakaolin (MK) was the subject of much research to see how it would affect its mechanical and durability properties (Landa-Sánchez et al., 2020). There is very little information about the durability of SCC with MK and fiber-reinforced concrete aggregate (FRCA), and CRCA, according to a literature review.

With the addition of FRCA and CRCA to SCC mixes, the results for strength and durability fall apart. It also makes the concrete more durable and strong because of its pozzolanic reaction with cement hydrates and ability to fill pores. This is why

MK is added to the concrete matrix. As a result, SCC combinations incorporating MK outperform SCC mixtures that do not contain MK.

In terms of compressive strength, SCC mixtures containing 50 wt% CRCA and 25% FRCA are comparable to control SCC. However, with the addition of MK, the strength reaches levels comparable to control mixtures with a 50% CRCA and 75% FRCA content, respectively. As a result, when 10% MK is added to the SCC mixture, the replacement of FRCA can be enhanced by 50 wt%.

At all maturation ages, reduced equipment and formwork maintenance costs, higher fleet utilization due to faster round-trip times for transmitting mix, and reduced environmental impact, SCC is more expensive than conventional concrete, and fine and coarse aggregates, as well as the binder and mineral additives, are rapidly accounting for a more significant portion of the total cost. Because natural aggregate supplies are becoming increasingly scarce, researchers worldwide are looking into alternatives, focusing on using RCA derived from C&DW.

In SCC mixes, the penetration of chloride ions and the formation of capillary suction are both accelerated by the addition of CRCA and FRCA, while MK was found to be very good at stopping chloride ions from getting into the body and forming capillary suction. MK-based mix total charge passed was lower even though FRCA was replaced entirely with FNA. The capillary suction findings were similar to the control SCC.

Even when the SCC mixture contained 50 wt% CRCA and 100 wt% FRCA, the UPV values decreased by less than 10%. When metakaolin (MK) is added to the SCC mixture while employing 100 wt% FRCA, the decrease becomes minor (Landa-Sánchez et al., 2020). As a result, based on the ultrasonic pulse velocity (UPV) values obtained from their SCC components, all SCC combinations combining CRCA, FRCA, and MK based composites can be classified as "excellent" as well.

3.4   RCA IN GREEN CONCRETE

Concrete will remain the most extensively utilized construction material in the built environment for the foreseeable future, and the weight of resource exploitation in its fabrication will continue unless better environmentally friendly and sustainable concrete production techniques are devised. In the creation of concrete, the principal component, cement, requires 4 GJ of energy to manufacture 1 ton of cement, resulting in 0.89–1.1 tons of CO_2 emissions.

Green concrete's use in sustainable infrastructure building has piqued the interest of structural design practitioners and other construction players. The worldwide population has expanded due to the global economic crisis, resulting in increased building waste output and innovative energy-efficient construction techniques for durable structures. The usage of regular cement concrete, including RCA, has shown considerable promise during the past decade in the field of concrete technology. Civil engineering specialists have gotten much attention since it represents a huge step forward in sustainability. According to published works (Gálvez-Martos et al., 2018), the developed world has undertaken a large portion of the work linked to the practical use of RCA-containing concrete, as well as the treatment and disposal of C&DW. The advancement of economics, the promotion of social justice, and the protection of the natural environment are the three pillars that support the concept of sustainability. As a result of successive summits, the world's nations have pledged to achieve these goals by conserving resources in the planning and building a sustainable built environment.

According to The Concrete Centre (2010a, b, and c), Portland cement is the part of the concrete industry that consumes tremendous energy, accounting for over 74% of the total energy consumed in the production process. The cement and ground granulated blast furnace slag (GGBS) industries have voluntarily agreed to participate in climate change agreements (CCA). Using SCMs, less cement will be required for concrete manufacturing (small-scale concrete mixing). These businesses have agreed to work with the U.K. government to reduce energy consumption, and if they fail, they will face financial penalties. The 14% greater-than-the-agreed-upon aim of 30%, according to Samad and Shah (2017); between 1990 and 2010, the cement industry improved its performance under the CCA. Between 1990 and 2010, the cement industry's CCA performance increased by 44.8%, above the agreed-upon target of 30%. Between 1999 and 2010, the GGBS sector achieved a 16% reduction in energy consumption by improving the grinding process. The SCM would continue to minimize the quantity of cement needed to produce concrete, with more studies being performed globally to standardize its specification and combine design techniques. With further advancements, binary cement, including SCMs, will continue to improve concrete's sustainability and environmental benefits.

Different SCMs are utilized to partially replace OPC, such as PFA, GGBS, silica fume, and rice husk ash (RHA). In addition, slag cement concrete has better flexural strengths and compressive than traditional Portland cement concrete. Compared to regular concrete, GGBS replacement provides reduced heat generated by hydration, improved robustness, particularly resistance to sulfate and chloride attack, and lower construction cost. In contrast, it helps protection the environment by minimizing the amount of cement required in concrete production, which helps to protect the environment. However, GGBS concrete's strength over time is more significant than OPC concrete, even though its strength development is slower than that of OPC concrete. These advantageous properties of GGBS concrete are subject to change based on changes in concrete mix proportions and curing conditions. When using GGBS, it is recommended to use a maximum replacement level of 50% and to cure at a temperature of at least 20°C. When compared to OPC concrete, the decrease of GGBS concrete is unaltered, and it is considerably easier to compact than OPC concrete. The quantity of entrapped air in slag concrete is decreased as a result of the improved workability of slag concrete (Alhozaimy et al., 2012).

Compressive strength and other technical features are linked. Concrete's impact on compressive strength is proportional to its impact on other technical characteristics, consistent with the prevailing design hypothesis for concrete in use today. Therefore, there is no need to alter design methodologies related to deformation or shear when employing GGBS and FA-containing concrete mixes. Although, for PFA, concrete is slower than the hydration responses, and they do not start up until about five days following the hydration reactions. As a result, when water curing temperatures are as low as 5°C because lower temperatures do not slow down pozzolanic processes any more than hydration reaction, the hardness of OPC concrete and concrete made with OPC/FA is quite comparable to one another (Alhozaimy et al., 1996). If the entire cementitious content of concrete is kept constant and the concrete is allowed to cure for a specified period, the concrete will not crack when compared to OPC concrete, the compressive strength and elastic modulus of FA-containing concrete are reduced (Kayali and Ahmed, 2013). A w/c of 0.48 is employed, and 20 wt% of OPC is substituted with PFA, resulting in an improvement in the concrete's flexural strength of 11%. This loss of strength becomes more pronounced as the substitution level is increased.

According to the Ayub et al. (2014), the elastic modulus of a FA concrete of the same grade is often equal to or slightly larger than that of OPC concrete. This assumption directly contradicts Kayali and Ahmed's findings (2013), which discovered that total cementitious content, curing period, and w/c all play an impact. Furthermore, FA was utilized to increase limit chloride diffusion, avoid alkali-silica reactions and sulphate resistance, and minimize heat generation in cement manufacturing. FA is utilized as a cementitious binder. It is possible to cut down on the amount of greenhouse gas emissions from building things, improve the structure's durability, and make it last longer. It takes 900 kg of CO_2 to make 1 ton of cement with each ton of FA) that is reused. Using fly ash in concrete production in the United Kingdom saves the environment around 250,000 tons of CO_2 each year (Rigamonti et al., 2012).

The usage of RHA is a concrete substitute material; it has demonstrated promising results in improving concrete qualities up to 10 wt% that are mechanical. Concrete's compressive and tensile strengths, on the other hand, decrease as the percent of replacement increases beyond 20 wt%. The water binder ratio (w/b) increases as the RHA material becomes finer, although this is countered by the use of high-range water reducers (HRWR).

GGBFS (as shown a typical composition compared to OPC in Tables 3.5 and 3.6)-containing RCA concrete's mechanical and physical qualities are developed at the 28-day-old standard cubic concrete specimens with a compressive strength of 47.6 MPa (Kapoor et al., 2021). It took 7 and 28 days to assess the UPV, water absorption, density of the reactive aggregate cement, and their mechanical properties (such as compressive strength, tension splitting strength, and binding strength). There are any relationships or a distinction between mechanical and physical characteristics. RCA (with and without ground GGBFS) was a sufficient substitute for conventional concrete in a wide range of applications because its contents ranged from 0–60 wt% (Tüfekçi and Çakır, 2017).

The influence that RCA and GGBFS have on the physicomechanical properties of concrete is shown below. After 28 days, the RCA used in the research had a compressive strength of 47.6 MPa, was made of concrete, and did not contain any pollutants. This finding is remarkable. Even though the RCA had been pre-soaked in water and the RAC had been made using SP, the RCA material's strong water absorption capabilities led to a significant slump loss at the 15-minute mark after mixing. This was the case despite the fact that the RAC had been created using SP (Landa-Sánchez et al., 2020). The compressive strength of concrete parepared by 25, 50, and 100 wt% RCA replacement decreased by 2.8%, 6.7%, and 6.9% compared to the potency of NAC.

TABLE 3.5

Chemical and Physical Properties of OPC and GGBFS (Tüfekçi and Çakır, 2017)

Chemical Composition Physical Properties	OPC	GGBFS
CaO	64.48	35.26
SiO_2	20.12	42.15
Al_2O_3	4.92	12.35
Fe_2O_3	3.57	2.35
MgO	1.23	5.23
SO_3	2.88	–
Cl^-	0.0425	0.0123
Na_2O/K_2O	0.24/0.89	0.21/1.25
Loss on ignition	1.72	0
Insoluble residue (max, %)	0.92	–
Specific gravity (g/cm^3)	3.14	2.91
Specific surface (cm^2/g)	3942	4130

TABLE 3.6

Chemical Composition and Physical and Mechanical Properties of the OPC (Paula Junior et al., 2021)

Component	OPC1	OPC2
MgO (%)	1.4	3.49
SO_3 (%)	3.2	1.63
Na_2O (%)	0.07	–
K_2O (%)	0.89	0.70
Na_2Oeq (%)	0.65	–
Insoluble Residue (IR) (%)	1.4	1.14
Properties	4617	4630
Blaine (cm^2/g)	157	200
Initial Set (min)	212	–
Final Set (min)		
Compressive Strength		
1 Day (MPa)	30	12.7
3 Days (MPa)	41.7	25.4
7 Days (MPa)	46.4	34.9
28 Days (MPa)	55.4	47.8

Compared to the compressive strength of 60 MPa (NS) after 28 days, the compressive strength of NS60 at 14 days is more significant yet weaker accordingly. In comparison to 60-MPa concrete, the compressive strength of RAC that is composed of 60 wt% GGBFS had increases of 2.8%, 7.4%, and 8.9%, or an increase of 60 wt%. The addition of GGBFS to RCA concrete results in a slight improvement in the tensile splitting strength, with the 25, 50, and 100 wt% RCA concrete having strengths of 2.9, 3.0, and 3.3 MPa, respectively, for the concrete samples. The addition of GGBFS to the 100 wt% replacement causes a significant increase in the tensile splitting strength of the concrete, and this increase is directly proportional to the concrete's compressive strength (Landa-Sánchez et al., 2020). Compared to its compressive strength, the concrete's tensile splitting strength is significantly increased when RCA is in sufficient quantities. The R100 series, on the other hand, had the worst correlation value (0.11) between tensile splitting strength and compressive strength. Additionally, after 7 days of exposure, the bond strengths of the 25, 50, and 100 wt% replacement rose by 2.8%, 4.2%, and 20.9%, respectively. As a direct result of the extended hydration time, the bond strength of the 25, 50, and 100 wt% replacement rose by 9.0, 5.1, and 25.6%, respectively, as compared to the bond strength of the NAC after 28 days. Compared to other combinations' binding strength, those containing 100 wt% RCA had a much greater value. The use of 60% GGBFS in the RAC led to the formation of bonds that were the strongest across the board for the 28-day RAC. The RAC specimens had the lowest densities and the highest water absorption, and the density to water absorption ratio was the highest. The density to water absorption ratio was inversely correlated (Landa-Sánchez et al., 2020).

Many practical and theoretical research have proved that using the C&DW is a realistic and beneficial option for conserving natural resources, safeguarding the environment, reducing pollution, and generating economic benefits. To maximize the proportion of RCA, Vivian et al. used the two-stage mixing strategy. The ideal quantities of RCA were approximately 25%–40% to improve the interfacial behavior as an intermediary layer between cement paste and aggregate (Arezoumandi et al., 2015). SF and sodium silicate were used in a method proposed by Bui et al. to enhance the mechanical characteristics of RCA (Arezoumandi et al., 2015). The proposed method can increase splitting tensile strength, compression strength, and elastic modulus according to the results—of Abreu et al. to explore the influence of multi-recycling on the mechanical behavior of recycled concrete aggregate. As a result of the research, the more recycling cycles a machine through, the less responsive it becomes mechanically; as the proportion of RCA integration increased, it decreased linearly. When Thomas et al. investigated the durability of RAC, they discovered that when the RCA replacement ratio increased, the durability of RAC deteriorated, and recycled RCA concrete's durability increased as the amount of cement in the increased blend.

When natural sand was entirely substituted by fine RCA (Table 3.7), the mechanical characteristics of the concrete improved significantly. It has been found that the compressive strength has gone up by 10% (Tamanna and Tuladhar, 2020). The aggregate with many fines (particles smaller than 75 m) has a lot more surface area, meaning it needs more water. However, the increased quantity of fine paticles

TABLE 3.7

Physical Properties of the Coarse and Fine Aggregate (Tamana et al., 2020)

Aggregate type	Apparent Density (kg/m³)	Loose Packing Density (kg/m³)	Dry-Rodded Density (kg/m³)	Water Absorption (wt.%)	Crush Index (%)	Void Ratio (%)
RCA	2,640	1,302	1,412	4.85	17.7	50.3
NCA	2,814	1,568	1,630	1.40	8.8	44.3
Sand	2,556	1,611	1,486	0.56	–	–

in the concrete mix improves to strengthen the aggregate skeleton by filling in the spaces between the larger particles. Cement mortar is found in high concentration in particles with sizes ranging between 125 and 500 μm. This could result in improved mechanical and permeability qualities for the concrete. It is also vital to note that fine and coarse RCAs can impact the formation of RAC.

Because RCA is not very good, it has only been used in a few areas. Steel fibers can be added to recycled coarse aggregate concrete (RCAC) to create a bridging effect. This is because steel fibers can help stop or slow the development of minor flaws in RCAC. They can also improve RCAC's mechanical properties and make it easier to control how it breaks. As already said, adding steel fibers to RCAC improves the material's technical properties. For example, the material is less likely to break, is more flexible and rigid, and can take more wear and tear. Also, steel fiber-reinforced recycled aggregate concrete is stronger and lasts longer than recycled aggregate concrete. Steel fibers and RCA also benefit the environment and the economy in a big way when used together. Because of this, steel fiber-reinforced recycled coarse aggregate concrete (SFRCAC) has much potential to be used in structural members.

Extensive research has been conducted around the globe on the properties of concretes that include both recycled coarse aggregates (RCA) and RFA. They are inferior to natural virgin aggregate-based concrete in every scenario that has been investigated. According to numerous pieces of study findings, several causes might be at play here. These factors include porous old mortar or cement paste sticking to the aggregates and weak interfacial transition zones between old and new mortar or cement paste connected to the aggregates. Recycled aggregates may be made better in several ways, including adding additional cementing materials to the mix and completely removing cement from the design. Kou and his colleagues carried out the research for the study. The effects of using class F fly ash as a partial replacement for cement, as an addition to cement, and in cement-free concretes were tested in concretes, including recycled coarse aggregates obtained from broken old concrete portions. The results of these tests were compared to the effects of using class F fly ash in any of these capacities. Cement made of fly ash was used instead of the 25 wt% FA. According to the study's findings, substituting FA for cement in concrete results in a decrease in the material's elastic modulus, compressive strength, and tensile strength. This effect is seen even after curing the concrete for 28 days.

When recycled aggregate concrete is combined with fly ash, the resulting mixture improves its strength and durability throughout its lifetime. On the other hand, after a treatment period of 28 days, such robust characteristics begin to emerge. In a separate piece of research, Berndt investigated the qualities of recycled aggregate concretes, which included using high-volume slag in the mix as a partial replacement for cement. The mechanical properties of the concrete, as well as its durability, show signs of significant improvement. Corinaldesi and Moriconi conducted an experiment in which they used recovered coarse aggregates from C&DW and investigated the impact of adding 30 wt% FA and 15 wt% SF to OPC (Corinaldesi et al., 2002). They discovered that adding them had no impact, which surprised them. Over time, it was discovered that RAC combined with FA and SF have superior mechanical properties. Corinaldesi and Moriconi discovered that adding 30 or 40 wt% FA cement replacement enhanced the qualities of concrete, including recycled coarse aggregates and fine RAC obtained from C&DW. They concluded after testing the effect of adding the cement substitute. It has been shown that replacing cement with FA of class C as a replacement for cement may also enhance the mechanical qualities of recycled aggregate concretes. For example, the author discovered that improving concrete's mechanical and durability properties by adding nano-silica to concretes prepared using C&DW as coarse aggregates had positive results.

Ultrafine FA (UFFA) has just been introduced as a form of fine pozzolanic material and is being used in the construction industry. It is generated using a patented separation technology that produces particles with an average particle diameter of 15 microns and a higher concentration of amorphous silica than conventional class F FA (more than 20%), as shown in Table 3.8. Compressive strength was increased when UFFA was used as a partial replacement for cement in concrete with a low w/c. UFFA concrete has superior strength and a greater tendency to decrease alkali-silica reaction expansion than conventional concrete (Ajmani, 2019).

In contrast to the impact of SF, Hosain et al. assessed the impact of UFFA on concrete. Compared to regular OPC concrete and SF concrete, the cracking

TABLE 3.8
Phase Compositions of Ultrafine Fly Ash Samples (Shaikh, 2016)

Phase (wt%)	Class F Fly Ash	Ultrafine Fly Ash
Hemitite	1.7	–
Maghemite-C	2.8	0.7
Mullite	16.8	6.0
Quartz	15.0	11.7
Amorphous content	63	81

resistance of concrete was enhanced when 12 wt% UFFA was substituted for cement. The UFFA concrete containing 8 and 12% cement substitution. The researchers determined that the compressive strength of 8% UFFA demolishes somewhat after one day but that this will not impede the long-term growth in strength. However, increasing the amount of UFFA to 12 wt% significantly enhanced the material's resistance to contraction cracking. Furthermore, the compressive strength of concrete increased with increasing FA fineness in the mixture. Before 7 days, the test mixture's compressive strength was lower than the control mixture's, but it increased after 14 days in which the fineness of the FA played a considerable effect in the changes in features and was the key factor influencing compressive strength (Hossain et al., 2007). Supit et al., in addition, evaluated the compressive strengths of cement mortars containing various amounts (5–15 wt%) of UFFA as partial cement replacement and determined that an UFFA cement replacement percentage of 8% is greatest for this application (Supit et al., 2014). Although extensive research has been conducted on the characteristics of concretes containing ultrafine fly ash as a partial cement substitute, there has been little research on the effectiveness of ultrafine fly ash in improving the characteristics of concretes containing recycled coarse aggregates derived from waste building and C&DW.

The compressive strength of concretes containing 25 or 50 wt% RCA was enhanced by up to 56 days, regardless of the RCA content, when UFFA was added to concretes containing 25 or 50 wt% RCA. This conclusion was generated using commonly available C&DW research. The improvement in compressive strength is more remarkable in concrete containing 25 wt% RCA than in concrete containing 50 wt% RCA, and it is even more significant in concrete containing 25 wt% RCA. After 56 days, the compressive strength of concrete containing 25 wt% RCA and 10 wt% UFFA had reached 94% of the compressive strength of control concrete. This difference can be minimized even further by curing over an extended period (Landa-Sanchez et al., 2020). The addition of UFFA to RAC caused a slight decrease in tensile strength. After 56 days of curing, both recycled aggregate concretes were ready for use and tested had tensile strengths of roughly 88% of the control concretes, and this gap can be closed with additional curing time. Concrete that had 25 wt% RCA and 10 wt% UFFA had a lower sorptivity at all ages than concrete with no RCA and no UFFA. In addition, it looks like adding 10 wt% UFFA to RACs could make them more resistant to chloride ions because it helps the concrete hydrate and closes the capillary gaps inside the concrete, which makes the concrete more durable. In concrete with 25 wt% RCA and 10 wt% UFFA, chloride ions penetrate less than in concrete with no RCA or UFFA (Landa-Sanchez et al., 2020).

Recycling waste materials and reusing them in the concrete industry can be an economically effective way to save natural resources while preservation the environment. However, before they can be used in practical applications, it is vital to study the qualities of recycled concrete. The consequences of concurrent use were determined in this study of waste rubber powder (WRP) and RCA, both of which recycled materials are, have a positive effect on the mechanical qualities and long-term strength of concrete. In addition, it should be mentioned that this research's limitation is using various cement types.

When RCA is used in concrete, the absence of proper coordination between the RCA and the concrete paste at the ITZ results in a decrease in mechanical qualities. The detrimental effects of WRP and RCA additions on the mechanical characteristics of concrete are exacerbated in concrete containing WRP and RCA as a result of the interaction between WRP and RCA. In comparison to the conventional concrete, the chloride ion migration rates in the WRP-concrete specimens are lower; as electrical insulators, WRP has a higher blocking capacity. The penetrating resistance of concrete is improved by 5% when the WRP rate is increased by 5%. As a result of the created concrete's lesser ITZ and the presence of the aged concrete's weak old ITZ, the use of RCA in concrete allows chloride ions to penetrate the concrete. The penetration rate increases dramatically by increasing the RCA from 25% to 50% (Wei et al., 2019). However, there are no statistically significant variations in terms of durability performance between concrete with WRP and RCA and the positive effects of WRP on sustainability outweigh the negative consequences of RCA.

Concrete mixtures have volumetric, durability, and mechanical qualities that contain a variety of RCA fractions in varying proportions. For example, concrete mixtures with RCA substitution have a lower density when combined with the coarse fraction. In contrast, large amounts of the binder are absorbed by the RCA coarse particles, but the RCA sand and filler fractions do not absorb significantly more binder than the control. The substitution of coarse RCA by the quantity of energy required for compaction increases dramatically as mass increases. It is slightly higher in the instance of RCA sand, whereas, in the instance of RCA filler, it is essentially unchanged.

RCA total filler replacement is implemented as a waste reduction strategy that simultaneously conserves natural resources. In addition to the cement-rich hydrated powder, recovered glass powders, and recycled mineral fibers examined in this study, additional recycled goods produced by C&DW, such as coarse and fine recycled aggregates, are also feasible possibilities. Researchers found that using more than 75 wt% of RCA from construction and demolition projects can produce the most environmentally friendly and long-lasting concrete.

Recycled coarse and fine aggregates can be used in place of natural coarse and fine aggregates without affecting the mechanical properties of the concrete. Revolutionary technology (ADR + HAS) makes it possible to increase the number of recycled aggregates used. As a result, when recycled aggregates completely replace fine and coarse aggregates, the behavior of recycled concrete changes in a big way. This presents a chance to transform the building industry into one that operates entirely in a circular fashion. The amount of cement paste-rich recycled powder that can be added is limited to just 5% of the total. The manufacturer says that adding more glass powder from C&DW to recycled concrete lowers its mechanical strength by a small amount. Calcium ion (Ca^{2+}), known to slow down the reaction, may not dissolve well in the early stages of the pozzolanic process. More research is needed to learn how glass powder changes over time. Thus, mineral fibers may increase recycled concrete's tensile strength and modulus of elasticity, but their presence at lower percentages makes the concrete harder to work with. Using such fibers in higher proportions can make it more difficult for recycled concrete to work and last longer. As a result, more recycled materials and products are added to the mix, and the density of the concrete decreases. This can help make buildings and other things made of cement lighter and better at insulating heat while maintaining mechanical performance.

Compared to typical high-performance concrete mixes, RCP compositions have more SCM. SF is the most common supplemental SCM in the green concrete. Some other SCMs can produce the granules denser. FA, blast furnace slag, glass powder, and the recycled powder of ceramic bricks are all examples of SCMs that can be used. These components can also be used to increase the performance of a combination in its fresh state and reduce the amount of SP used and the amount of cement used. Incorporating nanoparticles, such as those of nano silica and titanium dioxide, into green concrete performance is the subject of several areas of investigation to ensure that green concrete can be produced and used on a big scale in the real world, more research is needed into varied w/c and the volumetric replacement of both aggregates to understand the properties of green concrete and how it operates in different weather conditions. For example, fresh concrete attributes are essential for placement and durability; green concrete is impractical until it possesses sufficient hardened concrete qualities. Since a typical mix of green concrete can be created from FRP waste aggregate, as shown in Figure 3.4, RCA with properties shown in Table 3.9, or any mix of these materials may be achieved through the RCA.

(a)   (b)

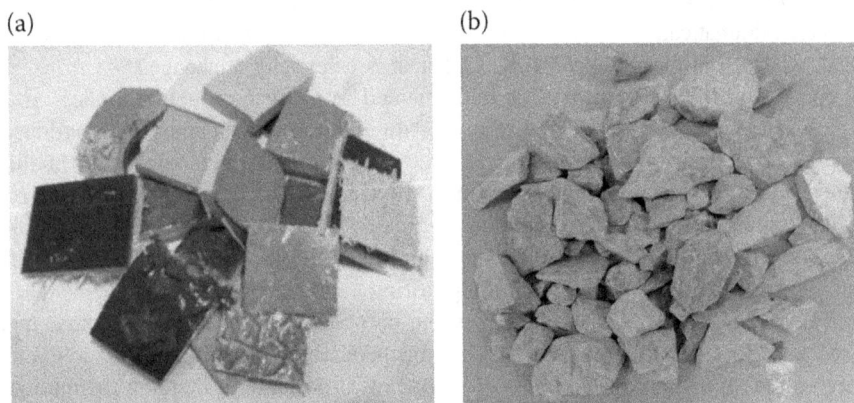

FIGURE 3.4 Aggregates in concrete: (a) fiber-reinforced polymer (FRP) scrap aggregate; (b) RCA (Shahria Alam et al., 2013).

TABLE 3.9
Properties of Coarse Aggregate (Ahahria Alam et al., 2013)

Aggregate type	Absorption Capacity (%)	Bulk Dry Specific Gravity	Bulk SSD Specific Gravity	Bulk Dry Density (kg/m³)	SSD Density (kg/m³)
NA	2.17	2.56	2.11	2,064	2,109
RCA	5.23	1.93	2.03	1,925	2,026
FSA	1.37	1.34	1.36	1,341	1,359

The behavior of RCA or fiber-reinforced polymer scrap aggregate (FSA) con-crete, as well as the behavior of FSA and RCA combined, are examined. Because of the presence of various mortars on the surface of the RCA, the concrete may have been rendered more absorbent and angular as a result. This rendered it less work-able, which resulted in a little lower slump.

In comparison to RCA concrete, FSA concrete is more malleable and workable. It is also lighter than RCA or control concrete, which could help reduce a structure's dead load in some cases. Moreover, in terms of mechanical qualities, the RCA concrete exhibited characteristics identical to the control concrete. Concrete having 25 and 50 wt% RCA substitution produced compressive strength very close to the control concrete and significantly higher than the design strength of the concrete in question (Shahria Alam et al., 2013). The fundamental difference between the two concrete aggregates can be attributed to RCA's substantially higher absorption capability.

Another thing to think about is the quality of the FSA that is being used, which could be very important. The mechanical qualities of the FSA concrete were superior to what was predicted. As long as the FSA concrete's 28-day compressive strength was not too low, it was fine. It was still in the range of 20–30 MPa, which was good enough. It is possible that the poor gradation of the FSA that is being used is to blame for the FSA concrete not being as strong as it should be. When FSA concrete is properly graded following the Canadian standards association (CSA) rules, it may have a higher strength overall (Shahria Alam et al., 2013). One is its characteristics were halfway between those of RCA concrete and pure FSA con-crete because it contained FSA and RCA. It was discovered that certain combi-nations of FSA and RCA might produce concrete that was as good as or better than the control mix. Due to the enormous disparities in shape and texture between the FSA and RCA, the combination batches could not attain compressive strengths much more significant than those of the concrete. There was no indication that combo batches were either ineffective or effective.

RCA and FSA in ready-mixed concrete are expected to save resources and minimize landfill waste (Shahria Alam et al., 2013). However, they are also ex-pected to stimulate additional research and development in Canada on how to design more ecologically friendly houses. Green concretes will need a lot more testing and study before they can be applied in the concrete industry due to the lengthy process involved in making concrete suitable for use in the field. However, this project's research and experiments will act as a springboard for introducing green concrete into the ready-mixed concrete industry. As long as environmental stewardship remains a high priority in engineering, it will pay huge dividends to devote more time and resources to investigating green concrete. Shortly, it will be critical for the construction industry to find cost-effective strategies to reduce the environmental impact of building materials like concrete. Even though the results of this study lead to crucial conclusions concerning the creation of green concrete in the ready mix industry utilizing FRP scrap and RCA together. In this point, addi-tional research is required to ensure that green concrete can be produced and used on a wide scale in the real world for read-mixing concrete production (as shown a typical plant in Figure 3.5). In order to better understand green concrete's properties

FIGURE 3.5   Ready-mixed concrete plant.

and how it operates in various weather conditions, more research is needed in varied w/c and volumetric replacement of both components.

In order to construct in a way that is more favorable to the environment, recycling concrete is critical. It also reduces the quantity of building and demolition trash generated, thereby reducing the environmental impact, which benefits to preserve natural resources from being used. A decrease in strength and resistance to deterioration can happen when RCAs are used instead of new concrete aggregates to fix these bad habits and get rid of the mortar paste stuck to recycled aggregates. The RAC properties are negatively impacted. Other additions, such as FA, SF, and fibers, can be added to concrete to make it more durable when used with the right mix design producing concrete more durable. As a result, if the suggested procedures are implemented, RCA can be a useful instrument for preserving the environment while still offering an appropriate level of structural performance in engineering applications.

For chloride resistance qualities, supplemental elements combined with cement impacted by RACs properties; their impacts were deemed substantial. It turns out that when Obla et al. conducted a test on concrete with slag and FA, the number of coulombs passed decreased as the w/c went up because more significant charges mean more chloride ions can get into the concrete. In their study, Kou and Poon found that the total porosity of RAC went down when they used 25% fly ash. Total porosity rose when they used 35% fly ash instead of cement, which they said was because the cement was diluted and fewer products of hydration formed in the first stages of the process (Demiss et al., 2018). Obla et al. reposted that establishing a schedule of a concrete mix that includes slag and FA decreases as the amount of cement in the mix increases. This is because the high-range water-reducing admixture (HRWRA) affects the setting time. Another big problem with RAC is that it can hold much water, which has caused many people to be worried about its durability, primarily when it is used in dangerous places. In these situations, RAC does not do as well as NCA (Demiss et al., 2018).

Most studies have found that adding coarse CCA to structural concrete might have a negative impact on the microstructure and water penetration. In CEM III/A concretes, adding GGBS reduces the negative impact. Using a CCA enables more efficient production. If it is essential to consider the 28-day characteristic strength, CEM I and natural aggregate (NA) should not be substituted with GGBS or CCA in proportions above 50 and 30 wt%, respectively. In unusual cases, the compressive cube strength of CEM III/A concretes can be improved by up to 60% if this condition is disregarded. At a later age, the concretes will be evaluated for conformity. This will make the concrete more environmentally friendly because it will have more CCA. More research is required to determine how superplasticizers influence early strength growth and endurance performance. In addition, CEM III/A concretes created with up to 100 wt% CCA are more durable than control CEM I concretes made with 100% NA. In addition, if it is possible to provide more cover to CEM III/A CCA concretes in the same manner as CEM I concrete, the potential for durability issues can be minimized even more (Landa-sanchez et al., 2020). This is due to the fact that around the ITZ apparent adverse effects may be attributed to other factors. Also, the use of sources of CCA is suggested. However, it is also recommended that they be checked for water absorption before being analyzed chemically and petrographically to detect water intrusion, probable contamination, and the composition of the original concrete mixture.

Green material development is seeing an uptick in demolition-derived RAC, causing a rising global population, fast urbanization, and poor economic conditions in developing nations. Building and demolition debris must be properly disposed of if we are to maintain a healthy ecosystem. Building and demolition trash can now be disposed of more environmentally friendly, reducing travel distances and increasing the amount of disposal space available. The material's flexibility is still one of RACs most conservative features, even though RAC has various downsides, such as low split tensile strength, high water absorption, and high porosity. The manufacture of Portland cement produces CO_2 emissions as a by-product. Geopolymer concrete is green concrete made from RCA to lessen the impact on the environment of the production of concrete. Silica fumes, ash, slag, and red mud are all examples of inorganic alumino-silicate polymers used to create the binder in geopolymer concrete (GPC) concrete. Bridges and other constructions are built with GPC concrete. Apart from that, a large amount of urban runoff and industrial wastewater is released into rivers and landfills.

Another factor is the increasing demand for alternative dumping methods at a reasonable cost due to rigorous environmental regulatory criteria and the repudiation of open dumps close to communities. Freshwater supply has been boosted due to the rapid population growth and increase in economic activity. A trillion gallons of water are consumed annually by concrete. It is the second most common material after wood. The amount of freshwater utilized in the construction industry and other businesses must be reduced to maintain a balance between demand and supply. According to the study, freshwater will deplete half of the world's population by 2050. The usefulness of effluent recycling, particularly in concrete, is increasingly recognized. By using wastewater to build concrete, it is possible to reduce the expense of treating it. Furthermore, this contaminated water negatively impacts the natural environment, human health, and well-being. As a result, effluent from concrete manufacturing could be used to mitigate the negative impacts on the climate and biodiversity of living organisms to some degree.

Aggregate substitution had a significant effect on how the concrete worked. GGBFS showed the most promise as a replacement for fine aggregate, followed by electric arc furnace dust (EAFD) and ceramic powder, all of which contributed to an improvement in the concrete's resistance to deterioration and its mechanical qualities. As long as using sand and incinerator sewage sludge ash (ISSA) for small and small-sized aggregates, easily use aggregates can be replaced with various alternative materials in most cases; nevertheless, significant substitution rates are frequently associated with adverse side effects in some situations.

Researchers Akhtaruzzaman and Hasnat (1983) found that the recycled clay brick material has a low density, resulting in the bulk density of the cured concrete having a lower value than regular concrete. This results in the concrete having a lower density than regular concrete. Because of this, a type of concrete with a lower density can be produced. A different study, which was conducted by Milicevic et al. (2015), discovered that concrete mixtures that contained clay brick and roof tile as aggregates fared better in terms of their physical and mechanical properties after subjecting to high temperatures. Bricks and tiles made of clay can be utilized as aggregates in concrete mixes for precast concrete beam and block flooring systems; Milicevic et al. (2015) said this was a good amount. In addition, several studies have demonstrated that recycled waste brick can be used as a proper replacement for fine aggregate in cement mortar and concrete. When the crushed brick was employed as reinforcement for the concrete, there were no significant differences observed in the long-term strength attributes of the concrete. The strength of the concrete was reduced by around 10% compared to conventional concrete and the usage of crushed clay brick material as fine aggregate should be limited to less than 50% of the time. On the other hand, broken tiles and bricks might be utilized to build precast concrete floor blocks that are superior in terms of their thermal and acoustic attributes by 50%. Milicevic et al. (2015) stated that one should consider the possibility of this happening. Gonzalez et al. (2017) discussed the use of recycled clay brick aggregate in structural concrete, which has the potential to be utilized in the production of precast prestressed beams for use in the flooring of buildings. The aggregate of recycled clay bricks can make up no more than 35% of the total material (Gonzalez et al., 2017).

According to several studies, crushed clay brick, which may be utilized as a fine aggregate in concrete, has many benefits and needs much water. This is one reason the material is not used as much as it could be. There was a lot of recycled brick material in the concrete mix, which made it difficult to work with and made the finished product less durable (Bektas et al., 2009). In addition, the aggregate's size significantly affects the compressive strength because crushed brick is used as coarse aggregate which is low compressive strength value of the concrete by two times more than if the coarse aggregate was made from the fine aggregate. Thus, the possibility of enhancing the performance of concrete produced from recycled brick aggregate by adding pozzolanic admixtures and fibers to the mixture. This substance was created from waste products from various industrial processes.

For instance, a study by Erdem et al. (2011) looked into how adding synthetic macro fibers to improve the impact resistance and microstructure of concrete could help it last longer such as using dense silica fumes could improve the performance of concrete made with recycled aggregate, adding rice husk ash changed the mechanical properties of concrete prepared with RCA. Hayles et al. (2018) developed an environmentally friendly RCA concrete mix for the structural application that made use of the equivalent mortar volume (EMV) approach to reduce the amount of cement used while simultaneously increasing the effectiveness of the binder for use in building structures. Additives such as calcined clay are effective in lowering carbon emissions and halting the depletion of resources, which is especially important in countries that do not have much money. Clay is found in many parts of the world, and when it is calcined at temperatures between 700 and 850°C, it turns out to be very pozzolanic (potassium-rich) (Fernandez et al., 2011). In addition to being good for the environment, calcined clay (Figure 3.6) can help cut greenhouse gas emissions by up to 30% when used as a pozzolan (Scrivener et al., 2018). In environmentally friendly high-strength concrete that can be utilized for structural applications, the issue of whether or not it is possible to utilize crushed clay bricks that have been produced from local waste as a replacement for up to

(a)   (b)

FIGURE 3.6 Cementitious materials: (a) OPC, (b) calcined clay (Olofinnade and Ogara, 2021).

50% of the fine aggregate. The possibility of adding calcined clay to concrete produces it more durable while maintaining the same rate of cement replacement (Fernandez et al., 2011).

Recycled clay brick aggregate could be used to produce high-strength, durable, eco-friendly concrete by mixing it with calcined clay and recycled clay bricks. Crushed clay brick aggregate (CCBA) (also known as crushed clay brick wastes) can be utilized in producing concrete that is kind to the natural world and durable enough to withstand the test of time. It can also be used to make concrete that is both strong enough to last for a long time and environmentally friendly. On the other hand, crushed brick clay is the ideal replacement level, at 10%, which makes a big difference in the strength of the building over the previous amount of replacement.

In order to produce concrete, it is not possible to work with it if they use natural sand instead of the mixture containing crushed clay brick aggregate. This is because the CCBA material can absorb much water. Additionally, the angular form of the CCBA particles renders the concrete mixture more durable and less fluid. As a result, a superplasticizer should be used. By substituting 10% of high-strength concrete with clay rather than by substituting a control concrete mixes with no clay. Micropores had been filled with a substance that had been squeezed extremely hard. Especially at 10% calcined clay, the concrete mixtures with 10, 20, and 30% CCBA replacement amounts have a significantly higher tensile strength than ordinary concrete. The regular concrete did not. Using CCBA instead of sand and adding 10% calcined clay made the concrete stronger after 28 days, for example. In addition, concrete mixtures with 10 wt% CCA were 5% stronger than normal concrete, while the mixtures with 50% CCA were about 20% weaker than normal concrete. Using CCBA aggregate in concrete had the same effect on its flexure strength as on its compressive strength, which is essential to using this aggregate due to hydration rate and a high concentration of CCBA particles in the concrete mix might also lead to deterioration in the strength qualities of the concrete with more CCBA in the concrete mix.

Clay bricks that have been crushed may be recycled as a pozzolanic component in structural concrete, an alternate disposal method to dumping them in landfills. Aside from that, the article suggests that crushed clay bricks can partially replace fine natural aggregate in the construction of high-strength concrete with pozzolanic admixture in the 40 MPa and higher, up to 50 wt% of the total amount of fine natural aggregate. This is in addition to the fact that crushed clay bricks can partially replace fine natural aggregate in the construction of high-strength concrete. In the production of high-strength concrete, the use of crushed clay bricks as a partial replacement for fine natural aggregate is recommended. This suggestion is given because the use of crushed clay bricks is possible. In addition, using these wastes continues to be the best approach for tackling the complicated problems of depletion of natural resources and pollution of the environment, as well as ensuring long-term profitability and sustainability.

In case of FA and RCA together affect the different properties of concrete, as well as the economic performance of the resulting concrete mix. By adding FA to the mix design, the adverse effects of RCA on the workability of concrete are reduced to a minimal level. FA behaves for the loss of workability that comes with adding RCA, but it also reduces the need for plasticizers to reach the workability in the final product. As FA content increases, the amount of plasticizer required to achieve the desired workability decreases proportionately. However, the harmful effects of RCA, and FA should be used together to lessen the adverse effects of RCA on hardened concrete properties, says the author. The more RCA in concrete, the less solid and dense it is, and the more it lets water in. When FA is used with RCA, it makes concrete more vital, especially at the 90-day mark when it is still wet. FA also makes concrete denser, which makes it more durable. The use of FA cuts down on the amount of time it takes the porosity of RCA concrete while simultaneously improving its chloride penetration resistance. If you take into account all of the investigated hardened qualities of the concrete, you will find that the concrete prepared with 50 wt% RCA and 20–30 wt% FA performs practically as well as the concrete made with standard materials. However, based on the fact that all concrete mixes have the same level of workability, RCA does not lower the total cost of building with concrete. Instead, concrete with this additive has a higher cost-to-strength ratio (CSR) than concrete without. As the need for plasticizers to keep the workable concrete increases, the total cost of the finished product does not go down much, even though RCA costs less. Thus, the cost of transportation determines the impact of FA on the concrete economy. The increase in transportation expenses does not result in a significant reduction in concrete's total cost. The gains in efficiency from using FA are proportional to the coal power plant to end-user location. It does not matter how far you have to go to your concrete. FA can help increase the CSR performance of RCA concrete (by causing significant improvements in strength). While the overall cost of production is reduced due to reduced shipping distance and distribution costs, it also reduces the amount of plasticizer needed.

According to current estimates, OPC accounts for 10% of global CO_2 emissions, with the likelihood of this amount growing to 15% in a while. Various techniques, such as novel alkali-activated polymers, have been presented as a potential alternative to this highly polluting binder, combining lower greenhouse gas emissions with noble corrosion resistance properties. There are many different types of these new binders, but some of the more common ones are: slags, metakaolin, SCBA, and RHA. Because SCBA and RHA are agricultural waste materials with corrosion characteristics equivalent to OPC, interest in them has recently surged. The SCBA has a pozzolanic activity after treatment, making it a viable alternative to OP). In terms of corrosion behavior, only a few studies have considered these innovative binders. However, despite the poorer workability of SCBA, as a result, substituting OPC in amounts of 10 to 30 wt% reduces both the SCBA diffusion coefficient of chloride and the SCBA permeability to water. As a result, there is little agreement on how well they will corrode, so they have been used only with SCMs as a safe bet. Because the new binders are made from recycled materials, this partial replacement of OPC is good for the environment and cheap.

Some cement is replaced with pozzolans, such as furnace coal BA, SF, and GGBS to reduce CO_2 emissions and waste steam in the green concrete industry. This technique has been widely used around the world to address these issues.

Using by-products instead of cement has been demonstrated to benefit the environment in some circumstances; thus, this should be included. Crossin showed that using 30 wt% GGBS instead of cement reduces greenhouse gas emissions by 47.5% compared to concrete built with the same amount of cement and the same amount of GGBS mixed in because as opposed to standard concrete reduced the number of respiratory inorganics, GWP, and nonrenewable energy, and acidity. In addition, the use of SF and GGBS, Kim and Lee, as well as Kadam and Patil, studied BA-based concretes (in which sand was substituted with BA in varying proportions) with a high binder concentration. Their compressive strength was comparable to that of conventional concrete. An experiment with high-strength concrete using BA was completed after 28 days, and Kim and Lee evaluated their findings in the range of 40 to 100 wt% (Cwirzen et al., 2008). The compressive strength was determined to be between 62 and 72 MPa. After 28 days, the compressive strength of normal-strength BA-based concretes measured 41.52–45.76 MPa when the total binder (cement and SF) was 8 wt%, according to Kadam and Patil's study (Cwirzen et al., 2008).

BA-based concretes with varying quantities of BA in place of sand are the subject of this investigation. Because these concretes showed compressive strengths comparable with ordinary concrete, their life cycle was examined. For greater strength, the concrete made using BA was supplemented with additional binders like SF and GGBS.

Green concrete combinations with SF as an extra binder and with SF and blast GGBS as an extra binder were found to have a more significant environmental impact than the mixtures with BA-based concretes, which were shown to have a lower environmental impact than standard concrete mixtures. There was a way for the study system to expand so that other industries' waste could be used in concrete. To make things even better, this analysis took into account the fact that there was no need to transport BA, SF, and GGBS and put them in a landfill. The need to make new steel was also avoided.

By-product-based concretes have been intensively studied to improve mechanical performance while simultaneously reducing the cost and environmental impact of this type of cement substitute. FA, FBA, and GGBS are all by-products of different industries that can be substituted for cement in the concrete sector. There have been a slew of positive outcomes reported in this area. Paul et al., for example, observed a 26.3% increase in the compressive strength of concrete with GGBS and FA utilized to replace 0% to 30% of the cement. For UHPCs, Meng et al. investigated the cost-effectiveness of cement replacement with Class C fly ash (FAC), GGBS, and SF, in addition to quartz sand replacement with ordinary concrete. Such combinations cost 4.1–4.5 $/m^3/MPa under average concrete curing conditions. In concretes, abiotic depletion, acidity, and eutrophication were reduced by 40%–70%, in which GBFS substituted 66 wt% of the OPC (Cwirzen et al., 2008).

A more pressing environmental concern is the loss of natural aggregates like sand and gravel. Researchers have studied the possibility of replacing sand in concrete with industrial by-products, including FA and furnace coal bottom ash (FBA), copper slag (CS), and quarry dust powder (QDP). FA and FBA are two industrial waste compounds studied. Even while sand generated less particle matter than FA or FBA, HPC created higher CO_2 emissions than standard cement (i.e., it significantly influenced climate change). It was shown that by replacing sand with QDP in lightweight foamed concrete, CO_2 emissions were reduced by up to 10%. Therefore, it is clear that additional research is needed in the long-term effects of replacing sand in concrete making with other materials.

3.5 RCA IN NANO CONCRETE

In recent years, there has been an increase in the use of nanotechnology in concrete projects. This is due to the considerable impact that nanotechnology has on structural stability and the longevity and strength of concrete materials. In addition, the use of nanoparticles in the formulation of concrete mixes, such as nano-silica, results in concrete that is friendlier to the surrounding ecosystem. Concrete may be improved in various ways, both chemically and physically, by using nanomaterials, the sizes of which range from 1 to 100 nanometers. The addition of nano-silica to concrete causes a response that results in forming calcium hydrate (C-H) crystals and calcium silicate hydrate (C-S-H) synthesis, for example. Both of these reactions improve the cement structure and fill any gaps in the concrete. Cementitious materials that contain suitable amounts of silica, such as microcrystalline silica, are added to the concrete. These materials then react with the C-H crystal to produce secondary C-S-H, decreasing porosity and permeability while increasing the compressive strength. Microcrystalline silica is one example of a cementitious material that contains suitable amounts of silica, as shown in Figures 3.7 and 3.8. Concrete may be strengthened with the help of nanoparticles in a number of different ways. Ca(OH)2-reacting nanoparticles create more calcium silicate-hydrate (C-S-H) gels than crystalline nanoparticles due to the amorphous form of the nanoparticles and their extraordinarily large surface area. As a result of their "filling function," they can also fill very thin holes. Compared to silica fume, the overall performance of

FIGURE 3.7 The nano-silica powder.

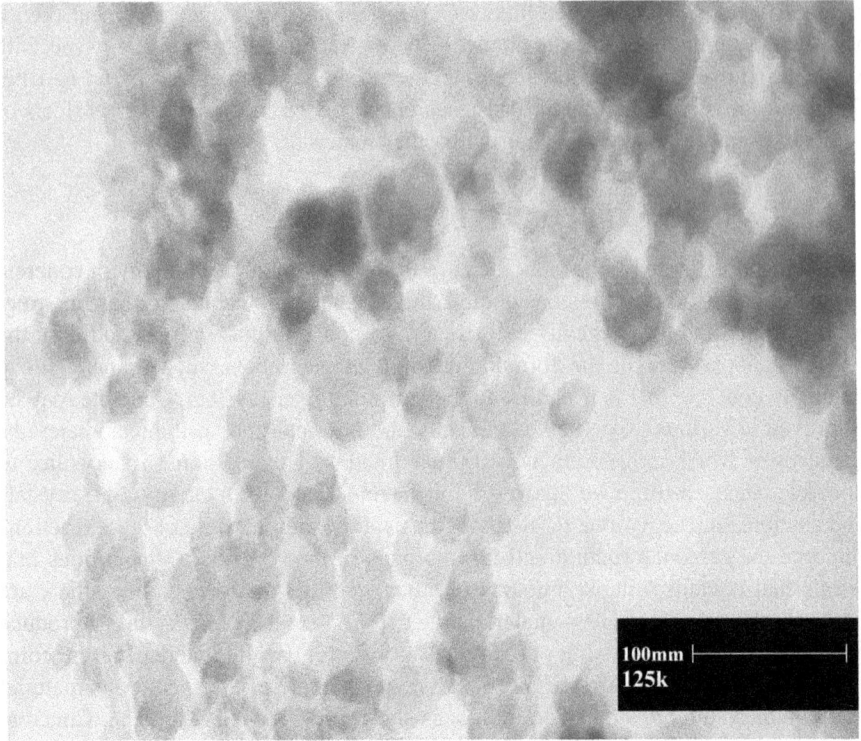

FIGURE 3.8 The scanning electron microscope (SEM) image of nano-silica (Provided by Iranian Nanomaterials Pioneers Company) (Shahbazpanahi et al., 2021).

nano-silica (NS) particles in concrete was superior to that of silica fume, as stated by Zhang and Islam (2012).

An additive in concrete is composed of extremely fine particles with sizes between 10 and 100 nm, which are ball-shaped and have a diameter of less than a millimeter. When calcium hydroxide is reacted with nano-silica compounds, the materials have the same pozzolanic effect as more significant concentrations of these materials. Nano-silica has better compressive strength than, say, nano-clay or similar nanomaterials and is more extensively utilized and environmentally benign than these other nanomaterials. When nano-silica is added to cement paste, the hydration acceleration of the cement accelerates, and calcium hydroxide is generated in the first few minutes due to the increased contact surface of the nano-silica with the water in the paste.

Nano-silica use in conventional concrete has been the substance of wide-ranging research, with encouraging results. In addition, numerous studies have seen a reduction in the concrete's slump in numerous studies involving the use of nano-silica with RCA in concrete. Additionally, nano-silica has recently piqued the interest of the construction sector to incorporate it into concrete to produce a more durable and long-lasting material. Furthermore, the volume of demolished concrete that contains nano-silica is expected to rise significantly shortly. The recycling of used nano-silica particles in the form of recovered concrete aggregate is yet primarily studied, but RCA-UNS in a building built of coarse aggregates containing nano-silica that is recycled once construction is complete may have another life cycle. Concrete has been made with recycled coarse aggregates containing nano-silica without any testing. As a result, there is no information about the quality of these recycled coarse aggregates, which include nano-silica, that has been used. This makes it even more important to reuse recycled coarse aggregates that have been used before with nano-silica.

In contrast, NS was incorporated into the concrete during the mixing process. Suppose you soak dry RCAs overnight in an evenly dispersed sodium sulfate solution, with salt deposits on the aggregate's surface and some even making it down to the aggregate subsurface. According to this notion, porous surfaces such as RCAs are effectively densified by this coating method, which uses pozzolanic reaction and filling effects to harden the surface rapidly. It is also possible to densify the porous ITZ between older mortar and older aggregates by using a nanoparticle layer placed on the aggregates' surface and acts as both impregnating filler and an adhesion promoter. A strong link between the recycled aggregate and the new mortar is predicted to form, increasing the qualities of this environmentally friendly concrete significantly.

Using crystalline additives in the mix may improve the concrete's physical, mechanical, and long-term qualities. The recycled masonry aggregate in the concrete is used to manufacture the concrete. Concrete's mechanical characteristics and durability suffer as a result of substituting recycled components for natural ones in its composition. Because recycled masonry aggregate concrete is porous and may soak up water, it cannot be used to make concrete that lasts as long. On the other hand, these issues may be resolved by including an appropriate addition into the concrete mix. Earlier studies investigated and ranked various strategies for enhancing admixtures' quality. These strategies were shown to be effective.

For this reason, a crystalline addition could be added to recycled masonry aggregate concrete to make it more durable. On the other hand, in the absence of the ameliorating influence of the crystalline admixture, the rate of immersion absorption was three times higher than was anticipated. In contrast to what most people believe, capillary water absorption demonstrated that the crystalline admixture had a beneficial impact; despite this, the value was still more than twice as high as the value at the beginning of the experiment. By enhancing the freeze-thaw resistance of recycled masonry aggregate concrete using crystalline admixtures, it is possible to meet the criteria of the Czech standard for frost resistance (Pavlů et al., 2022). Compared to conventional concrete, the carbonation depth of recycled masonry aggregate concrete (RMAC) was more than twice as deep as that of conventional concrete. Because crystalline additive was used, the carbonation depth of concrete mixed with RMAC was greatly enhanced. This study investigated the potential for crystalline additions to improve the strength of concrete. Because this mixture interacts with the water in the recycled masonry aggregate, it was hypothesized that the product that it would produce would be able to fill the voids that are present in the masonry aggregate. Because the outcomes of water absorption through immersion and water absorption through capillary are distinct from one another, it is impossible to demonstrate beyond a reasonable doubt the excellent effect of crystalline admixture on the water absorption capacity of recycled masonry aggregate concrete. On the other hand, it has been shown that the addition of crystalline admixtures may increase the durability of a substance, particularly with regard to its capacity to endure being frozen and thawed.

The RCA is available in 2/0.25, 0.125/2, and 0/0.125 mm parts were used in the experimental program to substitute virgin aggregates in a semi-dense asphalt (SDA) at 100% and 50% of the cost of coarse aggregates, sand, and filler, respectively. For every mixture, only one fraction was substituted, while the air voids and binder percentages were maintained at their original levels. Following that, the mixtures' volumetric characteristics, compactability, and mechanical qualities/durability were examined. Mechanical properties/durability was assessed in the wheel tracking test (WTT) using indirect tensile strength, moisture resistance, and rutting resistance. The major goal of the experiment is to see how each RCA fraction affects the volumetric and mechanical properties of the final combination, which to understand how to take advantage of the mechanical and economic benefits of grading RCA fractions. As a result, more successful RCA recycling in new materials would be possible, as well as enhanced acceptability of the material as a suitable concrete substitute. Two further modest experiments are the investigation of RCA in higher porosity concrete and the utilization of fracture energy—calculated using two different approaches—in determining how sensitive the mixtures are to moisture.

One of the modification strategies to strengthen RC's inherent resilience and tensile strength using admixtures and improving RCA through modification therapies. Using fly ash enhances RCA, as listed in Table 3.10, resistance to sulfate attack and freeze-thaw cycles. Mukharjee and Barai, during their research, concluded that the compressive strength of RC that had been treated with 3% nano-SiO_2 was equivalent to that of natural concrete (NC). When RA is immersed in a solution containing nanomaterials, specific nanomaterials, such as nano-SiO_2 and nano-$CaCO_3$, can cover the holes and gaps in adhering mortars, producing a more cohesive bond. In addition, specific nanomaterials, when combined with calcium hydroxide ($Ca(OH)_2$), can produce C-S-H gels. These gels have the potential to make RCA even more robust and long-lasting. Polymers, when coated with a hydrophobic layer, can be utilized to prevent water from entering porous objects by acting as a barrier. RCA was submerged in a polymer emulsion

TABLE 3.10

Physical Properties of Recycled Aggregate (Lei et al., 2020)

Apparent density (kg/m^3)	Water Absorption (%)	Crushing Index (%)	Size Distribution (mm)
2531	4.0	17.6	5–31.5

which the molecules of the polymer emulsion filled the holes of the porous ceramic mortar and the surfaces of the RCA particles, which resulted in the RCA particles becoming more stable. When the polyvinyl alcohol (PVA) emulsion concentration was more than 10%, the results were only slightly better than when the concentration was less than 10%. This led Kou and Poon to conclude that treatment with PVA emulsion could improve RCA's physical and mechanical properties. Furthermore, there is a chance that impregnation with PVA could make it easier for RCA and concrete to stick together.

3.6   RCA IN REPAIR CONCRETE

Recycling concrete has developed as a realistic solution that offers a number of advantages, including the reduction of expenses associated with landfills, the preservation of the environment, and the achievement of sustainable development goals. RCA is an abbreviation for recycled concrete aggregate, and it describes the particles produced as a result of activities such as crushing, screening, and grading old concrete. Fine RCA, medium RCA, and recycled aggregate (RA) are the three subcategories that fall within the category of medium size in the RCA classification system. The term "RCA" refers to any concrete that contains at least some recycled aggregate. Due to the low water absorption ratio of recycled fine aggregate, the usage of recycled fine aggregate in the manufacturing of RCA is still fraught with difficulties. Due to the fact that it is a coarse aggregate, it is frequently referred to as coarse RCA. Recently, there has been much interest in the concept of reusing re-covered coarse aggregate. The RCA has two fundamental problems that have not been fixed yet, restricting the range of applications for which it may be used. In order to attain a high water absorption ratio (RA), the old hydrated mortars are connected to the new particles. When new recycled concrete is contrasted with freshly recycled concrete, the fresh recycled concrete's workability is significantly inferior to that of new recycled concrete. The compressive strength of the RAC, as well as its durability, are both reduced as a result of the use of the same quantity of water in each batch of concrete. This is because the mixture did not include a sufficient amount of water. If RCA is forced to crumble and crush any residual concrete, it will become weaker and less durable due to this need. Before the creation of RCA, a number of studies have discovered that they can successfully enhance RA by striking and rubbing between particles or by pre-soaking RA in silica fume solution for a significant amount of time. This approach solves both of the problems that were discussed before. However, they are seldom used significantly because of the inconveniences and high costs connected with these pre-treatment methods. According to the results of experiments, the compressive strength and durability of RCA are increased not by pre-treating the material but rather by adding FA to it after it has been manufactured. There have only been a few studies that have investigated the influence that FA has on the microstructure of RAC as well as the damage caused by RCA. When using recycled concrete mixes that include FA or UFA, it is important to analyze the interfacial transition zone between the old mortar and the original aggregate. This will allow you to identify how the old ITZ's porosity structure and fracture characteristics alter.

For a special case, the range of the recycled aggregate's former ITZ may be determined to be, on average, roughly 4 microns thick. In FA and UFA, the percentage of particles with a size less than 4 micrometers is 34.42% and 80.73%. These particles can fill and shut pores and fissures and do an outstanding job. Because RCA can potentially be problematic, supplementing it with FA or UFA may be beneficial. The effectiveness of the repair may be evaluated by looking for changes in the structure of the pores, an increase in the length of the bonding zone, a reduction in the breadth of the ITZ, and a lessening in the number of fractures in the material. When compared to FA, UFA has a special mending effect. This is primarily because of the fine dispersion of the UFA, which was developed by UFA (2020).

3.7 RCA IN REINFORCED AND PRE-STRESSED CONCRETE

Reinforced concrete is typically long-lasting. Its versatility and low cost have led to its widespread application in constructing semi-permanent structures due to its durability and affordability. However, it has become increasingly clear that aggressive chemicals, such as chloride ions, are capable of causing damage, which results in corrosion of implanted steel, which is becoming more common. It is thought that corrosion of the steel reinforcement causes concrete to crack and shatter, and the concrete cover is delaminating and spalling. It also reduces the cross sections of concrete and reinforcement; they also lose their ability to stick together and become strong and flexible. Because of this, the concrete building will not last as long as it could.

The significant environmental impact of cement implies that its use should be reduced while its utilization should be increased. Optimizing structure shape depending on the qualities of the materials employed is one strategy that could be applied. For instance, concrete constructions can benefit from topological improvement. Reducing the amount of material used in concrete constructions makes it feasible to reduce cement use.

Occasionally, ingredients such as FA or slag sand can be substituted for hydraulic binder cement. Another idea is to employ high-performance materials in concrete construction instead of standard ones. Because of their great strength and stiffness capabilities, they can significantly reduce the amount of concrete used. Fiber-reinforced polymer (FRP) (Figure 3.9) has received much attention in past years, in using fiber reinforcement instead of steel reinforcement in research and the construction industry. Ehlig et al. (2012), for instance, provide a summary of fabric-reinforced concrete constructions that have been built. Construction components with fiberglass reinforcement promise to construct lightweight, long-lasting concrete structures (Kromoser et al., 2018). Compared to other fiber types such as basalt, glass, and aramid, CFRP is distinguished by having a high Young's modulus equal to steel. Because of its good stiffness attribute, CFRP reinforcement is the best solution for restricting building structural deformations to a bare minimum because of its superior rigidity feature, as necessary in serviceability restriction. In addition, CFRP reinforcements can have a tensile strength of up to 3,000 MPa or more, and a specific non-corrosive behavior is also an option here. It is now feasible

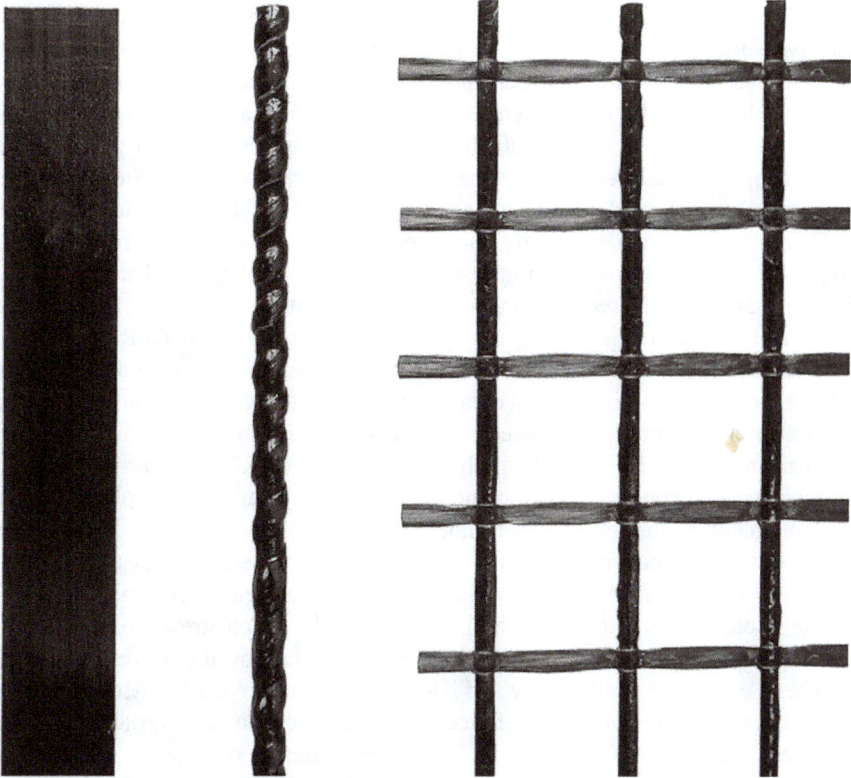

FIGURE 3.9 CFRP reinforcement products from left to right: laminate, rod with wrapping and grid (Stoiber et al., 2021).

to employ a concrete cover with a minimum thickness of 10 mm or less; this reduces the amount of material utilized in concrete buildings.

One of the most significant disadvantages of the material, in addition to the previously mentioned advantages, is the significant need for raw materials and energy during the manufacturing process of CFRP reinforcement. This need is significantly greater when compared to the typical requirement for steel reinforcement. It is often challenging to discover design values that enable you to look at how well CFRP reinforcement performs in the environment due to CFRP-reinforced structures. This is because it is a result of the fact that CFRP reinforced structures. BFRP reinforcement in concrete beams was the focus of investigation in the research carried out on this topic by Inman et al. (2017). This research was carried out as part of a study on this topic. Even if its stiffness attributes are less desirable than steel reinforcement, the investigations show that BFRP reinforcement is better for the environment than steel reinforcement. The LCA was performed with the assistance of software, and the results indicate that CFRP-reinforced concrete buildings provide a lower risk to both people and the ozone layer when compared to traditional concrete structures.

The environmental impact of CFRP reinforcement in concrete structures is frequently overlooked by individuals in the corresponding field. For the bridge design under discussion, the results suggest that CFRP reinforcement is an environmentally benign option to steel reinforcement, as long as its strength and stiffness qualities are used to their utmost extent and effectively in the bridge construction process. The prestressing of CFRP-reinforced structures is one strategy that might be used to increase this use. According to Hammerl and Kromoser (2021), the serviceability of a CFRP-reinforced beam is greatly improved by prestressing CFRP rebars. Their investigation revealed encouraging results. It should be highlighted that the illustrative application considered within this study does not allow for the development of a reasonable conclusion about the impact on the environment of CFRP-reinforced concrete structures. CFRP reinforcement in concrete structures must continue to be studied because the environmental data is constantly evolving and becoming more up to date, if not more transparent. In corrosive environments, such as chlorine salt, RCA replacement percent of 100% should be avoided. Instead, RCA replacement percent for high-strength concrete should be no higher than 30%, and RCA replacement for low- and medium-strength concrete should not be higher than 60 wt%.

In a reinforced concrete structure, the damage is caused by inelastic deformations. As a result, any damage variable used to represent a demand parameter contains some distortion. While the material is under much stress, cross-sectional curvature and regional damage condition is determined by the movement of the member's end. Specifically, story and inter-story moves are used to determine how much damage there is worldwide. Forces can be used as a severity variable to assess damage (i.e., members' resistances, base shears, and story shears). Inelastic reversed cyclic loadings can also be used as a damage metric on rigid composite members (RCM); energy is absorbed or released, as shown in Table 3.11 (Arjomandi et al., 2009; Heo and Kunnath, 2013; Zameeruddin and Sangle, 2021). Using a collapse mechanism allowed us to link the damage amount to a specific level of structural strength. This sequence of transitions followed plastic hinge formations and the transitions between construction quality levels. Different drift limitations are applied to calculate a collapse zone, enabling a faster seismic evaluation method than was previously possible.

TABLE 3.11

Behaviors of the Structure in the Entire Range of Nonlinear Action (Zameeruddin and Sangle, 2021)

Performance level	Behavior	Damage State
Operational (OP)	Elastic	No damage
Immediate occupancy (IO)	Strain hardening	Light damage
Life safety (LS)	Ultimate strength	Moderate damage
Collapse prevention (CP)	Strength reduction	Severe damage
Collapse	Imminent collapse	Extreme damage

Xiao and Zhang found when RCA contained up to 30 wt%, there was a more considerable reduction in residual compressive strength, while Vieira et al. reported that containing up to 100 wt% RCA behaved after being heated to high temperatures. Adding RCA to concrete does not have any relevant effect on the performance of the concrete. Reduced w/c ratios increase the fire resistance of reinforced concrete (RC), which already has a high base level of fire resistance, as seen by fewer cracks following exposure to high temperatures. The residual mechanical properties were higher based on their RCA experiments. RCA from super strength sources concrete is better than RA from natural aggregate concrete when using RCA. The use of cement-replacing materials (CRMs) can also improve the performance of RAC, as has been established at room temperature. Bui and colleagues discovered that the use of these CRMs improved the residual characteristics of RAC. The utilization of RCA did not have an appreciable effect on the performance of the concrete when it was subjected to high temperatures; for instance, the fire resistance of RAC is improved when the w/c ratio is reduced.

The concrete-filled steel tube, often known as the CFST, is a kind of composite construction that fully uses the mechanical capabilities of both steel and concrete to create a robust and lightweight structure. Numerous analytical and experimental research has been conducted about CFST, and substantial progress has been achieved. In order to encourage the use of recycled aggregate concrete-filled steel tubes (RACFST) in engineering, several academics are concentrating their research on the mechanical characteristics of these tubes. According to an investigation into the mechanical behavior of a RACFST column, the RACFST column has a high ultimate bearing capacity and excellent deformation performance. Additionally, the ultimate bearing capacity of the RACFST columns grew in tandem with the eccentricity of the columns as they increased (Demiss et al., 2018). Xu et al. experimented to determine how RACFST columns behave in a seismic environment. Based on the findings of this experiment, they reported that the final strain of the RACFST short column was greater than that of the natural concrete-filled steel tube (CFST) after being subjected to high temperatures. This was the case even though both types of columns were subjected to the same amount of stress. The research concluded that the local buckling of the metal tube was the root of the problem that led to the collapse of the RACFST columns. Regarding how they responded to earthquakes, the RACFST columns performed very well, and the RACFST structures themselves might be used in areas prone to earthquakes (Demiss et al., 2018).

Using RCA from crushed reinforced concrete constructions that have been demolished is a great way to save money. It has been used in concrete as a partial substitute for natural aggregates and extensively explored and tested in various real-world applications in numerous countries. According to the general understanding, the mechanical and durability properties of RCA have inferior mechanical and durability characteristics compared to natural aggregate concrete (Ntaryamira et al., 2017). Due to the bad qualities of recycled aggregates, such as increased porosity, the existence of old mortar, microcracks caused by crushing, and so on, this is the case. During their service life, reinforced concrete structures are subjected to fire, which causes the concrete to reach extremely high temperatures. The mechanical qualities of standard and high-strength concrete are drastically diminished when exposed to fire at varying temperatures. In addition, the research on mechanical features of concrete disintegration when exposed to high temperatures has required considerable time and effort.

In contrast, the behavior of RAC at elevated temperatures when temperatures exceeded a certain threshold, the mechanical qualities of RACs degraded and became comparable to those of other types of concrete. Kou and colleagues also report comparable outcomes. Xiao and colleagues disclosed that the residual mechanical properties of recycled aggregate concrete diminished when temperatures approached 300°C. This was followed by a pattern of temperatures decreasing between 400 and 600°C, then a significant drop to 800°C (Ntaryamira et al., 2017). Martins et al. conducted an additional study examining the residual mechanical properties of recycled aggregate concretes consisting of recycled ceramic coarse aggregates after exposure to increased temperatures up to 600°C. The estimated residual mechanical properties of recycled aggregate concrete deteriorate in reaction to a rise in temperature. In other research, residual compressive strength of recycled aggregate concretes decreases as their exposure to severe temperatures increases (Ntaryamira et al., 2017).

Reinforced concrete, with a weight replacement of RAC concrete of 43%, can save up to 37% of primary raw materials compared to conventional concrete. It is possible to save up to 50% on building expenses if the RC concrete is manufactured entirely from RCAs. Concrete recycling is critical in the building industry's efforts to reduce environmental impact. RC-concrete does not contribute to environmental protection on its own. This is due mainly to manufacturing cement, the most significant source of greenhouse gas emissions. The RC-concrete mixtures may need to be tuned to get the necessary physical and mechanical properties, depending on the quality of the RAC. This can be done by adding admixtures or increasing the cement used. However, if the concrete is built of OPC, the cement need for RC-concrete with an exposure rating could be decreased by 5 kg/m^3. As a result, the environmental implications of RC-concrete manufacturing should be assessed on a case-by-case basis rather than using a blanket approach.

Concrete using FRC has been found to have higher mechanical qualities than conventional concrete. Anti-cracking properties of composite materials can be improved via fiber-based interface bonding. With the proper selection of FRC may typically outperform conventional in terms of compressive and tensile strength, hinting that it may be a viable candidate for RAC enhancement. For fiber-reinforced RAC, however, it is necessary to consider fabrication costs and environmental impact from a practical aspect. Because of this, in terms of performance, BFs surpass other fibers, such as steel, carbon, and glass fiber. Regarding performance, BFs are less expensive and easier to disseminate in concrete than SF. BFs offer better heat, alkali, and thermostability resistance than other fibers because of their refined structure. Most importantly, basalt is classed as an inorganic silicate, implying that it has an appropriate consistency when mixed with cement.

RFA was not considered concurrently because a single component (the replacement of coarse materials) may be disregarded, and other components may also have an impact. Researchers conducted a study to compare the performance of various fibers, such as concrete-filled steel (CFS) and polypropylene fiber (PF), to fiber-reinforced composites made using BF and the polypropylene-basalt hybrid fibers (PPBHFs) because the elastic modulus is greater than the compressive strength, the compressive strength was found to be superior to the splitting tensile strength.

Fiber-reinforced RAC structures with varying fiber and volumetric percentage BF can have greater compressive and splitting tensile strengths than CC when the combination ratios are identical due to the water-absorbing capacity of recycled coarse aggregates, RAC. Fibers can have a beneficial or detrimental impact on the characteristics of RAC, depending on the fiber type utilized because of the non-uniform distribution of fibers within the concrete, the latter effect may result from this. On the other hand, several factors contribute to the RAC network's failure modes in various ways (Ntaryamira et al., 2017). As a result, compressive strength in the cube and axial compression, tensile strength in the split, and elastic modulus were all best achieved at volume fractions of 10% or less by using BF. In addition, the design codes can be used to create formulas for converting cube compressive strength to other mechanical parameters, in addition to the recommended strength and elastic modulus estimates and test results, yield predictions that are reasonably close to practical (Ntaryamira et al., 2017).

RC, the most prevalent building material on the world, is manufactured with rebars composed of AISI 1018 carbon steel (CS) (Arezoumandi et al., 2015). For a long time, hydraulic concrete was the planet's most widely used construction material. Reinforced concrete structures have a long service life and require little maintenance. As a result, each government must spend billions of dollars to repair and maintain rusted steel reinforcement in infrastructure such as bridges, bridge tunnels, highways, and ports, among other things. An electrochemical process leads to the corrosion of steel embedded in concrete. The cathode reduces oxygen while the anode oxidizes iron. Steel embedded in concrete corrodes because of this. Corrosion is caused by some factors that favor passivity breakdown, the most important of which are carbonation and hostile ion incursion. Sulfates are found in inorganic salts, and chlorides are found in marine settings. Ions that occur in groundwater and surface water are the most active depassivating ions. On the other hand, the concentration of unfriendly substances may vary significantly in specific environments. The presence of sulfates in contact with a cured cement paste enhance the solubility of matrix components and promote concrete degradation through leaching, lowering the degree of reinforcement protection. According to another research, galvanized reinforcements outperform standard carbon steel reinforcements not only in harsh situations but also when they come into touch with pollutants in the concrete mix.

Currently, RCA is mainly used in low-stress structures, like practicalities. Some of the concerns with RCA could be solved using RAC packed with steel tubes. You can make the RAC core more robust and durable by wrapping it with steel tubes. Some researchers have studied steel tube columns filled with RAC. Xu et al. (2020), for example, employed data mining to determine how well RAC-filled steel tubes would withstand axial and lateral loads.

Concrete mixing materials containing steel fiber are relatively new. There is no pelleting of steel fiber in the concrete, and it is uniformly dispersed throughout. It has become increasingly popular in buildings because it dramatically enhances the mechanical qualities of concrete, making it easier to build with, even though the volume fraction of SF can significantly impact workability. In order to compensate for this inadequacy in performance, SF might be added to RCA concrete mixes. Additionally, adding fibers to concrete, like PVA and polymeric fibers, has been shown to improve the strength and hardness of concrete and make it more durable and robust. Steel fiber-reinforced recycled aggregate concrete was investigated by Carneiro et al. for its compressive stress-strain behavior. Mohseni and colleagues discovered that adding steel or polypropylene fibers to RCA concrete improved the material's behavior. Gao et al. (2019) discovered that including sulfur in RAC improved the compressive and flexural characteristics of the structure.

Steel fibers, as shown in Figure 3.10, can help to reduce the friability and sessility of RAC by incorporating them into the material. Steel tubes can be improved by filling them with SF-reinforced concrete, as shown in the example above. However, there are only a few research reports on this type of composite structure. Steel fiber-reinforced RCA concrete-filled circular steel tubes are studied in this paper to widen the range of applications for RCA concrete.

Because recycled concrete will be used in place of natural concrete in the columns with SF, column-bearing capacity will be lowered. When there is a deficiency in either the steel fiber content or the strength grade of the concrete, the amount of RCA in the columns considerably affects the columns' structural performance. It is possible to increase the amount of weight that short columns can hold. As more

FIGURE 3.10 Steel fibers.

steel fibers are added to the material, the rate at which the columns can hold more weight decreased.

Concrete strength has a significant impact on column bearing capacity. It is possible to reduce the mass of concrete by increasing its strength significantly. Durability is most affected by concrete's strength, while steel fiber content has a negligible impact. The stiffness of the short column is greatly influenced by the steel fiber and concrete strength used in its construction. Especially in a beam with a 3% longitudinal steel ratio and 50 wt% RCA, there is no statistically significant difference in deflection or maximum shear strength between RAC and CC beams.

Choi et al. evaluated the shear strength of 20 beams with varying span-depth ratios (1.50% and 2.50%), longitudinal reinforcement ratios (0.53%, 0.833%, and 1.61%), and RCA replacement ratios to determine the shear strength of reinforced concrete beams a larger RCA replacement ratio resulted in lower shear strength. Addtionally, Schubert et al., who researched the behavior of 14 slabs made from 100 wt% RCA, found these slabs could be constructed using the same design equations as conventional concrete slabs. Xiao et al. investigated and analyzed 32 shears push-off specimens with varying percentages of recycled coarse aggregate replacement. No statistically significant differences were found between the RAC concrete and the CC specimens in terms of the shear stress-slip curves, fracture propagation pattern, or shear transfer capabilities across fractures. It was also shown that while the proportion of recycled aggregate replacement in the mix did not affect the ultimate shear load up to 30 wt%, a more significant percentage of RCA replacements did result in a decrease in the ultimate shear load. The shear strength of RAC was determined by testing full-scale beams without shear reinforcement. Standard indicates that 100 wt% RCA beams have a lower experimentally predicted to code-predicted capacity ratio than 50 wt% RCA and CC beams, although this is not necessarily true. Shear strength is directly related to reduced fracture energy, splitting strength, flexural strength, and other properties of RAC mixes when they are compared to their counterparts in cementitious composites.

There have also been several studies that have looked into the mechanical characteristics, mixture design, and structural performance of reinforced concrete beams constructed with recycled aggregate. Both of these aspects are essential. The flexural performance of beams produced by reinforced RAC has been examined extensively by a variety of different people, and their findings indicate that the flexural performance of beams made of RAC is equivalent to that of beams manufactured with conventional NAC. On the other hand, the maximum flexural strength of RAC has been found by some people to diminish when specimens are replaced with water that is composed of 100%, no matter what the water-cement ratio is. These findings show that using fine RCA in structural elements does not significantly affect how well they bend but that these materials make them more flexible. This means that they can better dissipate energy during the plastic phase.

The fact that these investigations of reinforced RAC beams were carried out in sterile environments devoid of corrosive substances does not diminish the significance of the findings. Locating in an acidic environment, such as seawater, is highly detrimental to any structure or component it may affect. Because of this, it may be difficult to use it safely. The individuals who carried out the experiments assert that the flexural strength of damaged concrete beams is significantly impacted by the various corrosion ratios of the longitudinal bars, even when the coarse aggregate is restored to its original level of 100%. Even if you do not use much steel, the sliding ability and final strength of RAC reinforced with rebar will diminish as more of the rebar corrodes. This will be the case even if you use much steel. On the other hand, the amount of RCA in the mixture does not significantly impact the strength of rebar that has not been corroded and RAC. The anti-permeability quality of RAC degrades directly to the amount of recycled aggregate utilized in its composition. It is essential to determine whether or not RAC can be utilized successfully as a structural beam in corrosive environments. Finding out the answer to this question is vital. This is because RCA can conserve resources and energy, protect the environment, and provide structural functions. Furthermore, the flexural strength of corroded reinforced concrete beams built of NAC to that of corroded reinforced concrete beams manufactured of RAC. As RACs anti-permeability properties are linked with how much RCA is used as coarse aggregate in concrete mixtures (RCA as a coarse aggregate replacement material of 0, 33, 66, and 100 wt%) and two different concrete strengths (C30 and C60), the RCA replacement and concrete strengths have a minimal bearing. Compared to concrete with the same strength, the ultimate bearing capacity of middle-strength and high-strength RAC decreases by 1.18% and 3.74%, respectively, when utilized. This demonstrates that the proportion of RCA replacement and the concrete's strength has no bearing on the concrete's ability to support the weight. When compared to non-RAC, the displacement ductility ratio for medium- and high-strength RAC decreases by 7.87% and 24.22%, respectively. The medium-strength concrete has a negligible impact, but it has an enormous impact on the quality of concrete (Kapoor et al., 2021).

4 Specifications, Production, and Applications of RAC

4.1 SPECIFICATIONS AND CODES OF RAC

The importance of sustainability in our society cannot be overstated. It contributes to preserving the environment by minimizing natural resources that cannot be replenished. Concrete, the most commonly manufactured material globally, has a vast range of uses and consumes a substantial quantity of nonrenewable resources in its production. By 2050, it is estimated that the demand for concrete would have increased to a yearly production of 18 billion tons. As a result of these discoveries, many researchers have investigated the use of recycled materials in concrete production, including fly ash recycled aggregate.

The Federal Highway Administration estimates that the United States produces 2.2 billion tons of new aggregate yearly. By 2022, the market is expected to have increased from its current 1.8 billion tons annually to 2.8 billion tons (2022). There is rising worry about the depletion of present natural aggregate sources and the availability of new aggregate sources to meet the anticipated great demand for new aggregates. Similarly, the quantity of trash generated during building projects in the USA is expected to rise. Demolition accounts for the vast majority of the estimated 12.3 million tons of yearly construction waste. Right now, landfills are the most common option for disposing of such types of waste.

Many states in the United States have learned that RCA used as aggregate in the production of freshly new concrete containing RCA is a more sustainable solution to the problem of the rising need for new aggregates and rising waste output. Although a substantial study has been conducted on the fresh and hardened properties of RAC (as shown in a typical RCA properties and concrete mixture in Tables 4.1 and 4.2 (Arezoumandi et al., 2015), the structural behavior of the material has received significantly less attention. Japan was the birthplace of structural performance research on RAC. For example, Maruyama et al. evaluated beams with longitudinal reinforcement rates ranging from 2.4% to 4.2%. Even though conventional concrete (CC) and reinforced concrete beams had identical fracture patterns and failure mechanisms, the shear strength of the RAC beams was 10% to 20% lower than that of the CC beams for beams with a 50 to 100 wt% RCA substitution rather than the NCA.

RCA properties inherent in the fabrication of concrete that is stable in the fresh state and capable of developing attributes comparable to equivalent natural aggregate concrete in the hardened state were taken into account in this study. For

TABLE 4.1

Aggregate Properties (Arezoumandi et al., 2015)

Property	NCA	RCA
Bulk specific gravity, oven-dry	2.72	2.35
Dry-rodded unit weight, kg/m^3	1,597.44	1,437.76
Absorption, %	0.98	4.56
LA abrasion, % loss	43	41

TABLE 4.2

Mixture Designs (per m^3) (Arezoumandi et al., 2015)

Composition	Conventional Concrete	50 wt% RCA	100 wt% RCA
OPC, kg	242.89	242.89	242.89
w/c	0.18	0.18	0.18
Natural coarse aggregate, kg	888.93	444.46	–
Recycled coarse aggregate, kg	–	384.08	749.10
Fine aggregate, kg	568.86	568.86	654.66
HRWR, mL	1.63	1.48	1.24
AE, mL	0.59	0.41	0.21

various applications, including foundations, pavements, and reinforced and pre-stressed concrete, RCA has already been tested for its appropriateness. Using coarse RCA in high-strength concrete, on the other hand, was found to be feasible according to the results of the research. Specifically, pervious concrete (PC) is a kind of with high porosity concrete, allowing it to have a high drainage capacity and, as a result, the ability to limit the flow of rainfall that collects on its surface. However, more significant and in-depth research is required before OPC concretes can be considered a solution to city flood-related problems. This lack of standardization persists, even though the American Institute of Concrete (ACI) PCs should be used in civil construction, according to a recommendation in the form of ACI 522R-10 (Arezoumandi et al., 2015).

Concerning the use of pervious concrete, the most important thing to bear in mind is that it is straightforward for things to get in and out. Measurement of this can be done using both the falling-head and constant-head tests. The American Society for Testing and Materials (ASTM) C1701 permeameter method and the NCA permeameter method can also be used to measure this on the ground (Yang et al., 2020). In the lab, PC's permeability cannot be accurately measured because there is no standard method. Nevertheless, the ACI stated that both tests should be done. Keep in mind that concrete's permeability and porosity are closely linked. More interconnected voids in the concrete mean improved water permeability. Even so, there is more to it than meets the eye. Concrete's coefficient may be affected by the aggregates used in its production. An investigation into how concrete permeability changes when sustainable aggregates made from manufacturing and industrial waste are utilized is required by the ACI recommendation.

The suggested replacement levels demonstrate the stringency of the preexisting British and European standards, particularly when structural concrete made with coarse crushed concrete aggregate (CCA) is subjected to chloride ion conditions. This can influence structural concrete design in a way that favors CCA requirements. Since the imposed constraints counter the conclusions that may be taken from publicly available data, this adds to the impression that broad worries and uncertainty in the business are being disregarded. It is noted that there is a lack of recent data on optimal procedures that might facilitate collaboration between the construction sector and academic institutions. For future projects committed to ethical sourcing, sustainable concrete may be utilized if it is crafted from a blended cement containing at least 50 wt% GGBS or EN 197-1 certified CEM III/A 42.5N (Nicoara et al., 2020), and a reliable and continuous supply of CCA is established. This study supports the widespread use of CCA in structural concrete and has positive implications. It is recommended that data from regional life cycle inventory be included in the LCA analysis, although there are a few caveats. Although the research draws on worldwide standards set by organizations like the ACI and the American Society for Testing and Materials (ASTM), it mainly employs Brazilian national standards in its effort to characterize Brazilian-made products (Nicoara et al., 2020). Portland concrete with RCAs may have different qualities depending on the county since RCA properties are well known to be highly dependent on the materials used in the buildings from which they were made.

Concrete aggregate made from RCA can be used as sustainable concrete buildings can be addressed by including RAC in the production process. For fresh concrete building purposes, less than 10 wt% of the RCA is used, while the vast majority of RCA is used for non-structural purposes such as backfilling, road base, and sub-base. Concerns about the structural application of RAC have been raised because of the technology's time-dependent behavior under service pressures. Because of the shrinkage and creeping caused by the remaining mortar associated with RCA, reinforced concrete structures suffer from large deflections, as previously discussed. Changes in RAC necessitate new analytical models for controlling deflection. It was found that the deflection behavior of RAC one-way slabs and beams differed significantly when the coarse RCA inclusion ratio exceeded 25 wt% by Toi's and Kurama.

The specifications that RCA produced have been used all around the world. The quality of the aggregate is a very important factor to take into account if one is contemplating the production of concrete for either its tenacity or its tenacity, respectively. It is vital to investigate the aggregate quality since there is a significant amount of variety in the waste concrete that may be acquired by recycling from several different RCA sources. This variation can be found in the waste concrete. Utilizing a classification technique is one of the critical problems that need to be handled because of the possibility that the strength qualities of concrete made with RCA might be affected by physical impurities. This is one of the primary concerns that need to be addressed. It was found that RCA concrete systems with waste brick as impurities had a strength that was 50%wt lower than that of systems with just RCA. This was established via testing. In addition, the experimental study that Peng and his colleagues carried out came to the conclusion that increasing the impurity content with waste brick lowered the whole strength of RAC by 15%–25% more than a control system that consisted only of RCA particles. This was determined by comparing the results to those of the study. Bricks are not the only object that has the capacity to contain impurities; gypsum, asphalt, and wood chips all have the possibility of doing so. Bricks are the most common example of this. These impurities can reduce the elastic modulus of RCA concrete systems, which, in turn, may negatively impact the concrete's strength. When using RCA as coarse aggregates, it is clear that impurity concentrations need to be regulated and controlled; therefore, this must be done. New classification schemes have been suggested by a wide range of countries, regions, and organizations based in different parts of the world.

In terms of RCA, there are a wide variety of acceptance conditions that have been established by a number of countries up until the present time (Hou et al., 2017). As shown in Table 4.3, RCA of class IA is suitably graded and contains no

TABLE 4.3
Acceptance Criteria Regarding RCA (Hou et al., 2017)

Country (Standard)	Recycled Aggregate Type	Oven-dry Density Criterion (kg/m³)	Absorption Ratio of Aggregate Criterion (%)
Australia (AS1141.6.2)	Class 1A	≥2,100	≤6
(AS 1996)	Class 1B	≥1,800	≤8
Germany (DIN 4226-100)	Type 1	≥2,000	≤10
(DIN 2002)	Type 2	≥2,000	≤15
	Type 3	≥1,800	≤20
	Type 4	≥1,500	No limit
Hong Kong (Works Bureau of Hong Kong 2002)	–	≥2,000	≤10
Japan (JIS A 5021, 5022,	Coarse—Class H	≥2,500	≤3
and 5023)	Coarse—Class H	≥2,500	≤3
(JIS 2011, 2012a, b)	Fine—Class H	≥2,500	≤3.5
	Coarse—Class M	≥2,300	≤5
	Fine—Class M	≥2,200	≤7
	Coarse—Class L	No limit	≤7
	Fine—Class L	No limit	≤13
Korea (KS F 2573)	Coarse	≥2,500	≤3
(KS 2002)	Fine	≥2,200	≤5
RILEM (1994)	Type 1	≥1,500	≤20
	Type 2	≥2,000	≤10
	Type 3	≥2,500	≤3
Spain (EHE 2000)	–	≥2,000	≤5

more than more than 0.5% brick; however, RCA of class IB is combined with no more than 30% crushed brick. These requirements are taken from the Australian standard from 1996 (AS 1996) (Hou et al., 2017). In Germany, according to DIN 2002 (Hou et al., 2017): chippings of type 1 concrete coupled with crusher sand; chippings of type 2 construction combined with crusher sand; chippings of type 3 masonry combined with crusher sand; and chippings of type 4 mixed materials combined with crusher sand. There are no restrictions put on the type and segment of the material for concrete and structures that have a nominal strength of 45 MPa or less and that fall within the Japan (JIS 2011) Class H standard (Nicoara et al., 2020). Pile members, underground beams, and concrete-filled steel tubes are examples of Class M members. These types of members are not subject to drying or freezing-and-thawing action. Backfill concrete, blinding concrete, and concrete-filled steel tubes are all examples of materials that belong to Class L. While the RILEM (1994): type 1 aggregates are derived from masonry rubble, type 2 aggregates are derived from concrete rubble, and type 3 aggregates are a mixture of natural (at least 80 wt%) and recycled (at most 20 wt%) material. Masonry rubble is the source of type 1 aggregates, while concrete rubble is the source of type 2 aggregates. In addition, the regulation on RCA is outlined in Annex G of the document (informative) providing guidance on how the different chemical components of aggregate impact the strength of concrete in which they are utilized as an element. Especially, G.3.2 The effect of alkali and silica on aggregates made from recycled materials (Nicoara et al., 2020).

As was discussed before, the use of RCA may have an impact on the application of the safety standards. In the event of RCAs, it will be required to make certain that the initial concrete did not include any form of reactive (or reacting) aggregate. In addition, when the amount of alkali contained in the new concrete (or the cement included inside it) is restricted, the amount of alkali present in the RCAs will need to be assessed and taken into consideration. In the case of RCAs generally, it will be fair to approach the material as a potentially reactive aggregate unless it has been expressly proved to be non-reactive. This must be done before the substance can be considered non-reactive. This is the case regardless of whether or not the substance has been shown to be non-reactive.

The criteria for RCA that must be fulfilled to comply with BS EN 12620 are outlined in the table that can be seen below (Hou et al., 2017). When the BS 8500-2:2015+A2:2019 is taken into consideration, as stated in Table 4.4, RCA is necessary; thus, the use of this material in concrete must be reviewed on a case-by-case basis while taking into consideration the concrete's unique composition. It is necessary to establish and consider the chloride and alkali content of RCA to calculate the maximum amount of chloride and alkali presence that can be found in concrete that contains RCA. Additionally, it is necessary to take into consideration the variability of that content. After being washed and processed, it is now suitable for use in concrete.

On the other hand, it is commonly known that some concrete components have been plastered with gypsum. This was done in order to protect them. When these components are crushed, the vast majority of the gypsum plaster is converted into RCA that can be used. It is well known that the use of gypsum plaster raises the

TABLE 4.4
BS 8500-2 Requirments for Coarse Crushed Recycled Aggregate[A] (Hou et al., 2017)

Properties	BS EN 12620 Size or Category	Description of Category
Aggregate size	d≥4 mm, D≥10 mm[B]	d/D
Maximum fines	F_4	≤4% by mass of particles passing the 0.063 mm sieve
Maximum acid-soluble sulfate (SO_3)	–[C]	By mass acid-soluble sulfate
Content of: concrete …	Rc_{NR}	No requirement
Content of: concrete … hydraulically bound aggregate	Rcu_{NR}	No requirement
Content of: Clay masonry units …	R_{NR}	No requirement
Content of bituminous materials	Ra_{10-}	≤10% by mass
Content of other materials … gypsum plaster, and glass	XRg_{1-}	≤1% by mass
Floating material by volume	FL_{2-}	≤2 cm^3/kg

Notes

[A] When the material to be used is obtained by crushing hardened concrete of known composition that has not been used, such as an overabundance of precast units or returned fresh concrete, and when the material has not been contaminated by storage or processing, the only requirements are for aggregate size, fines content, drying shrinkage, and resistance to fragmentation. This is the case when the material to be used obtained by crushing hardened concrete of known composition that has not been used.

[B] The designation accepts that some particles would remain on the top sieve (oversize) while others will flow through the lower sieve (undersize). According to BS EN 12620 (Hou et al., 2017), the aggregate sizes that should be used for single-size coarse aggregate with a specified maximum aggregate size of 40, 20, 14, and 10 mm are, respectively, 20/40, 10/20, 6.3/14, and 4/10.

[C] Each situation requires its unique analysis to identify the acceptable limit and testing procedure.

level of sulfate, which in turn raises the potential for delayed ettringite production (Wardeh et al., 2014). Because there are so many distinct sources of sulfate in the environment, it is difficult to identify isolated areas with high sulfate concentrations. This further complicates the detection process. In addition, there is no information on the potential separation of the stockpile into other categories. As a result of these aspects, the choice about whether or not to use acceptable coarse RCA is left up to the project specification, which might consider the particular source of coarse RCA. However, some fine RCA may also be suitable for use in concrete; however, it is not feasible to establish general criteria at this time because of the extensive range of compositions and the unavailability of detailed information. In addition, the presence of appropriate levels of gypsum in coarse RCA is also significant to specific sources of fine RCA.

In accordance with the Hong Kong Civil Engineering Specifications, the use of RCA in freshly manufactured concrete was first authorized in 2001 (Yang et al., 2020). Those works related to the "Specification facilitating the use of recycled aggregates" included in Bureau Technical Circular No.12/200281 (Yang et al., 2020) have two applications for RCA that are used in concrete production. These applications are as follows: 1) for applications involving lower grade concrete, and 2) for applications involving higher grade concrete. Both of these applications are utilized in the production of concrete. It is acceptable to make use of lower-grade applications using concrete that is made up completely of RCA. However, using recycled fine aggregates in concrete is not authorized because the material may lose some of its hardening and durability with time. The target strength has been determined to be 20 MPa, and some of the possible uses for the concrete include stools, benches, planter walls, and concrete mass walls, in addition to a variety of other small concrete structures. In higher concrete grade applications with strength requirements of less than 35 MPa, the current standard allows for the use of RCA at a replacement level of up to 20 wt%, which is the maximum allowable amount. On the other hand, using this replacement level in any concrete application connected to water retention structures is strictly forbidden. In order for RCA to be successful in Hong Kong, the conditions that must be followed are outlined in Table 4.5 (Yang et al., 2020).

In 1998, RILEM established a standard for RCA in the European Union. RCA may be categorized as "Class I," that derived from waste concrete rubble as "Class II," and that formed from a mixture of RCA and natural aggregates as "Class III." Twenty wt % is the maximum proportion of RCA that may be replaced with Class I brick, and no more than 10 wt% of the brick may include impurities. In contrast, Class I and Class II replacements may be employed in their entirety. Table 4.6 summarizes the RILEM criteria for each of these classes, as well as the compliance standards for strength and durability (Rao et al., 2019).

TABLE 4.5
RCA Specification Requirements Prescribed in Hong Kong (Yang et al., 2020)

Requirements	Limitation
Minimum dry particle density	2,000 kg/m^3
Maximum water absorption	10%
The maximum content of wood and other materials less dense than in water	0.5%
The maximum content of other foreign materials (e.g., metals, clay lumps, asphalt, glass, tar, etc.)	1%
The maximum content of sulfate	1%
Maximum contents of finer material (<4 mm)	5%
Maximum chloride content	0.05% (by mass of iron chloride of combined aggregate)
Flakiness index	40%

TABLE 4.6

RCA Particle Requirements Prescribed in RILEM Standards (Rao et al., 2019)

Requirement	Class I	Class II	Class III
Saturated dry density	1,500 kg/m^3	2,000 kg/m^3	2,400 kg/m^3
Maximum content of material with SSD < 2,200 kg/m^3	–	10%	10%
Maximum content of material with SSD < 1,800 kg/m^3	10%	1%	1%
Maximum content of material with SSD < 1,000 kg/m^3	1%	0.5%	0.5%
Water absorption	20%	10%	3%
Maximum content of foreign materials (metals, glass, soft materials, bitumen)	5% (by volume)	1% (by volume)	1% (by volume)
Maximum content of metals	1% (by mass)	1% (by mass)	1% (by mass)
Maximum sulfate content	1% (by mass)	1% (by mass)	1% (by mass)

The Building Contractors Society of Japan has released and made accessible to the general public a document about RCA classification. This standard establishes a maximum and minimum oven-dry density threshold for RCA. The quality of recycled aggregates plays an important role in selecting the kind of recycled aggregates used, influencing the maximum RCA design strength. Class L aggregates are of worse quality compared to other categories, and they are often employed for screed concrete, which is used for backfilling, filling, and leveling concrete. Class H aggregates are considered good quality and may be used in normal concrete applications. The JIS A 502383 (Wardeh et al., 2014) and JIS A 502184 (Wardeh et al., 2014) documents provide the standard specifications for the L and H classes of RCA, respectively. Table 4.7 depicts the RCA use constraints in JIS A 502184. The constraints identified in Table 4.7 are comparable to those used for conventional aggregates, and these limitations comprise the Japanese standard for coarse aggregates of higher quality (Wardeh et al., 2014).

TABLE 4.7

RCA Requirements Prescribed in Japanese Standards JIS A 502184 (Wardeh et al., 2014)

Requirements	Coarse aggregate
Oven-dry density	$\leq$2.5 g/cm^3
Water absorption	$\leq$3%
LA abrasion	$\leq$35%
Amount of material passing No. 200 sieve	$\leq$1%
Chloride content	$\leq$0.04%

TABLE 4.8

RCA Particle Requirements Prescribed in Chinese Standards (Wardeh et al., 2014)

Requirement	Class I	Class II	Class III
Water absorption	<3%	<5%	<8%
Apparent density	>2,450 kg/m^3	>2,350 kg/m^3	>2,250 kg/m^3
Porosity	<47%	<50%	<53%
Content of clay by mass	<1%	<2%	<3%
Content of clay lumps by mass	<0.5%	<0.7%	<1.0%
Content of elongated and flaky particles		<10%	
Content of organic		Standard	
Sulfide and sulfate by mass		<2%	
Chloride by mass		<0.06%	
Other impurities		<1%	
Mass loss	<5%	<10%	<15%
Crushing Index	<12%	<20%	<30%

As shown in Table 4.8, the Chinese specification for RCA classification bases includes three distinct groupings. The aggregate classes are defined and measured based on several aggregate properties, such as chemical impurities, porosity, shape, and the presence of related foreign materials. Class I aggregates are of the highest quality and may be used in structural applications, while Class III aggregates are of poorer quality and cannot be utilized in concrete structural applications. Class I aggregates are used for structural purposes (Wardeh et al., 2014).

In the United States, the most current version of the ACI building code does not contain specific concrete standards. This is the case throughout the coast of the United States (ACI 318-14) (Gebremariam et al., 2021). However, the ACI technical committee 555 and their current state-of-the-art report handle the majority of the responsibilities for the rules for employing RCA materials. The RCA may be establishd in Table 4.9, and the remainder of this report discusses the exposed impurity levels.

According to the ACI 221R (Gebremariam et al., 2021), the choices on the use of recycled concrete aggregate must take into account the results of trial batches, detailed testing, chemical, and petrographic studies. Aggregates manufactured from

TABLE 4.9

RCA Particle Requirements Prescribed in ACI 555-01R (Gebremariam et al., 2021)

Impurities	Lime Plaster	Soil	Wood	Hydrated Gypsum	Asphalt	Paint-Made Vinyl Acetate
Percentage of aggregate by volume	7%	5%	4%	3%	2%	0.2%

municipal or industrial wastes (slags not created by an iron blast furnace), recycled materials, or marginal resources can feature a wide variety of undesirable physical and chemical qualities. Waste from buildings may include hazardous quantities of brick, glass, or gypsum, and recycled concrete may contain reactive or low-quality materials in addition to high chloride concentrations. Building debris may also contain gypsum. All of these different aspects are important considerations throughout the decision-making process. This is the primary justification for why it is essential to carry out comprehensive testing. In addition, it highlights that recycled materials should be characterized and assessed in accordance with ASTM C33 (Gebremariam et al., 2021), except for circumstances in which the composition indicates the requirement for specific additional criteria.

When fine RCAs are used for concrete that will be exposed to abrasion, the gradation limitations may rise from 0%–3% to 0%–5% of material finer than the No. 200 sieve (Gebremariam et al., 2021). This is because recycled fine aggregates are typically finer than fine natural aggregates. This is owing to the fact that recycled fine aggregates may include dust from fractures as well as clay, which might result in the existence of finer materials. If the concrete is not exposed to abrasion, the gradation constraints may be stretched further, from 0%–5% to 0%–7%. In addition, this specification suggests using appropriate test methods that are in place at the time of use at the local, state, and federal levels in order to undertake evaluations of environmental elements such as air quality, water quality, and storage conditions. The ASTM C33 (Gebremariam et al., 2021), on the other hand, does not provide any additional rules for categorizing or directing RCA in specific concrete applications.

The American Association of State Highway and Transportation Officials (AASHTO) M80 standard (Rong et al., 2021) does allow for the use of crushed concrete as aggregates, even though it does not include many limitations on the usage of RCA. It is important to note that for any crushed aggregates to be used, they need to comply with the standards of the American Association of State Highway and Transportation Officials (AASHTO) M80 standard (Rong et al., 2021). However, before this provision was included in the American Association of State Highway and Transportation Officials (AASHTO) M80 standard (Rong et al., 2021), the AASHTO M1688 standard (Rong et al., 2021) was responsible for regulating the use of RCA in concrete. The primary focus of AASHTO MP16 standard (Rong et al., 2021) was on aggregates that were used for uses that were not structural. AASHTO MP16 standard (Rong et al., 2021) distinguished RCA into three classes: Class A for aggregates subjected to severe exposure, Class B for aggregates subjected to moderate exposure, and Class C for aggregates subjected to insignificant exposure. In MP16 (Rong et al., 2021) of AASHTO, the standard practice standards for RCA have been compiled and are shown in Table 4.10. However, the limited usage of RCAs in Canada is not primarily related to the product's misleading definition in national, provincial, or municipal building regulations and standards. This is true despite this description being found in all three levels of Canadian building regulation. For instance, the Canadian Standards Association (CSA) has a more permissive stance on RCA than other organizations. It is important to pay particular attention to dangerous chemicals, durability, alkali-aggregate reactivity, workability, and physical qualities, as stated in a remark in the

TABLE 4.10

Standard Specification for RCA Recommended in AASHTO MP 16 (Rong et al., 2021)

Requirements	Limitation
Maximum LA abrasion loss	50%
Soundness loss	12% (under sodium sulfate)
	18 (under magnesium sulfate)
Amount of material passing No. 200 sieve	1.5%
Chlorite ion content	0.6 Ib/yd^3 of concrete

CSA A23.1-clause 0990s (Rong et al., 2021). Depending on the RCA source and its inherent unpredictability, a testing plan that includes daily checks can be necessary. They will permit the use of RCA so long as the finished product is able to satisfy the performance criteria set out by the CSA. A new Annex to this Canadian Standard was published in 2014 (Rong et al., 2021), and it specifies the manufacturing process as well as the quality of aggregates created from recycled concrete. These aggregates may be used in hydraulic cement concrete.

The German standards DIN, 42226-100 (Rong et al., 2021), experienced their most recent revision in 2002. As a result of this revision, the standard now allows the use of RCA in freshly mixed concrete so long as it satisfies the requirements relating to a particular aggregate class. These requirements can be found in Table 4.11. The various impurity levels required of aggregates result in the

TABLE 4.11

RCA Particle Requirements Prescribed in German Standards (Rong et al., 2021)

Requirement	Constituent by Mass [%]			
	Type 1	Type 2	Type 3	Type 4
Concrete and natural aggregates	≥90	≥70	≤20	≥80
Clinker, no porous clay bricks	≤10	≤30	≥80	
Calcium silicate bricks			≤5	
Other mineral materials (e.g., porous brick, lightweight concrete, plaster, mortar, porous slag)	≤2	≤3	≤5	≤20
Asphalt	≤10	≤30	≤1	
Foreign substances (e.g., glass, plastic, metal, wood)	≤0.2	≤0.5	≤0.5	≤1
Oven-dry density	≥2,000 kg/m^3	≥2,000 kg/m^3	≥1,800 kg/m^3	≥1,500 kg/m^3
Maximum water absorption (in 10 mins)	10%	15%	20%	No limit

classification of these materials into a large number of distinct groups. A variety of different sources determines these levels. As the RCA types go from Type 1 to Type 4, the quality of the aggregates deteriorates, and the use of these aggregates in structural concrete becomes progressively limited. Type 1 RCA aggregates have the highest quality. Type 4 RCA aggregates have the lowest quality. Consequently, it is highly recommended that RCA of Type 3 or a higher grade be applied in non-structural applications such as pavement, curbs, and pathways.

Regarding the criteria that are relevant to Australia, VicRoads in Victoria, the Department of Transport, Energy and Infrastructure in South Australia, the Institute of Public Works Engineering Australia in New South Wales (NSW), the Roads and Traffic Authority (RTA) in New South Wales, and Main Roads in Western Australia are some of the organizations that have contributed to the development of the Australian specifications (MRWA). The standard standards for employing RCA by VicRoads both as a subbase and base layer were released in 1993 in Victoria. These specifications were based on traffic design. In March of 2009, an updated version of the specs was made available to the general public. According to the standard, RCA applications might fall into three categories: light-duty base, heavy-duty upper, or heavy-duty lower sub-base. Heavy load traffic consists of more than 5×106 equivalent standard axles, whereas light duty traffic consists of less than 5×105 equivalent standard axles. Table 4.12, which has a nominal size of 20 mm and presents an overview of the RCA material standard, has a size designation of 20 mm. Crushed brick may be utilized more significantly than specified in Table 4.13 as long as the sub-base is created using a "registered crushed concrete mix design." There is a chance that Class 3 sub-bases will have a complete crushed brick content of 15 wt%, whereas Class 4 subbases will have a maximum crushed brick content of 50 wt%. For example, in New South Wales, numerous road agencies and local government bodies in New South Wales have successfully included RCA into their local road specifications after formulating their own particular needs. Participation in the usage of RCA has been witnessed by a range of agencies, such as the Institute of Public Works Engineering Australia (IPWEA), the Southern Sydney Region of Councils, and the RTA (Babalola et al., 2020).

Many countries in Europe, to mention just a few of these countries, have given their stamp of approval for the use of RCA in the construction of bases and sub-bases. These countries include Finland, Sweden, Denmark, the Netherlands, and Portugal. Knowledge gained via construction field trials and case studies in a variety of European countries was used to aid in the development of the great majority of the criteria. The first country is an example, and it is important to note the requirements for using RCA as an introductory course in the Netherlands. When designing the specification for the material, the gradation of the material, as well as its stiffness, were taken into consideration. Table 4.14 presents an overview of various significant properties of the 20 mm RCA and is formatted in a manner compliant with Dutch regulations. Australian regulations require the sieve sizes and percentages of material that must pass through that are not materially different from those used in other nations. In addition, the kinds of raw materials used and the technical characteristics required for the base and sub-base layers are used to divide the sub-base and base layers into four distinct groups in Finland.

TABLE 4.12

Summary of VicRoads Specification for 20 mm RCA

Particle Size Distribution	Light Duty Base Class 2	Heavy Duty Upper Sub-Base Class 3	Heavy Duty Lower Sub-Base Class 4
Sieve Size, mm	Grading Limits, %	Grading Limits, %	Grading Limits, %
26.5	100	100	100
19	95–100	95–100	...
13.2	78–92	75–95	...
9.5	63–83	60–90	...
4.75	44–64	42–76	42–76
2.36	30–48	28–60	...
0.425	13–21	10–28	10–28
0.075	5–9	2–10	2–14
Atterberg and strength and LAA limits	Base (Class 2)	Upper sub-base (Class 3)	Lower sub-base (Class 4)
Liquid limit, % (max)	35	35	40
Plasticity index, % (max)	6	10	20
CBR 4 days soaked, % (min)	100	80	20
Los Angeles Abrasion, % (max)	35	40	45
Foreign material limits			
High-density materials such as metal, glass, and brick, % (max)	2	3	3
Low-density materials such as plastic, rubber, and plaster, % (max)	0.5	1	3
Wood and other vegetable matter, % (max)	0.1	0.2	0.5

There was a lack of clarity regarding how to apply each of the many different grades properly. Table 4.15 comprehensively summarizes the primary attributes that distinguish the RCA materials. A plate load test or a falling weight deflectometer was used to ascertain RCA's "design bearing capacity," which was then represented in terms of the material's stiffness. In addition, the National Portuguese Laboratory for Civil Engineering (LNEC) in Portugal developed guidelines for using RCA in the base and subbase pavement layers. These guidelines may be found in Portugal. This standard was developed using the criteria described in European Standards EN 13242 (Hamad and Dawi, 2017) and 13285 (Hamad and Dawi, 2017) as its primary reference point throughout the process. The materials that go into recycled goods are produced to determine whether those goods are given a B or C rating (refer to Table 4.9, which is based partly on EN 933-11 (Hamad and Dawi, 2017)). For instance, according to Table 4.16, the parts that make up Class B may contain 85% crushed concrete, 10% clay masonry, and 5% recycled asphalt.

TABLE 4.13

Summary of Dutch Specification for 20 mm RCA Materials (After) (Hamad and Dawi, 2017)

Particle Size Distribution	Base (0–20)
Sieve size	Retained, %
C 31.5	0
C 22.4	0–10
C 16	…
C 8	15–45
C 4	…
2 mm	45–70
63 μm	92–100
CBR after preparing, % (min)	50
Crushing factor	0.65
Foreign material limits	
Crushed concrete content, % (min)	80
Asphalt, % (max)	5
Other broken crushed stone, dry density > 2.1 tons/cubic meter, % (max)	10
Other broken, crushed stone, dry density> 1.6 tons/cubic meter such as light concrete, glass, slag, etc., % (max)	10
Organic materials such as wood, rope, paper, etc., % (max)	0.1
Gypsum, metals, and plastics, % max	1.0

It was determined that the recycled crushed asphalt (RCA) materials could be separated into four unique groups according to the characteristics that differentiated them from one another: categories 1, 2, 3, and 4. This was discovered in the country of Sweden. It is possible to use base or subbase material from either Class 1 or Class 2 in areas subjected to mild traffic loadings, such as those encountered by pedestrian and bicycle lanes. On the other hand, Class 3 material may be used for capping layers, while Class 4 material can be used as fill material. Before the concrete was crushed, the quality of each class was defined by the qualities of the concrete itself, which is an interesting fact to keep in mind. These qualities included compressive strength, assessed in accordance with EN 12390-3 (Hamad and Dawi, 2017), and abrasion, measured in accordance with EN 1097-1 (Hamad and Dawi, 2017) using either the LAA or Micro-Deval test. Both of these tests were conducted in the same manner.

Regarding F's, the minimum criteria for Class 1 were 30 MPa, whereas the requirements for Class 2 were 20 MPa. In addition, the Danish Road Institute developed national requirements for using RCA as a road basis. These standards may be obtained in Denmark. When assigning the material to one of the three categories, A, B, or C, the back-calculated modulus, also known as E, and the abrasion resistance as determined by the Los Angeles Abrasion tests purity of the material were all taken into consideration. On the other hand, a base of Class C can only be deployed on certain types of roads, but bases of Classes A and B may be utilized on any road.

TABLE 4.14

Summary of Finland Specification for RCA Materials (After) (Hamad and Dawi, 2017)

Property	I	II	III	IV	In General
Grain size, mm	0–50	0–50	0–50	Varies	...
Optimum moisture content, %	8–10	8–12	...	...	8–12
Maximum dry density, kN/m^3	18–20	17.5–20.5	...	...	...
Specific gravity	...	...	...	...	2.55–2.65
UCS at 7 days, MPa	1.2–1.3	0.3–1.1	...	...	...
UCS at 28 days, MPa	2.0–2.1	0.6–1.3	...	...	...
CBR	...	...	...	...	90–140
Design E-modulus, MPa	700	500	280	200	...
Los Angeles Abrasion, %	23	28	...	...	...
Friction angle (φ),°	...	...	...	...	40
Permeability, m/s	$(1–7) \times 10^{-5}$	...	...	...	...
pH	12.7–12.9	...	...	...	≥11
Capillarity, m	0.25	0.2	...	...	...
Foreign material limits					
Brick content, % (max)	0	10	10	30	...
Other materials such as wood, plastics, etc., % (max)	0.5	1.0	1.0	1	...

TABLE 4.15

Classification of RCA Following the Nature of the Constituents of Coarse Fraction (After) (Hamad and Dawi, 2017)

	Constituents According to EN 13242 (Hamad and Dawi, 2017)				
Class	$R_C^a + R_u^b + R_G^c$	R_B^d	R_A^C	$FL_s^f + FL_{NS}^g$	R_C^A
B	≥ 90%	≤ 10%	≤ 5%	≤ 1%	≤ 0.2%
C	≥ 50%	≤ 50%	≤ 30%	≤ 1%	≤ 0.2%

4.2 PRODUCTION OF RAC

In the current effort, a national initiative known as eco-construction with concrete recycling (ECOREB), deals with recycling issues in construction waste for a future sustainable city and recycled aggregates, has significantly impacted both the physical and mechanical qualities of concrete. National and local governments are in the process of adopting a number of significant regulations, including the "code for re-cycling construction and demolition waste" and the "technical code for recycled concrete applications." These are just two examples of the numerous significant requirements currently being adopted. When these requirements are used in practical

TABLE 4.16

Physical Properties of the Coarse and Fine Aggregate (Gao et al., 2019)

Aggregate Type	Apparent Density (kg/m^3)	Loose Packing Density (kg/m^3)	Dry-Rodded Density (kg/m^3)	Water Absorption (wt%)	Crush Index (%)	Void Ratio (%)
RCA	2,640	1,302	1,412	4.85	17.7	50.3
NCA	2,814	1,568	1,630	1.40	8.8	44.3
Sand	2,556	1,611	1,486	0.56	–	–

engineering applications, the quality of recycled aggregates and the mixed RAC are able to meet them successfully. In the future, RAC-specific standards will make its use more uncomplicated and straightforward. Because of extensive study and a tremendous amount of experience working with RAC, it is recommended that both new standards and changes to existing standards be implemented.

The properties of RAC in both its fresh and hardened stages will be compared in order to describe the characteristics of RAC in their fresh and hardened phases as a function of the replacement ratio. This will be done by comparing the properties of RAC in both its fresh and hardened states. During this inquiry, developing and analyzing two different kinds of concrete have been carried out. These include NAC suitable for control operations, in addition to three distinct types of RAC that each has a workability rating of S4 and compressive strengths close to 35 MPa. In this research, the correlations found in Eurocode 2 (EC2) (Nicoara et al., 2020) are investigated, along with the application of such relationships to concrete made from recycled materials. The elastic modulus, the ultimate strain, and the connection between stress and strain may all be calculated using these equations, which can be derived from a fundamental comprehension of compressive strength.

The American Concrete Institute's American Concrete Institute (ACI)-555R provides guidelines for proportioning concrete mixtures containing recycled concrete (Nicoara et al., 2020). However, neither it nor any other source provides a step-by-step process for making steel fiber–reinforced recycled coarse aggregate concrete (SFRCAC) that satisfies the proper fresh and hardened quality requirements. Its primary objective is to devise a method of mixture design for SFRCAC that is both specific and efficient; this will ensure that the material possesses the same fresh and hardened properties as conventional natural coarse aggregate concrete (NCAC), regardless of the replacement ratios of recycled RACs, as shown by the typical properties in Table 4.17, that are used. This will be accomplished by developing a method of mixture design that is specific and efficient.

A more intriguing code is the one that was developed for the New York City Building Code governing the amounts of concrete mixtures that were made accessible for usage. In order to effectively compute the concrete ratios, one must refer to the guidelines provided in Sections 1905.2.1 through 1905. 2.3. 1905.2.5. 1905.2.5. The minimum amount of materials that have been recycled in all concrete mixes for cast-in-place concrete (Gao et al., 2019) that requires a compressive

TABLE 4.17

Volume Percentages of Impurities[*] (Hansen, 1986)

Impurities	Lime Plaster	Soil	Wood (Japanese Cypress)	Hydrated Gypsum	Asphalt	Paint Made of Vinyl Acetate
% of aggregate by volume	7	5	4	3	2	0.2

Note

[*] Resulting in a 15% or more significant reduction of compressive strength as compared to control concrete.

strength of 4,000 pounds per square inch or less, recycled concrete must make up at least 10 wt% of the aggregate. This requirement is in place to protect the environment. These criteria must be satisfied. This criterion, which is now 10% of the overall, will be raised to 15% of the whole, based on weight, effective in 2013. The aggregate used in the production of concrete should have a diameter of no more than 19.0 mm at its maximum, and there should be no more than 1% hazardous materials. Buildings designed to store water, sewage, or other liquids will not be subject to the regulations outlined in Section 1905.2.5 (Gao et al., 2019), nor will they be allowed to be utilized to transport these substances.

Both concrete and the components that make up the base course are required to include a specific amount of recycled material. a. Concrete mixes purchased by any government agency and need a compressive strength of 27.6 MPa or less are not allowed to include the concrete aggregate of less than 10% recycled concrete as measured by weight (Omary et al, 2017). This rule applies to both new and recycled concrete. This guideline applies to freshly produce as well as previously used concrete. After the first of July in 2013, it will be illegal for certain kinds of concrete mixes to include concrete aggregate that is made of less than 15% recycled concrete as measured by weight. This restriction goes into effect in the United States and Canada. The maximum diameter of any concrete aggregate cannot be more than 19 mm, and the overall quantity of harmful material cannot be greater than 1%. In spite of what has been said above, the requirements of this subdivision do not necessarily have to apply to any concrete mixes intended to be utilized in the construction of buildings or other structures.

Recycled concrete is a technically competent aggregate and base course. It improves concrete and asphalt compaction and constructability over equivalent virgin aggregates. ASTM and AASHTO recognize it as a concrete aggregate. It encounters state and federal standards. Recycled concrete weighs 10%–15% less than quarry goods, reducing material and transportation costs (Hamad and Dawi, 2017).

It is possible to begin concrete production directly on the job site with RCA. The demolition site is the first location for the manufacture of recycled concrete. A massive hydraulic hammer is used to initially break the concrete pavements into pieces that are 600 millimeters in size. After that, the pavements are broken up using a big hooked instrument known as a rhino horn, which is installed in place of the typical bucket on a rubber-tired loader or excavator. The slabs are broken apart, and most reinforcement is removed when the hook is dragged through the pavement. In the process of demolishing buildings or other types of structures, impact hammers that are mounted on vehicles are often used. After a pile of debris has been accumulated on the ground, hydraulic breakers may be used to further process the rubble by reducing its size to one that is more manageable. The subsequent process involves removing 90–95 wt% of the reinforcement using hydraulic shears and torches. The leftover steel is extracted from the concrete at the unit that performs the final processing. It has been shown that jaw-type crushers are adequate for crushing materials containing steel (Hamad and Dawi, 2017).

Additionally, the steel material itself is recyclable. Concrete that has been demolished is loaded onto trucks and sent to a factory where it will undergo additional processing. This facility may either be directly related to the selective demolition and removal construction, or it may be a different organization established to dispose of the concrete, comparable to facilities for recycling aluminum cans. Alternatively, the plant may be a combination of the two. When, in the development of plant operations when the concrete arrives at the factory, it will either go straight into the crushing and reprocessing procedure, or it will be further broken down using hydraulic breakers that are placed on tracked or wheeled excavators. The majority of different types of excavators are compatible with a variety of standard attachments. Putting the main waste feed through a jaw crusher is the recommended course of action to reach the desired size reduction. When the product has been discharged from the initial crushing operation in the plant, a self-cleaning magnet may be used to remove any leftover reinforcing steel from the product quickly. The untreated material, typically between 300 and 400 mm in size, is cut down to between 64 and 76 mm. A smaller jaw crusher, cone crusher, or impact crusher is used on the product to create a final product with a top size of between 19 and 25 millimeters. The final aggregate generated has a moisture content of less than 2% and can generally pass through a sieve sized No. 200 (0.75 mm) (Hansen, 1986). After the removal of the reinforcement, the pavement concrete is left in a pretty clean state. However, this is not the case with all the concrete that has been dismantled. Even once the steel has been removed from the concrete, there is no assurance that the material created will be of a high enough quality to be used as aggregates in new concrete. Other pollutants may still be present in the material; they should be eliminated. Plaster, wood, plastic, oil droplets, and other nonmetallic construction materials might be considered examples of these pollutants. It is not a good idea to utilize RCA from pavement concrete polluted with salt to expose reinforced concrete in a damp environment. Larger particles may be removed by hand or mechanical means, but further processing, either dry or wet, will be required to eliminate any and all possible pollutants (Hamad and Dawi, 2017).

ACI 221R covers the consequences of processing, handling, and beneficiation. 1.3 Plant design—The units used to treat recycled concrete are quite similar to the machines used to process concrete made from virgin resources (Hamad and Dawi, 2017). Almost the same equipment is used to eliminate impurities not often present in conventional aggregate deposits, with only minor adjustments. Depending on the circumstances, the aggregate processing facility might have an open or closed system. Because it enables more precise control over the largest particle size that

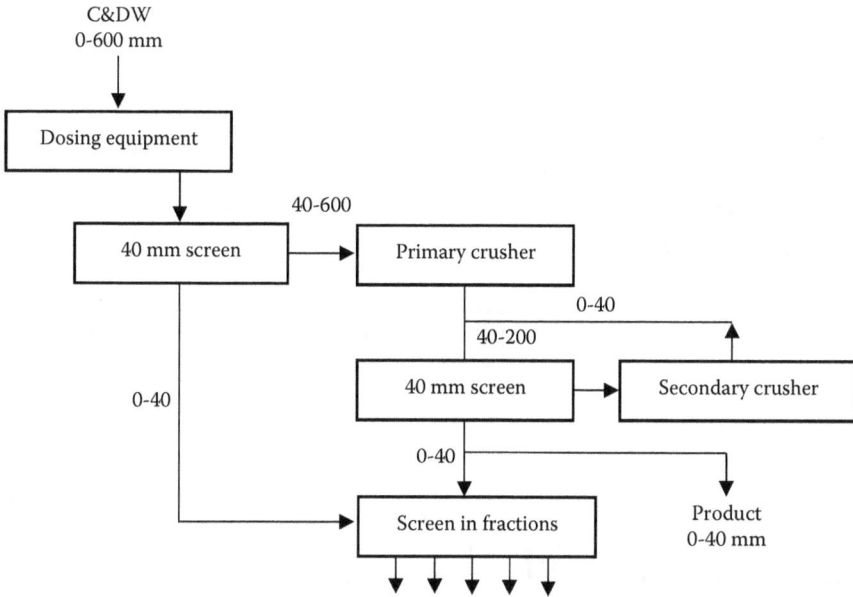

FIGURE 4.1   Flowchart of typical plant for closed-system production of RAC (Rao et al., 2019).

may be generated, the closed system is the one that should be used. The result has a greater degree of consistency as a result of this (Hansen, 1986). Figures 4.1 and 4.2 illustrate typical workflows for the closed- and open-system processing of aggregate materials. These pictures illustrate what is often known as "first-generation plants." Plants of the second generation can process waste and remove items that are not fit for use. Figure 4.3 depicts the flow diagram for a typical second-generation plant. Crushers–jaw crushers produce the particle-size distribution of RCA necessary for high-quality manufacturing concrete. Cone crushers can successfully process concrete with a maximum feed size of 200 mm. Swing hammer mills are not used very often. Impact crushers provide a better particle size distribution for use in road building and are less susceptible to materials that cannot be crushed, such as reinforcing bars. This makes impact crushers a more desirable choice overall (Commissie voor Uitvoering van Research, 1983). Impact crushers shatter not only the concrete but also the aggregate particles as they smash the material (Building Contractors Society of Japan, 1978).

It is well known that the aggregate qualities primary account for the size of the aggregate, once the materials have been processed, the next step is to size them appropriately so they may be used correctly. Hansen (1986) published his findings from an analysis of data gathered from two investigations in which coarse aggregates were created. The products created after screening using a No. 4 (4.75 mm) screen are depicted. Hansen (1986) concluded that it is pretty simple to create suitable coarse aggregates from recycled concrete and that both types of aggregates can fulfill the specification range if the opening of the crusher is set appropriately.

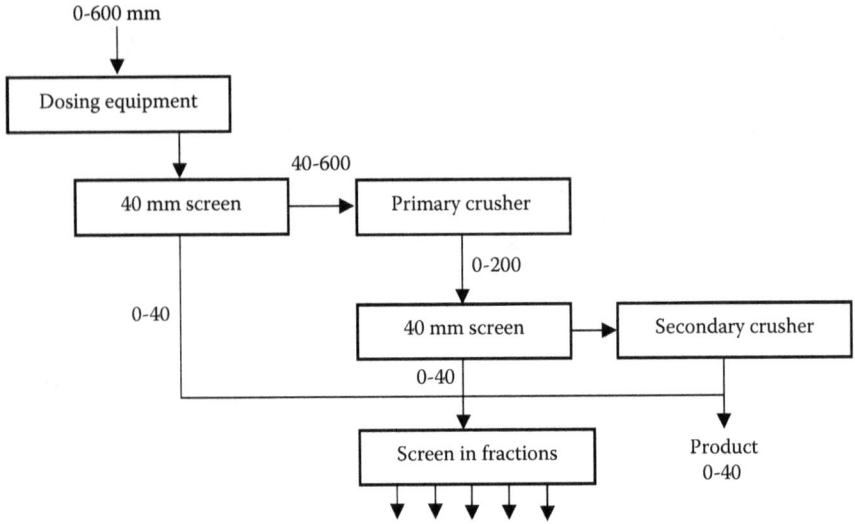

FIGURE 4.2 Flowchart of typical plant for open-system production of RCA (Rao et al., 2019).

Combining recycled sand with a small amount of natural blending sand that is of a finer grain may result in the manufacture of materials that have properties that are acceptable for use in the manufacturing of concrete (Hansen, 1986). The production of fine aggregate was the subject of research, and the findings revealed that the data from three different experiments created materials that fit inside the region illustrated in the study. These materials have a grain that is more substantial than what is required by the C 33 standard of the American Society for Testing and Materials. The bulk of the conventional aggregate that goes into concrete production has a grain size far less than that of their aggregate, which is significantly larger. To begin, the mass per unit area, the densities of recycled aggregates were only marginally lower than the densities of the initial materials used. This is due to the fact that the cement mortar that successfully bound the aggregates to itself had a low density (Building Contractors Society of Japan, 1978; Hansen, 1986). It would seem that very slight changes in the proportions of water to cement in concrete do not significantly impact the densities of the finished products (Hansen, 1986). The process of absorbing water comes in at number two. One of the physical properties that most obviously differentiate recycled aggregates from new aggregates is that recycled aggregates have a higher water absorption rate. The Building Contractors Society of Japan (1978) and Hansen (1986) concluded that the higher water absorption of coarse aggregates is due to the absorption of old cement mortar that is linked to the aggregate particles. This was the cause of the higher water absorption of coarse aggregates. Third, the amount of material lost due to abrasion in Los Angeles According to standard C 33 (Hansen, 1986) of the ASTM, aggregates that are going to be used in the production of concrete should have an abrasion loss of less than 40% for the crushed stone that will be used beneath pavements and less

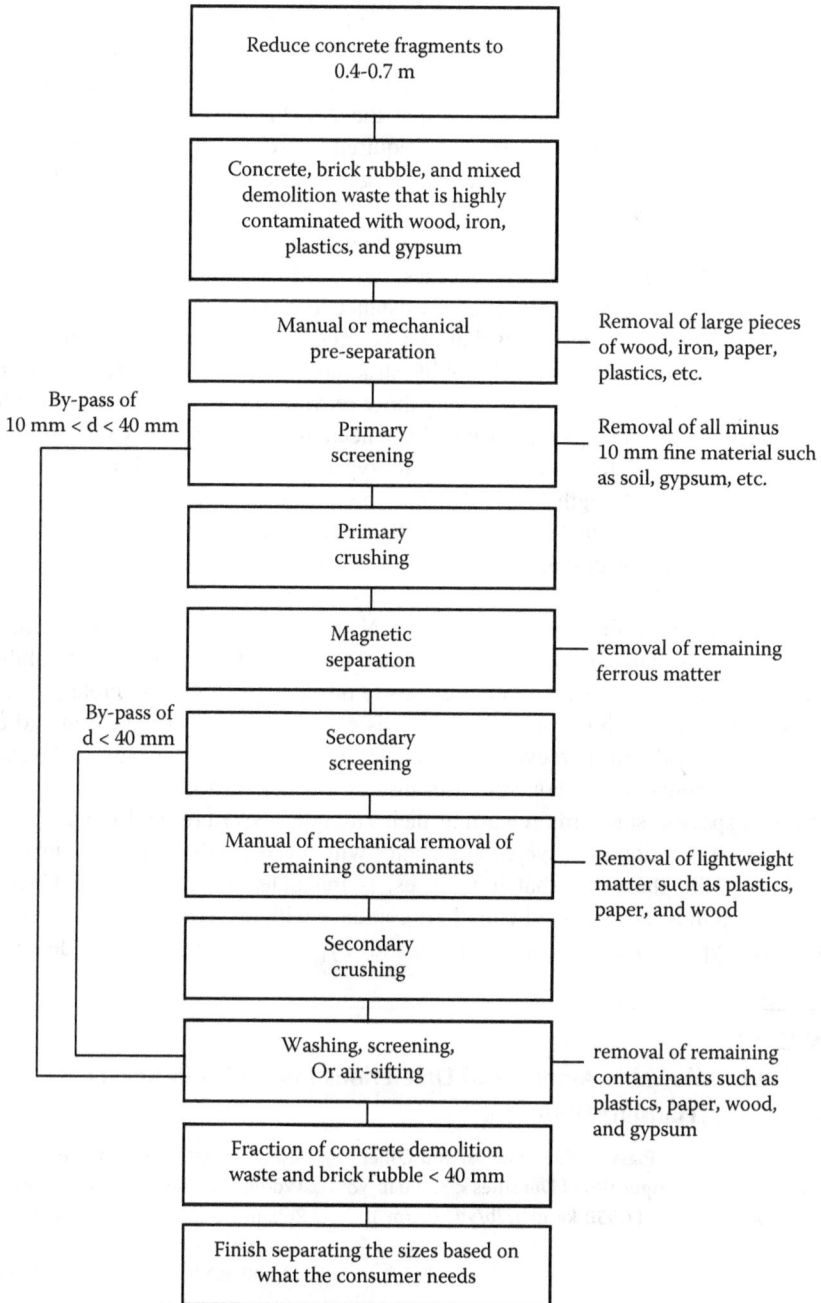

FIGURE 4.3 Processing procedure for C&DW (Rao et al., 2019).

than 50% for general construction. Hansen (1986) came to a conclusion, based on the data that was easily accessible at the time, that recycled concrete aggregates produced from recycled concrete of all but the lowest quality should be able to pass the standards for concrete aggregates that the ASTM. Fourth, sulfate soundness tests (ASTM C 88) (Hansen, 1986) are required by ASTM C 33 (Hansen, 1986), and recycled concrete fine and coarse aggregates may be tested by ASTM C 88 (Hansen, 1986) to ensure that appropriate resistance to freezing and thawing of the recycled aggregates. This is done to ensure that the recycled aggregates are not negatively affected by the freezing. These tests are carried out to guarantee that recycled aggregates have the necessary resistance to freezing and thawing. Fifth, the many different types of pollutants that may be present in recycled aggregates due to the deconstruction of previously built structures can significantly lessen the strengths of the concrete produced using those recycled aggregates. This possibility exists because recycled aggregates may have been used. Material may take the form of anything from plaster to soil to wood to gypsum to asphalt to plastic to rubber. The breakdown of strength losses that may be attributed to pollutants in recycled aggregates can be seen in Tables 4.18–4.20. These tables also define the acceptable amounts of hazardous contaminants that can be present RCA.

In general, RAC needs a substantially greater parameter when compared to NAC, which may influence its endurance. NAC does not have this requirement. Although this scenario is not always connected to a negative influence on durability, it can offer significant issues when it comes to producing durable concrete that is in precise agreement with the specification. It is expected that the production and use of concretes made from recycled aggregates would not display any differences compared to conventional concrete. Because of this notion, RCA in Germany must adhere to specific standards regarding their strength, exposure, and moisture content. The exploitation of recycled aggregates with a quality that meets or surpasses DIN 4226-100, the norm that it replaces, is mandated by the DAfStb Code of Practice and must be complied with. In accordance with this code of practice, coarse RCA of Type 1 (concrete chippings) and Type 2 (chippings from destroyed

TABLE 4.18

Maximum Allowable Amounts of Deleterious Impurities in Recycled Aggregates (Hansen, 1986)

Type of aggregate	Plasters, Clay Lumps, and Other Impurities of Densities < 3,300 lb/yd^3 (1,950 kg/m^3), lb/yd^3 (kg/m^3)	Asphalt, Plastics, Paints, Cloth, Paper, Wood, and Similar Material Particles Retained on a 0.047 in. (1.2 mm) sieve (also Other Impurities of Densities < 2,000 lb/yd^3 [1,200 kg/m^3], lb/yd^3 (kg/m^3)
Recycled coarse	17 (10)	3 (2)
Recycled fine	17 (10)	3 (2)

TABLE 4.19
JIS Overview of RCA for Concrete (Hansen, 1986)

		Recycled Aggregate for Concrete-Class H	Recycled Aggregate Concrete-Class M	Recycled Aggregate Concrete-Class L
Water absorption of aggregate	Coarse	3% or Less	5% or Less	7% or Less
	Fine	3.6% or Less	7% or Less	13% or Less
Main applications		There is no particular limitation, general purpose concrete	Concrete for piles, foundation beams, steel pip filling	Concrete does not require high strength and durability, such as leveling concrete
JIS Standard		JIS A 5021:2018 Recycled aggregate for concrete-Class H	JIS A 5022:2018 Recycled aggregate concrete-Class M	JIS A 5023:2018 Recycled aggregate concrete-Class L
Established date of the standard		March 20, 2005	March 20, 2007	March 20, 2006
Purpose of JIS standard		The standard for recycled aggregate used for concrete for general use which improved quality as aggregate by advanced treatment such as crushing and abrasion of concrete waste	The standard for recycled aggregate is produced by a comparatively simple method such as crushing and abrasion of concrete waste for concrete which is hardly affected by drying, shrinkage and freezing, and thawing.	The standard for the recycled aggregate of relatively low-strength concrete using recycled aggregate produced by crushing concrete waste

structures) may be used to produce concretes with a strength class of up to C30/37 based on the standards, specifications, and quality controls for the use of aggregates as shown in Table 4.21.

When producing aggregate that satisfies these criteria, the producer must, at the time of production, comply with all of the requirements of an IS EN aggregates standard appropriate for the use for which the aggregate is destined. Only then will the aggregate be produced in a manner that satisfies these criteria. You may discover information on the product and the uses to which it is ultimately put, as well as the IS EN Standards, the specifications, and the quality controls associated with aggregate manufacturing, is referred to in Table 4.22.

4.3   RAC FOR BUILDINGS

To promote environmentally responsible construction practices over the past few decades, research has been conducted on the use of crushed and RCA materials as coarse aggregates. The compressive strength of recycled aggregate concrete is inherently lower than that of traditional concrete with the same water-to-cement

TABLE 4.20

Standards, Specifications, and Quality Controls for the Use of Aggregates (Hansen, 1986)

Product and use	Standard	Specification	Quality Control
1 Unbound recycled aggregate: Pipe bedding Drainage	BS EN 13242: Aggregates for unbound and hydraulically bound materials for use in civil engineering work and road construction	Highways Agency Specification for Highway Works (SHW): Series 500 Highway Authorities and Utilities Committee (HAUC): Specification for the reinstatement of openings in highways (SROH)	BS EN 13242: Level 4 Attestation Evaluation of Conformity to BS EN 16236* SHW: Quality Control procedures per the Quality Protocol for the production of aggregates from inert waste SROH: Compliance with SHW
2 Unbound recycled aggregate: Granular fill General fill capping	BS EN 13242: Aggregates for unbound and hydraulically bound materials for use in civil engineering work and road construction	Highways Agency Specification for Highway Works: Series 600 HAUC: Specification for the reinstatement of openings in highways BS EN 13285: Unbound mixtures: Specifications	BS EN 13242: Level 4 Attestation Evaluation of Conformity to BS EN 16236* SHW: Quality Control procedures per the Quality Protocol for the production of aggregates from inert waste SROH: Compliance with SHW
3 Unbound recycled aggregate: sub-base	BS EN 13242: Aggregates for unbound and hydraulically bound materials for use in civil engineering work and road construction	Highways Agency Specification for Highway Works: Series 800 HAUC: Specification for the reinstatement of openings in highways BS EN 13285: Unbound mixtures: Specifications	BS EN 13242: Level 4 Attestation Evaluation of Conformity to BS EN 16236* SHW: Quality Control procedures following the Quality Protocol for the production of aggregates from inert waste SROH: Compliance with SHW
4 Recycled aggregate for concrete	BS EN 12620: Aggregates for concrete	Highways Agency Specification for Highway Works: Series 1000 BS 8500-2: Concrete	BS EN 12620: Level 4 Attestation Evaluation of Conformity to BS EN 16236* SHW: Quality Control procedures following the Quality Protocol for the production of aggregates from inert waste

TABLE 4.20 (Continued)

Standards, Specifications, and Quality Controls for the Use of Aggregates (Hansen, 1986)

Product and use	Standard	Specification	Quality Control
5 Recycled aggregate for asphalt	BS EN 13043: Aggregates for bituminous mixtures and surface treatments for roads, airfields, and other trafficked areas	Highways Agency Specification for Highway Works: Series 900 HAUC: Specification for the reinstatement of openings in highways	BS EN 13043: Level 4 Attestation Evaluation of Conformity to BS EN 16236* SHW: Quality Control procedures following the Quality Protocol for the production of aggregates from inert waste SROH: Compliance with SHW
6 Recycled aggregate for hydraulically bound mixtures	BS EN 13242: Aggregates for unbound and hydraulically bound materials for use in civil engineering work and road construction	Highways Agency Specification for Highway Works: Series 800 HAUC: Specification for the reinstatement of openings in highways BS EN 14227-1 to 5 Hydraulically Bound Mixtures: Specifications	BS EN 13242: Level 4 Attestation Evaluation of Conformity to BS EN 16236* SHW: Quality Control Procedures following the Quality Protocol for the production of aggregates from inert waste SROH: Compliance with SHW
7 Reclaimed asphalt for use in bituminous mixtures	BS EN 13108-8 Bituminous mixtures – Material specifications – Part 8: Reclaimed asphalt.	Highways Agency Specification for Highway Works: Series 900 BS EN 13108-1 to 5 Bituminous mixtures – Material specifications	BS EN 13108-8 NHSS Sector Scheme 14 SHW: Quality Control procedures following the Quality Protocol for the production of aggregates from inert waste SROH: Compliance with SHW

ratio. This is due to the fact that RAC breaks in a different way than traditional concrete. The recycled concrete aggregate (RCA)-cement interfacial zone of normal strength recycled concrete aggregate (RCA) was composed primarily of loose and porous hydrates, whereas the recycled concrete aggregate (RCA)-cement interfacial zone of high-performance recycled concrete aggregate (RCA) was composed of compact hydrates. The same is true for the strength of RAC, which is between 1% and 15% lower than that of natural aggregate (NA)-formed concrete. The same is true for the modulus of elasticity, which is between 13% and 18% lower, and

TABLE 4.21

Standards, Specifications, and Quality Controls for the Use of Aggregates (Hansen, 1986)

Product and Use	Standard	Specification	Quality Control
Unbound recycled aggregate: Piper bedding Drainage	IS EN 13242: Aggregates for unbound and hydraulically bound materials for use in civil engineering work and road construction	As required by the customer	Independent Audit of the Factory Production Controls
Unbound recycled aggregate: Granular Fill General Fill Capping	IS EN 13242: Aggregates for unbound and hydraulically bound materials for use in civil engineering work and road construction	As required by the customer IS EN 13285: Unbound Mixtures: Specifications	Independent Audit of the Factory Production Controls
Unbound recycled aggregate: sub base	IS EN 13242: Aggregates for unbound and hydraulically bound materials for use in civil engineering work and road construction	As required by the customer IS EN 13285: Unbound Mixtures: Specifications	Independent Audit of the Factory Production Controls
Recycled aggregate for hydraulically bound mixtures	IS EN 13242: Aggregates for unbound and hydraulically bound materials for use in civil engineering work and road construction	As required by the customer IS EN 14227-1 to 5 Hydraulically Bound Mixtures: Specifications	Independent Audit of the Factory Production Controls

TABLE 4.22

The Physical Properties of the Produced CRFA (Yang et al., 2020)

Content of fine powder (%)	9.8	Crush index (%)	21
Apparent density (kg/m^3)	2,470	Saturated-surface-dry water absorption (%)	6.4
Bulk density (loose condition) (kg/m^3)	1,350	Void content (loose condition) (%)	45
Grading zone	11	Fineness modulus	2.2

fractures energy, which is between 27% and 45% lower. RCA is produced due to the devastation caused by the collapse of a wide variety of distinct types of buildings and other structures. The properties of aggregates that have been recycled from these constructions vary, depending on the properties of the RAC that they were recycled from.

As a result of the fast growth of urbanization, there has been a massive rise in the number of construction materials, particularly concrete, used by the building industry. As a direct result of this, a much more considerable amount of construction and demolition waste (also known as C&DW) has been produced, severely affecting the surrounding ecosystem. Controlling and disposing of construction and demolition waste (C&DW) has always been challenging, and using RCA has been highlighted as one of the most promising options. It is mainly made out of concrete and bricks and other elements such as porcelain and glass. When concrete is crushed and sorted, it creates coarse RCA and fine RCA.

The pace of economic growth on a global scale is accelerating at an accelerating rate, and the breadth of infrastructure development is becoming an increasingly crucial aspect of the global economy. According to the International Building Code, each year, there must be a need for billions of tons worth of natural coarse aggregate to be used in the construction of new buildings. Meanwhile, the destruction of several old buildings generates a substantial amount of construction garbage. Additionally, many countries create about more than 2,000 billion tons of urban construction rubbish each year via building or destroying structures. This garbage accounts for approximately 40% of total urban waste (Liu et al., 2017). However, due to the low rate at which construction waste is used, most of it is buried or stacked. This consumes a significant amount of land and space and results in secondary environmental catastrophes. Recycling is done using waste from construction. As a consequence, recycling garbage from building projects has many positive effects on the environment and the economy.

In order to evaluate how glowing RCA made from crushed concrete functions as a construction material, this research looked at many different cement replacements weight ratios. The goal was to assess how the addition of RCA affected the concrete's appearance and strength. In addition, flexural tests are being performed on reinforced concrete beams with RCA in varying replacement ratios (0, 15, 30, and 50 wt%) to determine how they affect the flexural behavior of concrete beams in a practical range of tensile reinforcement ratios (0.5% to 1.8%) and to determine the optimal RCA replacement ratio for a given application. These tests aim to determine how they affect the flexural behavior of concrete beams in a practical range of tensile reinforcement. Even though there are very few structural tests of reinforced concrete beams with weak areas, there are still relatively few of them. The next step is to create a database that contains information on previous and ongoing studies with beams made from natural aggregates. This will assist us in determining how the moment capabilities of these beams constructed using RCA are related to the NA used in their construction.

The fast urbanization in developing countirs is causing a significant amount of concrete and other building materials to be wasted, with the former being the more prevalent of the two. RCA may be discovered in abandoned buildings or concrete structures that have been demolished. These recycled aggregates can be used to manufacture recycled concrete, which is better for the environment than virgin aggregates. However, the recycling process requires careful attention. It is possible to replace all or the majority of the coarse aggregate in a concrete mix with aggregate derived from abandoned buildings or concrete structures that have been destroyed. Because of this, RAC may be derived from RCA. The mechanical properties of RAC have been the subject of a significant amount of study in the past. Investigating recycled concrete members and structures is only one of the many things that have been done as part of this. Because of the inherent limitations of recycled aggregate, it has been established that RCA may primarily be used in applications for constructions subjected to lesser degrees of stress or strain. Because of this, RAC and concrete-filled steel tubular (CFST) constructions, which have been recommended by academics and have received much attention from real-world applications, make buildings lighter and less expensive to construct (Xiao and Ding, 2013). This technology not only makes the structures lighter but also reduces the cost of the concrete components. This is accomplished by filling the steel tubes with RAC or concrete formed from fragments of destroyed concrete. The use of recycled aggregate makes RAC structures more flexible, which makes them a viable option for buildings that need to be able to survive powerful earthquakes or are located in an area where earthquakes occur often. Meanwhile, it is generally believed that recycled aggregate concrete has significant and obvious economic advantages and a possible future for its usage in various applications. To fill concrete poured within steel tubular columns, it is technically conceivable to use RCA generated from waste building materials. This would make the process more environmentally friendly. It is possible that this is a workable plan for dealing with waste materials that have a high potential value and could be repurposed in the construction of load-bearing components for buildings. The RCA process is efficient and cost-effective in managing trash from buildings and demolition. Additionally, it assists in the conservation of resources, the preservation of the environment, and the promotion of sustained economic development within the building sector. Successful building projects in China that have used RAC as a structural material. This is because, due to the recent rapid expansion of China's construction industry and the occurrence of numerous natural disasters, the country has seen an increase in the amount of waste concrete being produced. It is essential to find new applications for leftover concrete in order to both protect the natural world and advance the cause of sustainable development. In addition, the seismic performance of RAC structures allows for the construction of cast-in-situ constructions, RAC frame buildings with a maximum height of six stories, and RAC block masonry structures in areas that have been damaged by earthquakes or are in the process of being rebuilt after an earthquake. Alterations are necessary regardless of whether or not a precast structural system is employed. It is recommended that the junction between the precast beam and column be designed to distribute the energy more effectively. In light of the growing utilization of RAC techniques in actual engineering projects, it is possible to deduce that the number of RAC building structures is increasing as a direct result of the development of RAC techniques.

Regarding the disposal of construction and demolition waste (also known as C&DW), the RAC technique is a very successful strategy. In addition, the possibility of recycling waste concrete is not only a technical obstacle but also a management challenge that calls for the involvement of international stakeholders. A significant number of important shifts have taken place in many countries with regard to RAC research and application. Nevertheless, more definitive study and investigation are still required in some RAC applications, notably those in which there is a lack of technology, standards, government backing, and public understanding. It is feasible to utilize RAC as a building material that does not adversely affect the surrounding environment if the structure's design and construction are carried out suitably. Because of this, it reduces the amount of waste created from natural aggregates and makes it possible to reuse waste from buildings and demolition, making the exploitation of RAC an environmentally responsible practice.

Extensive studes have been carried out on fine RCA, with the primary emphasis on determining how well the various recycled concrete components interact. The great majority of academics are of the view that the presence of connected mortar on the surface of the aggregates is the most crucial problem with RCA when it comes to its usage in structural applications. This is the perspective held by the vast majority of academics. However, a number of people feel that RCA may be used for structural applications in specific settings while still maintaining its effectiveness. On the other hand, fine recycled aggregates, sometimes referred to as fine RCA, are not the kind of material advised for use in concrete applications. The decreasing utility of recycled fine aggregates may be attributed, in part, to the increased number of linked cement pastes on recycled fine aggregates, as well as their angular form and high water absorption.

Additionally, it has an excellent capacity for absorbing water. Because of the issues discussed previously, recycled concrete suffers from a decrease in strength and durability. This is a direct effect of those factors. The CO_2 treatment had a more substantial impact on the lifespan of RAC than any other treatment, enabling it to increase RCA properties. When creating anything, it is vital to consider the inherent qualities of recycled aggregates, such as their high permeability and high water absorption capacity. This strategy helps make up for specific characteristics that need to be considered. In addition, Cartuxo et al. stated that the utilization of superplasticizers would improve the workability and porosity of concrete made from fine RCA, which is an exciting idea, and they discovered that the strength of the concrete increased as a result of it. This is a very promising finding. People have suggested that mineral admixtures, like fly ash, may be added to RAC in order to make it simpler to deal with and more substantial for construction purposes. Another strategy suggested for enhancing the mechanical properties of RCA is to include sodium silicate and silica fume in the mixture. This has been done in order to achieve this goal. However, most of the methods for improving the properties or strength of concrete may not be appreciated by the concrete industries because they increase the amount of time required to make concrete, which is linked to the cost of making things. Concrete industries may find this to be an issue.

There has been some thought given to the possibility of producing cement-based items using ultrafine recycled materials that are sourced from C&DW in order to make the most of the resources offered by C&DW while simultaneously reducing the carbon footprint of a product. This would accomplish both of these goals. In recycling EoL concrete debris into smaller pieces, the process produces ultrafine particles of recycled concrete as a consequence of the crushing step. Most of these particles are composed of wet cement products and silica particles. They might be used either as a supplement to cement or as a partial substitute for it due to the fact that they are formed of small particles and have a solid structure. According to the results of many studies, the amount of unhydrated cement and C_2S attached to recycled fine aggregates is directly related to whether or not the aggregates have the potential to cement themselves independently. This amount of cement and C_2S varies depending on the conventional concrete's age, the concrete's grade, and the number of cementitious components included inside the concrete. Despite this, it was discovered that the recycled ultrafine product had around 24% of unhydrated cement, which increased its hydraulic activity when it was utilized again (Hou et al., 2017). On the other hand, it has been shown that recycled concrete ultrafine goods include less than 4% unhydrated cement (Hou et al., 2017), indicating that the product has poor hydraulic qualities. According to the findings of the same research project, recycled concrete ultrafine products have the potential to successfully replace up to 25% of the cement in mortar without reducing its overall strength (Hou et al., 2017). In addition, utilizing recycled concrete ultrafine goods as a replacement for limestone filler could also be a good idea. This is because recycled concrete ultrafine products are outstanding. It has been shown that it is feasible to employ recycled concrete ultrafine particles in new eco-products, and it has also been demonstrated that these particles may be exploited to create a solution for the environment that is both technically and ecologically viable. It has been demonstrated that the material possesses pozzolanic properties when utilized as supplemental cementitious materials when ultrafine derived from C&DW are included in newly manufactured cement or added to the material as an additional SCMs. This was accomplished by including the ultrafine in the newly manufactured cement or the material itself. It has been established by Moreno-Juez et al. (2020) that it is possible to use ultrafine recycled concrete particles that have been processed by heating air classification system (HAS) technology to replace up to 5% or more of the cement that is used in the manufacturing of concrete. This replacement can be accomplished by employing ultrafine recycled concrete particles processed by heating air classification system (HAS) technology. The researchers uncovered these findings. This procedure results in a reduction of the amount of clinker that is contained in cement, as well as a reduction in the amount of time it takes for cement to cure and set, improved mechanical qualities at a younger age, and a reduction of between 5% and 7% in the amount of clinker that is contained in cement. A limited group of researchers has also looked at the possibility of using ultrafine recycled concrete products as a source of the component raw material in the production of concrete. In order to manufacture clinker with a mineral composition that is much more equivalent to that of commercial clinker.

The fine RAC is challenging to utilize in the construction industry since there is not yet a suitable mix design technique that can be used with it. This makes its use problematic. This study provides a novel packing density strategy that considers the fine RAC characteristic and achieves a good balance between a low vacancy ratio and a low specific surface area. The goal of this work is to find a solution to the problem. Research into the method of mix design is required to understand the properties of a new form of concrete and to encourage its application in the construction industry. In terms of its production method, the generation of a recycled concrete mix is comparable to that of a standard concrete mix. However, in order to get equivalent workability, extra water is necessary. This is because coarse RCA has a large capacity for absorbing water. In addition, the performance of recycled concrete can be comparable to or even superior to that of new concrete when a low water-cement ratio is used, when the concrete is constructed of medium to high strength, and when coarse recycled concrete aggregate is integrated into the concrete to a weight percentage of 30% or less of the total weight (Marie and Mujalli, 2019). In recent years, direct weight replacement, comparable mortar replacement, and direct volume replacement have all been used in combination with these tactics. Direct weight replacement was the first method to be developed. When recycled aggregate concrete is used, the modified equivalent mortar volume (EMV) mix design technique will not produce a low elastic modulus. This is because EMV stands for equivalent mortar volume. According to Kim and Sadowski, recycled aggregate concrete exhibits mechanical properties and durability equivalent to natural aggregate concrete (Marie and Mujalli, 2019). Gupta et al. discovered a revolutionary way to gain the same mechanical performance from RCA without using any of the three strategies that enhance performance. By altering the coarse RCA ratios, the researchers could boost the ratio by up to 50 wt% when utilizing RCA in conjunction with single strength source concrete (Marie and Mujalli, 2019).

The great majority of academics are of the view that the presence of connected mortar on the surface of the aggregates is the most important problem with RCA when it comes to its usage in structural applications. This is the perspective held by the vast majority of academics. In some circumstances, however, RCA is capable of serving as effective when applied to structural applications. On the other hand, fine RCA is not recommended for use in concrete applications as much as the other sizes. The number of cement pastes linked to recycled fine aggregates, the angular form of recycled fine aggregates, and the high water absorption rate of recycled fine aggregates are the aspects that lead to the lower usefulness of recycled fine aggregates. Because of the issues discussed previously, recycled concrete suffers from a decrease in strength and durability. This is a direct effect of those factors.

While creating the structures, it is vital to consider the intrinsic characteristics of RCAs. These characteristics include a high capacity for absorbing water and high permeability, both of which are amenable to being compensated for using this approach. In addition to the use of superplasticizers in order to improve the workability and porosity of concrete that was produced from fine RCAs, an increase in the compressive strength of the concrete with recycled mineral admixtures such as fly ash can be used to produce recycled concrete that is both more workable and more durable. This type of recycled concrete can be made by using recycled mineral admixtures that can be used to produce recycled concrete.

It has been looked into whether or not it would be possible to make cement-based goods that were created using ultrafine components supplied from C&DW and include them into the overall composition of a product. In this way, it is possible to reduce the carbon footprint of a product while simultaneously increasing the number of resources that are used that are generated from C&DW. It is mainly made up of recycled concrete ultrafine particles, by-products of hydrated cement, and silica particles, created when end-of-life (EoL) concrete waste is crushed into small pieces. Both of these types of particles come from recycled concrete. Because they are quite small and have a solid composition, there has been some recent discussion over the feasibility of using them as a substitute for cement or as an additive to concrete. According to many studies, the quantity of unhydrated cement and C_2S that adheres to a product is directly proportional to the amount of recycled fine aggregates with self-cementing properties. This relationship is determined by the original concrete's age and the quality and quantity of the cementitious components. In addition to that, using recycled concrete ultrafine materials for limestone filler could also be a good option. This is due to the very fine nature of items made from ultrafine recycled concrete. Utilizing recycled concrete ultrafine particles in the production of new items that are kind to the environment is a clever idea that is effective. This suggests that they may be included in conceiving an efficient solution for the environment. The ultrafine that C&DW produces either be included in already produced cement or utilized to create a new kind of cement. It would seem that the presence of CCA in structural concrete in an applied electrical field has a negative impact on the number of chloride ions that penetrate the concrete. Using GGBS to manufacture structural CEM III/A concretes decreases the negative consequences by enabling a more significant percentage, of course, CCA, to be used up to 100 wt%, outperforming CEM I concrete controls using 100 wt% natural aggregates. This is accomplished by allowing a higher proportion of CCA to be used. In addition, durability performance problems may be reduced even further by increasing the cover depth of CEM III/A concretes to a level equivalent to that of CEM I concrete. This will bring the danger of durability performance problems down to an even lower level. In order to keep the same structural concrete mixtures and coarse CCA sources, the percentages of GGBS and CCA replacements should be capped at 50 and 60 wt%, respectively.

4.4 MIX DESIGN AND QUALITY CONTROL OF RAC

In order to design a concrete structure that is both the most sustainable and the least harmful to the environment, it is necessary to use as many reclaimed materials from construction and demolition work as is practically possible. Not only does this product cut down on the use of natural resources, but it also cuts down on the CO_2 emissions linked with concrete production.

According to the "European Green Deal" and the "Circular Economy Action Plan" published by the European Commission, this is a significant milestone in the efforts of the European Union to combat the extraction of natural resources, as well as to eliminate potentially valuable C&DW, and to align the goals of sustainability and climate change. This study demonstrated the use of recycled aggregates in a structural concrete mix and other recycled products from C&DW processes, such as cement-rich hydrated powder, recovered glass powder, and recycled mineral fibers. Additionally, this study demonstrated the use of recycled mineral fibers. Concrete that is beneficial to the environment and has a long shelf life might be manufactured by researchers using more than 75 wt% of recycled materials reclaimed from building construction and demolition operations.

It is possible to replace coarse and fine natural particles in concrete with synthetic equivalents without the end product suffering any degradation in its inherent mechanical properties. The application of revolutionary technology (advanced dry recovery (ADR) + heating air classification system (HAS)) improves the performance of recycled aggregates and, as a result, the behavior of recycled concrete. This occurs when all coarse and fine particles are replaced with recycled aggregates. Because of this, there is an opportunity to turn the construction sector into one that functions completely based on a circular model. In addition, the amount of cement paste-rich recycled powder that may be included in the product is limited to a maximum of 5% of the total. The mechanical strength of recycled concrete is only marginally impacted when more glass powder generated from C&DW is added to the mix. Due to the ease with which Ca^{2+} ions dissolve in water, they are able to inhibit the pozzolanic process at very early ages. It would be beneficial to conduct more studies to get a deeper and more comprehensive understanding of how the glass powder develops throughout time. In addition, adding mineral fibers to recycled concrete can boost both its tensile strength and its modulus of elasticity. Recycled concrete may be more challenging to work with and may not last as long if it has a higher concentration of certain types of fibers. However, recycled concrete becomes more challenging to work with when added in lower volumes because of the additional recycled material.

Concrete is made by mixing natural coarse aggregate (also known as RCA), recycled fine aggregate (also known as RFA), and natural cementitious material together with cement. Because of this, the term "recovered aggregate" is often used when referring to recycled concrete debris. This word refers to the several varieties of recycled concrete waste that include "RFA," "RP," and "RCA," among other expressions. The term "recycled aggregate" refers to a category that encompasses all three types of recycled materials: RCA, RFA, and recycled powder. The original aggregate and the mortar poured on top of it make up the "basis" of the structure (Rong et al., 2021). These two components, when put together, constitute the basis.

Due to the porous nature of the mortar that is used to bind it together, it has a lower density than that of natural aggregate, making it easier for water to penetrate the material. This, in turn, makes it simpler for water damage to occur. This indicates that the quality of the mortar used to keep the recycled aggregate together plays a crucial role in determining the grade when recycled aggregate is used.

The standard approach to recycling RCA is extracting the bigger recycled aggregate particles from the crushed concrete and then repurposing those larger recycled aggregate particles as RCA. The acronym RFA stands for recycled fine aggregate, simply smaller pieces of recycled aggregate. In this procedure, the quantity of adhering mortar rises as the average particle size of the recycled aggregate falls, which, according to a number of studies, indicates that RFA are capable of holding more water than RCA. Because RCA, which has the same mechanical characteristics as natural aggregate, is the most frequent approach to utilize leftover concrete, RFAs are used as little as possible. This is because RCA has the same mechanical properties as natural aggregate. Despite this, academics are becoming more worried about the shortage of natural fine aggregate, and they have conducted a significant amount of studies on recycled fine aggregate. For instance, Kou and Poon claimed that RFAs might make concrete less dense and cause it to shrink more when it cures. According to a few pieces of research findings, the mechanical strength and durability of concrete are unaffected by the incorporation of RFA at percentages lower than 30 wt% (Yang et al., 2020). When the percentage of RFA in newly mixed concrete paste is more than 50%, one's capacity to swiftly work with the material is severely hindered. The JGJ/T240-2011 standard (Yang et al., 2020) from China, according to the document titled "Technical Specification for the Application of Recycled Aggregate," recycled fine aggregates (RFA) should not be used more than 50% of the time, nor should they be used in concrete with a strength grade that is higher than C40. This is done to avoid the adverse effects of their use. The letter "C" denotes that the concrete in question has high compressive strength, and the phrase "characteristic value of cubic concrete compressive strength" is used here.

According to Yang et al. (2020), increasing the percentage of recycled concrete debris used improves the properties of RFA. This process completely transforms the discarded concrete into RFA without leaving any residual RCA. The application of this technique, which is referred to as fully recycled fine aggregate (CRFA) technology, has been given a patent in China.

Natural coarse aggregate, which makes up 40% of all C&DW, is often sieved into RCA rather than RFA, as is customarily done. This is because RCA can better simulate the properties of natural coarse aggregate. As a result of the increased frequency with which the mortar is used in RFA, the percentage of the mortar used in RFA has increased. Another thing to take into account is the possibility that fully recycled fine aggregate, also known as CRFA, would employ this natural coarse material. This is because, throughout the production process, it pulverizes all of the concrete rubble into fine aggregate. When it comes to crushed natural coarse aggregate, on the other hand, CRFA has a lower mortar ratio than typical recycled fine aggregates. This is because CRFA is made entirely from recycled materials.

In addition, recycled fine aggregate, also known as CRFA, gives off the impression of being denser and more porous than ordinary RFA. In addition, it seems to have a greater density and porosity than the typical RFA, making it better for the human body. Because of its many advantages, completely recycled fine aggregate, also known as CRFA, may successfully substitute for all-natural fine aggregate in concrete building projects. The same conclusion was arrived at by Fan et al. (2016). They claim that by crushing all waste concrete into RFA, it is possible to make RFA with a smoother surface than was previously thought. A roller sand washer was used to extract the bulk of powder (particles smaller than 150 μm) from the resulting RFA, as opposed to the technique that was used to manufacture the fully recycled fine aggregate (CRFA), as demonstrated by Fan et al. (2016). Because it is so expensive to remove the small-particle powder from the RFAs that were purchased, CRFA keeps and uses this portion of the RFA as a fine aggregate, in addition to using the RFA as the coarse aggregate. This is in addition to the RFA being used as the coarse aggregate.

At the Concrete Technology Unit (Yang et al., 2020) of the University of Dundee, comprehensive research and development efforts are now under way to overcome the practical and technological difficulties associated with using RCA in manufacturing concrete and how the use of coarse RCA impacts the performance of high-strength concrete with a C50 strength or greater, as well as the obstacles that come with putting it into practice. Porosity was present in the cement paste because, according to the RCA, it had a relative density of 7% to 9% lower and a water absorption rate double that of organic matter when it was in a wet condition on the surface. In addition, the mechanical qualities of RCA were on par with the limitations set by the British Standard (BS) (Yang et al., 2020), although they were somewhat inferior to those of the natural aggregates employed. In addition, it was discovered that the RCA coarseness did not affect the ceiling strength of the concrete up to a coarseness content of 30 wt%; nevertheless, this dropped when the coarseness of the RCA content grew above that threshold. In addition, coarse RCA may be used to achieve outstanding technical qualities in various high-strength concrete combinations. These attributes include compressive strength, flexibility, and modulus of elasticity. Nevertheless, when the RCA level rose, the concrete saw an increase in both its shrinkage and its creep stresses. In conclusion, the acceptability of RCA was determined by many different aspects; nonetheless, it will need to be reviewed separately. In addition, it is of the utmost importance to acknowledge the need for new criteria for RCA and the necessity to show that the materials may be used suitably and practical.

4.5   TEST METHODS ON RAC STRUCTURES

Serviceability is an essential quality in the realm of civil engineering, which deals with the construction of buildings and infrastructure. On the other hand, what exactly defines a secure infrastructural assessment has not been established as of yet since it is not generally possible to quantify how reliable the infrastructures are. This is essential in building the vast majority of substantially laden facilities, including bridges, highways, airports, dams, and offshore projects. Because these buildings are often exposed to loads, it is vital to perform structural inspections regularly to evaluate their health levels and prevent catastrophic failures. They are not necessarily expensive to correct but may cause injuries and other problems. Concerning the problems associated with fatigue in concrete buildings, the Rilem Committee indicated that the loading range, a variety of factors, and the fatigue performance of concrete buildings were affected by various factors. These factors included load frequency, stress level, cycle count, and stress ratios, which offered fundamental support for the methods that are currently in use. In addition to the findings of scanning electron microscopy (SEM), the process of crushing RA results in the presence of microcracks in the material. In addition, the interactions between RA and the cement matrix of the studied concretes are of very high quality.

The ultrasonic pulse velocity (UPV) test is one of the non-destructive inspection assessment techniques that is used on a widespread basis. It is used in various settings to assess the quality of the concrete that is already in place as well as the consistency of the concrete utilized in the many different structural components. It assesses whether or not there have been alterations to the properties of a concrete component due to its long exposure to high temperatures. This is done so that any changes may be accounted for. RAC has a higher UPV than natural aggregate concrete when subjected to temperatures at ambient and elevated levels. Because RA has a lower density than coarse natural aggregate. As the original equipment manufacturer (OEM) recommended, including cement-replacing materials (CRMs) into RAC boosts residual UPV by boosting density and improving binding between new mortar and RCA in the transition zone.

The UPV tool may be used to determine how strong concrete contains RCA. This is primarily because of the fact that it expresses variations in strength in a suggestive manner. It was used in the whole mix planning process. Numerous investigations have been carried out to determine what happens to the UPV of concrete when it is heated, particularly concerning concrete that is not very strong. It is vulnerable to such temperatures and grows worse regardless of the amount of concrete since the exposed temperature rises with all types of concrete. This is because the exposed temperature increases with each kind of concrete. Because of the UPV data, it is feasible to establish a correlation that can be used to calculate the residual strength. This correlation may be used to determine the value of the UPV.

5 Good Practices in RCA Concrete

5.1 ASSESSING THE USE OF RAC IN PRACTICE

The RAC can be processed further to produce unbound and RCAs for use in concrete production. Even though manufacturing RAC using RAC is becoming an essential concern in the construction industry, reinforced concrete (RC)-concrete quality remains a significant concern. There is a widespread belief that NAC are exchanged for RAC, the physical and mechanical qualities of the concrete will be diminished.

Using RAC when making new concrete structures can help save natural aggregates, which are essential for making concrete. Wood, glass, metal, and plastic may be recycled in various ways; nonetheless, RCAs from crushed concrete are relatively uncommon in new construction. Processes for converting demolished concrete into RCAs are now widespread in the building industry. However, aggregates are usually solely utilized for unconstrained purposes, such as road construction. It takes much energy to break down and treat concrete waste. It may also need to use more water depending on how it is treated, in any case. If more cement is needed for RC-concrete manufacturing, greenhouse gas (GHG) emissions may grow. In examining comparative footprints, using LCA techniques and tools, potential trade-offs could be detected, and issue shifting avoided.

By assessing the climate and resource footprints of concrete during the production phase of end-of-life (EoL) and recycled concrete (RC) products with those of business as usual (BAU) concrete. The computation is based on data from a real-life case study acquired during the deconstruction, waste disposal, and manufacturing sites. The RAC was converted into recyclable aggregates in a mobile unit. The new structure was built with RC, obtained through an urban mining approach. The literature was consulted to validate the data and to simulate various scenarios; an example involves the treatment of concrete at a fixed location (Hu et al., 2016). Additionally, a sensitivity analysis of different factors was done to find possible ways to reduce climate change and better use resources (Chua et al., 2016).

There are a few problems; most cost data comes from the Netherlands. If raw materials become more available and cheaper in other European Union (EU) countries, secondary materials' competitiveness and market share will be affected. This study's data were collected in a lab for the second time. Future research will look at how well the heating air classification system (HAS) works at a more advanced stage (e.g., on a pilot scale and an industrial scale). Third, like the proper location of recycling facilities, the cost of moving items and raw materials to their next destination, the fact that some recycling processes are not always the same, and the fact that it is hard to figure out what impact category indicators mean.

It is common knowledge that construction made of concrete must comply with the safety standards relevant to their industry. Even though the behavior of concrete after being heated to high temperatures has been the subject of a significant amount of research, the many different types of concrete that are available, such as SCC and RAC, have received a comparatively small amount of attention in the works that are pertinent to the topic. It is a well-established fact that concrete's residual mechanical properties decrease with increasing exposure temperatures; thus, this is also the case for SCC and RAC. The resistance of SCC to high temperatures led researchers to the conclusion that, because of the dense microstructure of SCC, the material is less resistant to high temperatures and has a greater susceptibility to spalling than vibrated concrete does. This is partly owing to the dense microstructure of SCC, which adds to the issue and plays a role in its development. Similarly, because the coarse RCA and the fresh mortar in RAC have such a strong interaction, RAC functions better after being subjected to high temperatures than NAC does, and its tendency to spall is significantly reduced because the interaction between these two components seems to be so powerful.

Cracking is common in RC structures; however, it is not always destructive to the structure. Many constructions are built to crack when subjected to service loads. When meeting the requirements for durability and aesthetics, cracking must be strictly controlled. The current rules, for example, include restrictions on the breadth of cracks and minimum amounts of reinforcing, as seen in. Plastic settling or shrinkage cracks in newly poured concrete to those caused by live loads on the structure are all cracks; several variables contribute to concrete member cracking. It is also possible for cracks to develop in concrete constructions as a result of internal or external restrictions. For example, temperature differences between different areas of a structure can result in varied deformation requirements. The occurrence of cracking is possible if those deformations are prevented. However, even if design requirements are strictly followed, inevitable cracks may expand beyond the permitted limitations. More importantly, cracking can harm the longevity and aesthetics of RC members and affect a RC member's ductility and strength. Depending on the location and shape of the crack, as well as the way it broke, the effect on toughness, and ultimate capacity could be both positive and negative. During the structural evaluation, you should consider how a crack affects toughness, capacity, and failure mode. In order to account for this, increased assessments of reinforced concrete infrastructure are required.

It is one of the most critical factors for engineers and practitioners to evaluate current reinforced concrete structures and infrastructures. Creep and shrinkage, time-dependent processes (such as pre-stressing steel relaxation), deterioration (like corrosion owing to carbonation and/or chloride penetration), and other environmental impacts and an increase in the magnitude of traffic loads are all factors that existing bridges and viaducts must contend with. Furthermore, the appraisal of existing structures is critical to national authorities and focuses on older bridges and viaducts due to their expensive maintenance, intervention, and upgrading expenses (Abed and de Brito, 2020). Engineering and practitioners have found that the partial factor method (PFM) applied under the most efficient methodology for designing and assessing structures subjected to static and dynamic stresses is the semi-probabilistic approach (i.e., level I) (Abed and de Brito, 2020). Partially safe materials and activities are used with partial safety factors to ensure that the goal reliability levels are met under a focus on service life, economics, and human safety.

The partial safety factors must be appropriately updated in order to carry out the evaluation using the EN 1990-defined technique of limited semiprobabilistic states (Amiri et al., 2021), taking into account the following factors: remaining service life, amended targets for extant structures' dependability, the structure's actual reality, and the degree of variable operations (for example, material characterization based on in situ and laboratory test results, environmental monitoring and assessments, and traffic loads for road bridges).

Because of the factors mentioned earlier, The International Federation for Structural Concrete (fib) Bulletin 8011 (Martinez-Arguelles et al., 2019) presents sophisticated and efficient approaches focused to develop target dependability levels for evaluating current structures while considering human safety (i.e., individual and group risk) and economic optimization criteria. Partially safe structures should have their partial safety factors recalculated to ensure that they are compliant EN1990 describes a semi-probabilistic framework for this (Martinez-Arguelles et al., 2019).

The Fifa Bulletin 8011 proposes two methods for redefining part safety factors for existing structures: the design value method (DVM) and the adjusted partial factor method (APFM). Both techniques may evaluate part safety factors for existing buildings based on leftover service life and newly specified target reliability standards. Following the two techniques established by fib Bulletin, the current work focuses on inspecting a concrete road bridge that already exists near the city of Avigliana in northern Italy (Piedmont, Turin) erected in 1990. Along the connection between the two highways is a bridge across the Dora Riparia, which connects the A32 Torino Bardonecchia highway with the SS25 motorway. The bridge is composed of pre-stressed concrete pre-cast box sections. The balanced cantilever technique is used to construct the three spans (30, 60, and 90 m), as shown Figure 5.1.

Creating new partial variables that take into account the remaining service life, data from in situ and laboratory tests, quantifying the actions of variables, and lowering goal reliability levels to meet economic and human safety needs will be done in order to use the guidelines found in Fib Bulletin 80 to evaluate existing reinforced concrete structures (Amiri et al., 2021). This will be done by creating new partial variables that take into account the remaining service life. An existing pre-cast box section pre-stressed reinforced concrete bridge erected in 1990 in northern Italy was subjected to an evaluation using the procedures indicated in FIB Bulletin 80. These methods included DVM and APFM. The findings were compared to those obtained from the EN1990 test (Anike et al., 2020), which demonstrated the advantages of DVM and APFM compared to the need to calibrate the component safety factor. Several restrictions are placed on how the FIB Bulletin 80 (Anike et al., 2020) may be used, and there are also some issues with the manufacturing process. There are no probability models for modifying partial safety factors for pre-stressing or induced deformations in Bulletin 80 of the F.I.B. (i.e., there will be foundation settlements and thermal actions).

FIGURE 5.1 A bridge across the Dora Riparia, which connects the A32 Torino Bardonecchia highway (1990).

According to EN1990, the partial safety factor for the action of pre-tensioning has been established at 1.00. This has been indicated in the standard. It is recommended that the procedures described in Fib Bulletin 80 be used. This is a safe assumption to make. This presumption needs to be taken seriously (because lower levels of dependability in new structures may result in values less than one). In addition, several safety considerations for settlements and thermal activities have been included at the same levels as those established by EN1990 (Anike et al., 2020).

In comparison to EN1990 (Anike et al., 2020), it is possible that adopting lower reliability target levels for existing structures will cause the partial safety factors linked to this condition to have significantly lower values. This outcome would result from adopting lower reliability target levels for existing structures. It could be used to avoid costly and ineffective quick fixes by, for instance, shortening the remaining service life of a bridge that the authorities have approved (i.e., setting a lower reliability goal that can be met in a shorter remaining service life) and planning for long-term maintenance, upgrading, and/or tearing down the structure and constructing something new in its place.

The current concrete construction in the Beibu Gulf port (China) as shown in Figure 5.2 had a probabilistic corrosion start time assessment. After conducting a field survey, it is found that the lognormal distribution best described the diffusion coefficient (DC). The mean was 23.6 mm^2/year with a standard deviation of 0.11

FIGURE 5.2 Beibu Gulf port (China) (1990).

and a correlation coefficient of 0.11, which means a normal distribution found for concrete cover depths. The mean is 59.5 mm, and the standard deviation is 3.93 mm. The chloride diffusion coefficient's time dependence should be considered when determining the time required for corrosion to begin. The corrosion initiation period of the structure was 40 years based on the target probability value of Pd = 10.0% provided. When the age factor (m) was 0, 0.2, 0.4, and 0.6, the corrosion start time was 19, 26, 40, and 98 years, respectively.

Numerous surveys have evaluated the fatigue safety of road and railway bridges using the probabilistic fatigue reliability framework, and it has proven to be effective. On the other hand, these investigations were restricted to RC bridges. Because of the possibility of fatigue damage over time, alongside the amount and frequency of increases in axle loads, RC bridges frequently require reinforcing to maintain the requisite degree of reliability while continuing to use the infrastructure (Kurylowicz-Cudowska et al., 2020). An evaluation is required prior to any intervention. However, assessing the intrinsic safety of RC current bridge structures or for the goal of extending their service life is typically more complex than building new bridges. During the entire service life of the structure, inspections, structural interventions, and monitoring campaigns may be utilized to update any uncertainty on both the action and resistance sides of the equation. Taking into consideration the weights of vehicles and their places within the road width of deck slabs, vehicle speed, and the width of the road; the number of cars passing via the bridge in each traffic direction; uncertainties associated with temperature; as well as temperature-related stresses, are all examples of action-side uncertainties. Uncertainties on the resistance side reveal a high dispersion in fatigue test data, even though the test campaigns are carried out under similar conditions. Uncertainty in resistance includes structural responses to these activities, represented as variations in the impact of the actions. The uncertainty in concrete compressive strength is connected to the rise in concrete strength over time due to the cement's continuous hydration. The probabilistic reliability technique used to evaluate the fatigue resistance of structures is effective in dealing with such a wide range of uncertainties as these.

The construction industry is one of the most dependent on non-renewable resources on a global scale, with construction accounting for around one-quarter of all such resources. Because of the country's aging infrastructure, the United States generates over 200 million tons of garbage from demolition each year, with roughly half of this waste consisting of OPC debris (Anike et al., 2020). Aggregates are the material used most often in civil engineering projects. Any strategy to reuse aggregates from concrete buildings that have been destroyed should lessen both the negative effect that development has on the environment and the expenses associated with construction. The term "debris" refers to any waste that results from the building, redevelopment, or destruction of any infrastructure or building. Several different kinds of materials and technology that have already reached the end of their useful life were included in the building process in some capacity. The scope of C&DW is expanded to encompass architectural and structural components. C&DW comprises a wide variety of materials because they include such a large variety of components. These components include concrete, masonry, wood, metal, and polymers. Additional components may include other materials. Crushed concrete aggregate, or RCA, is a waste product resulting from demolishing and crushing concrete blocks that have reached the end of their usable life. It has lately gained attention due to its improved advantages in terms of its greater usability and friendliness to the environment. To reuse RCA in an economical and efficient manner, academics and industry professionals must investigate the mechanical and technical challenges it presents. Concrete has been widely used in civil engineering projects for a significant number of years due to the fact that it can be produced at a low cost and has excellent mechanical characteristics. When constructing with concrete, tearing down buildings made of concrete, or manufacturing concrete, there is always a significant quantity of C&DW debris created, which has a detrimental impact on the environment. Concrete production is anticipated to release approximately 1.35 billion tons of CO_2 each year, which is a sizeable amount of CO_2 (Ding et al., 2020). In addition, the production of C&DW raises concerns about the occupation of land and water contamination in the surrounding area. As a consequence of this, it has been decided that action has to be taken in order to deal with this kind of waste.

It is well known that the global quarry methods for the production of coarse aggregates have had a significant impact on equilibrium in the ecosystem. This represents the quantity of aggregates produced per capita in 39 countries (Ding et al., 2020), expressed in tons of aggregates. To ensure the long-term viability of our world, it is critical to discover alternatives for virgin resources that are gathered to manufacture the binder and aggregates that the construction industry requires. End-of-life products are thrown in open fields as landfill at the same time. One type of trash that falls into this category is building and demolition debris, which has the potential to be turned into a viable recycled aggregate product. In practically all industrialized and emerging countries, RCA can be found because old buildings and structures. Such structures have become unusable and of little worth, nonetheless they have enormous potential as raw materials suppliers for rebuilding projects. Current practices provide for disposing of a considerable amount of radioactive waste as zero-value garbage. As a result, RCA has the potential to play a significant role in sustainable development. RCA has emerged as a significant domain in the building industry, serving as an alternative for natural aggregates and raw resources. Adding RCA to concrete would change the hardness of the finished product. Several investigations have shown that the substitution of RCA for aggregates in concrete at 100 wt% is unsatisfactory due to a considerable reduction in hardened strength ranging from 15 to 25 wt%, depending on the study.

Because concrete is one of the most important material making with natural resources, it significantly affects the demand for aggregates if there is much demand for aggregates in the construction business. This presents a significant obstacle for the mineral aggregates business in Europe, which, to fulfill the requirements of the concrete industry, is required to generate a total of 2,700 million tons of aggregates annually (Hahladakis et al., 2020). Excessive aggregate exploitation is becoming both a concern for the environment and an economic disaster on a global scale due to the depletion of raw resources and the energy-intensive extraction process required to obtain these elements. Despite this, new approaches to waste management are required because there has been a rise in the amount of waste generated from C&DW. In this type of economy, the maximization of resource recovery and reuse is encouraged. Therefore, debris left behind after a construction project has been demolished and can be recycled into aggregate for use in the production of concrete if it is first mechanically treated and cleaned (Hahladakis et al., 2020). To use the cut-off principle of RCA's environmental impact, there is no effect on the primary aggregate production process, and only the waste management procedures relevant to C&DW are considered in this way. Typically, the evaluation process does not include the destruction phase. However, there is also research in which demolition is considered the first phase in manufacturing RCA, prior to transferring garbage to a processing plant or a landfill and the processes employed for mechanical treatment.

Despite what has been stated above, once the period during which the structure was technically suitable for use has passed, the structure is demolished, regardless of whether or not it is possible to rebuild the structure or reuse or recycle the materials used in its construction. The demolition process is typically thought of as occurring at the end of the stage of the building's life cycle that is encompassed within the system boundary. This is when the standard concept of the building's life cycle is considered. As a result, the inclusion of this stage in examining environmental issues brought on by the production of RCA may rise to some controversy within the industry. However, due to the widespread assumption that transportation, waste processing, and disposal are all included in the previous system, the validity of examining the environmental implications associated with creating recycled aggregate may be called into question.

The use of RCA in the building industry is an important step toward achieving the goal of sustainable development. Additionally, it is a strategy for improving the resource effectiveness of construction projects. According to the Ministry of the Environment of Singapore, the majority of the country's solid waste consists of C&DW, which is produced as a byproduct of activities such as the demolition of buildings, concrete, and roadways (Ntaryamira et al., 2017). It is predicted that more than 1.2 million metric tons of waste from construction and demolition work was created in 2009, and this quantity is anticipated to increase even higher as the rate of development remains the same. On the other hand, C&DW has been used primarily in the construction industry as a short-term access road and in the construction of highway structures as a sub-base course (Ho et al., 2008). C&DW for more beneficial uses is becoming more relevant as a consequence of the rising demand for, as well as the expense of, natural aggregates. and a growing trend demonstrating how rubbish can be used more effectively.

RCs are the most prevalent and beneficial building materials in the construction business. They have also played an essential role in civilization growth during the previous century. On the other hand, construction facilities necessitate a significant quantity of natural resources to generate cement and aggregate. The acquisition of these natural resources has significantly impacted the natural environment, resulting in significant environmental issues. Sustainable waste management is a critical problem that countries worldwide are currently handling. Regarding environmental conservation and steady, sustainable growth, recycling building trash is a practical and substantial means of accomplishing both. Landfills and roadbed applications are two of the most frequent methods of reducing waste generation and disposal. The land is required for construction waste management, processing, and subsequent processing into a RAC for use in the construction sector, which is both expensive and time-consuming. When considered in this light, concrete recycling is a notion that is not only environmentally friendly but also cost-effective.

Germany is the world's biggest producer of RCA, according to the European aggregate association yearly assessment, followed by the United Kingdom (Gao et al., 2019). Germany produces over 60 million tons of RCA annually, with the United Kingdom producing approximately 49 million tons (Gao et al., 2019). In the European Union, RCA accounts for 5% of total aggregate output. A minimum recycling rate of at least 70% by weight of demolition debris must be attained by the year 2020, under the modified European regulation. Recent studies on the production of recycled aggregate (RA) have been undertaken in Turkey. Consequently, the number of research on the recycling of concrete is now fairly limited. According to the study conducted, there are worries about RCA. Several parameters of recycled aggregate concrete, including density, compressive strength, modulus of elasticity, and toughness value, are observed to decrease as the recycled aggregate concrete's waste concrete aggregate replacement ratio increases. According to Topcu and Güncan (Gao et al., 2019), RAC has many characteristics, including a decrease in compressive strength. Prior to using RCA in the production of concrete, its properties must be evaluated. When producing high-strength concrete, the abrasion resistance and approximate strength of waste concrete aggregate are crucial concerns.

In the possibilities of employing RCA in concrete applications in various countries, Corinaldesi discovered that RAC with a strength class of C32/40 could be made with 70 wt% NA and 30 wt% RCA. In addition, Pereira et al. (Gao et al., 2019) blended two distinct types of superplasticizers with finely recycled aggregate. In relative efficiency, they determined that RAC structures are less effective. Sheen et al. (Gao et al., 2019) indicated that RCA affected the compressive strength of RAC because small portions make the RAC less strong when it is compressed and a large amount of water absorption reduces the strength of RAC. In RAC, fine RCA can be partially or entirely substituted with a quantity of up to 30 wt%, which is appropriate. When it comes to durability, for example, SF and metakaolin both improve mechanical properties and durability as well as GGBFS and FA.

Each year, over 850 million metric tons of C&DW is produced, which accounts for approximately one-third of the entire quantity of waste produced worldwide (Abed and de Brito, 2020). There has been a significant rise in the speed at which the construction sector is growing because of more people living in cities, more people moving to cities, and better economic circumstances in developing nations. As a result, it is important to tear down old buildings to build new ones. An example of a recyclable waste product from demolishing historic structures is aged concrete. RCA refers to recycled material that comprises at least 95% aggregate; otherwise, it is known as recycled aggregate. In concrete mixtures, RCA can be used in place of natural aggregates in some cases, hence reducing the need for natural stone and the environmental and social difficulties associated with its exploitation. It is possible that the lower quality of, instead of natural aggregates, RCA is generated by lateral building materials such as wood chips, ceramics, bricks, or shards of reinforcing steel connected to them. As a result of this difficulty, the use of RCA in concrete production is restricted. Due to the fact that RCA, like natural aggregates, has the potential to be utilized in the production of new concretes, it is essential to analyze the properties of concrete that have been produced with the addition of RCA.

The composition of this waste is very close to that of new building material. Given this activity's negative environmental and economic consequences, it is possible to place a monetary value on C&DW due to the shortage of raw materials. The recovery of waste is one of the most important acts. To characterize innovative materials containing RCA for the building, several studies in Europe have shown that injecting crushed recycled concrete produces positive long-term significances (Abed and de Brito, 2020). It has been determined that this study is important since construction waste in Morocco will unavoidably differ in properties from prior investigations; RCA was utilized various types of aggregates and sand result in varying consistencies. More importantly, in this investigation, RCA replaces only a small percentage of the sand, which maintains the particle size distribution of the sand. For the recovered material, it will be critical to ensure that conventional concrete performs as well as possible. Based on the characteristics that will allow recycled concrete to be successfully integrated into the construction industry, a system for recycling concrete for every type of concrete anywhere in the world should be developed. For instance, RCA from a pavement (RCAP) displayed the best behavior and could be utilized as a viable substitute for NA in concrete pavement applications. It is necessary to do additional study to determine whether the HMA is effective when the recycled concrete aggregate from a pavement (RCAP) component is partially substituted (Ding et al., 2020).

It is impossible to think of a society without concrete foundations or walls; for the foreseeable future, humans will need to construct using flexible materials such as wood. The mixed concrete consists of several components, including cement and other elements that make it difficult to decompose. Also included are aggregates and water in this process. Cement is the most frequent material used as a binder in the production of concrete. The aggregates used to produce concrete, which account for 75% by weight of the volume of the concrete mixture, consist of fine-grade particles with a maximum diameter of 4.75 mm and coarse-grade particles with a maximum diameter of 20 mm. In recent years, various investigations have been devoted to evaluating the physical and mechanical properties and durability indicators of concrete mixtures, including RCA. The difference in density and water absorption between recycled and natural aggregates is mainly attributable to the legacy of an earlier adhering mortar, which has a lower density and more water absorption than the present hanging mortar. According to a study done by Ravindrarajah and Tam, building concrete using a blend of coarse recycled gravel and natural sand uses 5% more water than using solely natural aggregate. The use of recovered sand increases water use by 15%. This may need more mixing water and initial workability, especially when recycled aggregates are used in dry condition.

A growing body of evidence supports the notion of a circular economy (CE) as a response to how we use limited resources. It has garnered substantial attention from governments, and academic institutions, in the last few years (Dumlao-Tan and Halog, 2017; Ellen MacArthur Foundation, 2016; Ranta et al., 2018). Consumer electronics (CE) goods are made to last as long as possible and to be recycled with the least amount of energy at the end of their useful lives, which reduces waste and its effects on the world (Iacovidou et al., 2017; Velenturf and Purnell, 2017). The European Commission's (EC) Sustainable Development Goals (SDGs) were introduced by the United Nations (UN) in 2015, and the CE action plan connects the essential legislative steps with these objectives. The CE action plan explicitly addresses the objective of Goal 12, which focuses on sustainable consumption and production (which was approved in 2015). The EU did this in 2016: One of the five areas where the European Commission wants to see more progress in a "circular economy" is construction and demolition waste (C&DW). This is one of the five areas where the European Commission wants to see more progress in a "circular economy" (Circular economy action plan 2023). C&DW in the manufacturing sector contributes the most to the generation of trash and uses the most resources in today's society. It uses up more than 40%, on average, of the total amount of raw materials that are mined across the world, generating about 35% of all solid waste and contributes about 33% of all CO_2 and GHG emissions (Peng, 2016).

C&DW is not very well defined, and it varies a lot because there are so many different materials and construction methods out there now (European Commission, 2018). Most of it comprises crushed concrete, gypsum, tiles, and a small number of other materials. In many countries, the practice of environmentally responsible handling of waste from construction and demolition projects is yet in its infancy, with landfills being the most common place for C&DW to be conducted (Marzouk and Azab, 2014). C&DW has been recycled many times in the European Union (EU) by using it for lower-quality applications, such as the production of aggregates used in road building as base and infill material because the Waste Framework Directive lacks consistently recycled materials quality standards (WFD 2008/98) (Rutz et al., 2013).

It is common in the construction and demolition industry to use C&DW as a source of material for producing RCAs, which can be used in many different ways and are better for the environment because they do not have to be thrown away or deplete natural aggregate (NA) resources. Most inorganic building elements used to make recycled aggregate (RA) are crushed and processed. Bricks, tiles, metals, glass, paper, plastic, wood, and other debris can also be found in RCA, as can paper, plastic, wood, and other waste. RCA is mainly made by crushing and processing inorganic construction materials that have been used before. RCA is often produced by crushing previously used concrete structural parts that have been segregated from the rest of the structure. In general, NA is thought to be better because people are not very familiar with RCA. Looking at things from a more holistic perspective, the environmental effects of RCA are much lower than those of NA (Kurda et al., 2018).

Recent estimates reveal that C&DW accounts for more than 75% of Qatar's total waste production (Reid et al., 2016). In addition, Qatar is still in the process of implementing its Vision 2030 (Tan et al., 2014). Even though Qatar Construction Standards (QCS) had only permitted the use of RCA since 2014, when the Qatar Transport Research Laboratory (TRL) was established, the Qatari TRL has investigated inventive methods to employ RCA in construction. It occurred in 2014. In preparation for the 2022 FIFA World Cup, aggregates will be in great demand; thus, using as many RCAs as possible is imperative. Therefore, using recycled materials might reduce the environmental effect of importing primary NAs by as much as 50% (Clarke et al., 2017).

CCAs are replacing the more costly virgin raw materials, formerly known as RCA, in increasing numbers. This is due to a rising emphasis on purchasing ecologically friendly items. The United Kingdom produced 13.3, 18.8, and 21.2 million tons of complicated demolition debris yearly between 2015 and 2022. The quantity is anticipated to increase annually in the future. The United Kingdom's most complicated demolition debris is used as general fill, sub-base material, or low-grade concrete component. This is because the aggregate quality criteria for these applications are often less stringent than those for other applications. However, a lack of trust in the structural performance of recycled aggregates has restricted their use.

On the other hand, manufacturers of recycled aggregate always seek to enhance the quality and performance of CCA to increase its future use. The Waste and Resources Action Programme (WRAP) in the United Kingdom provides manufacturing criteria for CCA that may be applied to structural concrete (Pavlu, Koči, and Hajek, 2019). The availability of natural aggregate (NA in the United Kingdom NA discourages designers and builders from adopting CCA for concrete structural applications. If a project or client CCA or improved project sustainability credentials need its usage, it might function as a viable substitute material, on-site access to a high-quality, consistent source of CCA, and/or a NA shortage.

5.2   MIX PROPORTION DESIGNATION OF RCA IN CONCRETE

Over the last few years, sustainable construction has gained significant attention in various countries. With 80%, 75%, and 66%, Denmark, the Netherlands, and Japan are some countries with the highest recycling rates in the world (Tam, V. W., & Tam, C. M. 2006). The exceptional quality of the finished new product generated from recycled C&DW is one of the primary reasons many recycle. A strategy must be devised and implemented for any country to achieve the same recycling rate as the United States. Making a lot of construction and demolition debris poses a challenge in terms of repurposing or recycling it to create new goods or services that benefit both the environment and the people in the area (Yeheyis et al., 2013). Concrete is the world's most widely used building material, accounting for 52% of all C&DW (Tam et al., 2005). Instead of coarse aggregate, researchers in the Okanagan Valley hope to produce concrete with two components commonly found in local landfills (British Columbia, Canada). These alternative aggregates will make the concrete more substantial and environmentally friendly.

The slump is the most common and useful way to measure how concrete behaves in terms of its rheology. The slump is used to judge the consistency of a mixture and show how it flows, including how easy it is to work with and use new concrete. Changes in a slump can be caused by things like the amount of water used and how the aggregates are made, such as how they are spread out, how they look, and how much water they can hold. With the development of new concretes and new ways to install them, like pumping and projection, in the building industry, managing the rheological behavior of fresh concrete has become an essential part of making it easier to use these new materials and technologies. As well as naming the primary factors that affect a material's rheological behavior; it is also important to name the most common behavior models. There are two main groups of rheological behaviors: Newtonian fluids and non-Newtonian fluids. These two groups are further divided into subfamilies. The first of the three parameters is how consistent the concrete is. This shows how well it flows. The ability to stop people from separating is a sign of long-term stability. Only in the last few decades have more accurate ways to figure out numbers based on the Bingham parameters been found (Tam et al., 2005).

When fresh concrete is combined with RCA, the attributes of the new concrete may vary depending on the strength of the old concrete used to get the RCA, the process utilized to generate the RCA, and the moisture content of the RCA. Consequently, there is disagreement among the authors on the various properties of concrete produced using RCA. Using 50 wt% or more RCA will have a detrimental effect on the mechanical properties of concrete, if not worse. However, recent research has demonstrated that using RCA in concrete does not necessarily result in undesirable properties if the proportions are selected with care. Material properties such as the modulus of elasticity, drying shrinkage, water absorption, total pore volume, and carbonation are unaltered when about 20% of the new aggregate is replaced with RCA. Using RCA from concrete sources of better quality may result in compressive strength equivalent to or slightly greater than that of virgin material. Due to the presence of ancient mortars and cement pastes, fine RCA may enhance time-dependent stresses, such as creep and drying shrinkage.

In recent years, numerous productive and in-depth experiments on the mechanical characteristics of RCA-containing concrete have already been described (Xiao et al., 2018). The mechanical and tensile parameters of concrete mixtures using recycled aggregates varied considerably from 20 to 30 wt% replacements of NCA. This is nearly identical to the characteristics of concrete mixtures using natural particles at a 0 wt% replacement percentage, concrete with no RCA. As well as, for instance, demolished buildings in Rabat, Morocco, around 50 years old, have been reduced to sand-sized particles by crushing the concrete. The new concrete's recycled sand percentage ranges from 0 to 20 wt%. As more recycled fraction is used in place of sand, the compressive strength improves, particularly as the percentage of recycled fraction increases from 15 to 25 wt%. There is no difference in the workability of new concrete containing or not containing RCA. It is still necessary to determine the remaining requirements and conduct a series of acceptable tests.

The proposed method for RAC mix design is based on strength. RAC has a wide range of applications in the actual engineering structure, and the proposed approach proposes a practical and simple mixed procedure. Several aspects, including the amount of cement, the w/c, the quantity of old paste, the replacement ratio, and the grade of recycled aggregates that are still attached to the new paste, influence mechanical qualities in concrete that contains RCAs. The design of RAC mix necessitates the inclusion of additional water in order to accomplish equal workability to that of NAC, and this adjustment may have an impact on the mechanical qualities of RAC. When working with new concrete, incorporating water-conserving admixtures (superplasticizers) can aid in reaching the desired workability while having no detrimental effect on the completed hardened concrete's properties. In some circumstances, the high porosity structure of RAC is associated with the material's decreased workability. In terms of concrete microstructure, it is said that RACs porous nature is to blame for the material's inferiority to NAC. Providing RCA is used up to 30 wt% of the time, the strength of the resulting concrete is not adversely affected. Researchers have found that RAC that contains at least 30% RCA is weaker than RAC that does not contain any RCA. Compared to concrete built only with RCA, the compressive strength decreased by up to 30% when all of the former was substituted with NCA. When it comes to compressive strength, the quality of concrete paste and where it is applied to the surface play a significant role; this fluctuation is primarily because it is very different.

In order to considerably lessen the pressure on natural aggregates, this research is being conducted (coarse and fine). The amount of cement gained using the analyzed mix design determined in this study approach is particularly important in the quest for environmentally friendly construction materials. As a result of the mixture design, several observations may be made about the results. First, even though NAC and coarse RCA are planned in accordance with the American Concrete Institute (ACI) standard (2020) and have there was a significant difference in aggregate requirements between the two projects, even if the w/c was the same. This is most likely due to RCA's increased ability to absorb water despite having a lower specific gravity due to the presence of dry mortar. This indicates that coarse RCA mixes prepared according to ACI and British standards (2020) have the same aggregate composition. However, the latter requires a substantial increase in cement utilized. Third, because RAC was designed in compliance with equivalent mortar volume (EMV) requirements, it required far less cement; thus, it is the most budget-friendly choice. At long last, a tenfold increase in superplasticizer was required to attain slump values that were on par with the other mixes, according to the EMV standard. According to Fathifazl et al.'s research, this is the case. Additionally, the superplasticizer helps concrete workability by dissolving flocks and dispersing small particles throughout the mix, minimizing the wall effect between larger particles and boosting the concrete's strength.

Traditional and unorthodox mix design approaches were tested in a laboratory to examine how RAC would perform. This publication explains the experiment's findings. ACI and the Department of the Environment's (DoE) recommendations were among the traditional approaches examined. An unorthodox method, called "Equivalent mortar volume," was also looked at by Fathifazl et al. (2011). It took more cement to make RAC mixes with the ACI and DoE methods rather than the EMV mix design method. The concrete weights of 143 kg/m^3 and 206 kg/m^3 were used to make RAC mixes with the ACI and DoE procedures. As a result, the reference mixes, which were made with only natural virgin aggregate and the old way, were much weaker than the new mix and included more cement per cubic meter of concrete than the EMV mix, with a cement content of more than 143 kg/m^3 of concrete, as shown in Tables 5.1 and 5.2. The use of RCA in structural concrete uses a recommended acceptable methodology that considers RCA's impacts (Martínez-Lage et al., 2020). When it came to meeting the essential needs of coarse aggregate in concrete, practitioners looked into the qualities of RCA with w/c in the range from 0.35 to 0.65. RCA was employed in place of NCA, with up to 100 wt% RCA used to replace NCA.

5.2.1 FIBER-REINFORCED RAC

RAC that contains fibrous material that strengthens its structural integrity is referred to as fiber-reinforced RAC. It is composed of a variety of short, distinct fibers that are evenly dispersed and orientated randomly. Steel, glass, synthetic, and natural fibers are all examples of fibers, and each kind of fiber contributes a unique set of characteristics to the RAC. Fiber-reinforced RAC, sometimes known as FRC, is a relatively new building material that is gaining popularity. A significant number of RAC engineering features are enhanced by using fiber reinforcement in a discrete form. RAC that contains fibrous material that strengthens its structural stability is referred to as fiber-reinforced RAC, or FRC for short. It is composed of a variety of short, distinct fibers that are evenly dispersed and orientated randomly. Steel, glass, synthetic, and natural fibers are examples of different types of fibers. The properties of fiber-reinforced RAC shift depending on the RAC used, the kind of fibers used, their geometries, distributions, orientations, and densities, among other factors.

TABLE 5.1
Chemical Compositions and Physical Properties of Type 1 Portland Cements and FA

Chemical Compositions (% by Mass)	SiO$_2$	Al$_2$O$_3$	Fe$_2$O$_3$	CaO	MgO	Na$_2$O	SO$_3$	Free CaO
Type 1 Portland Cement	20.84	5.22	3.20	66.28	1.24	0.10	2.41	0.99
FA	42.10	21.80	11.22	13.56	2.41	2.90	1.88	1.44

Physical Properties	Type 1 Portland Cement	FA
Loss on Ignition (%)	0.96	2.33
Moisture Content (%)	0.19	1.50
Blaine Surface Area (cm^2/g)		
Fineness (Particle Size, % Retained)	3,200	2,850
− ≥ 75 μm	0.50	0.56
− 75 μm	5.25	8.25
− 45 μm	3.60	4.76
− ≤ 36 μm	90.62	86.43
Fineness (Retained) on 45 Micron		
(No. 325)	5.75	4.90
Water Requirement (%)	100	97
Bulk Density (kg/l)	1.03	0.51
Specific Gravity	3.15	2.13

The performance of any fiber reinforcement is highly reliant on the ability to achieve a uniform distribution of the fibers throughout the RAC, as well as on the casting and spraying processes, as well as the fibers' ability to interact well with the cement. The workability of RAC often suffers due to the addition of a higher proportion of fibers, particularly fibers with tiny diameters. As a consequence, the mix design has to be altered in order to accommodate the new characteristics of the RAC. This is due to the fact that fibers with extremely tiny diameters have a total surface area that is much higher. This need for the addition of more water and cement or admixtures will eventually result in a significant influence on the workability of the RAC. This will ultimately result in an increase in expenses.

The word "compressive strength" refers to the level of resistance a RAC block must meet before it can be said to have achieved its full potential as a material capable of withstanding a compressive load. Brandt (2008) proposed that plastic fiber–reinforced RAC breaks when there are several tiny fissures on the material's surface. This contrasts with the catastrophic failure of plain RAC during the compression testing. S. Spadea et al. (2015) researched why adding extremely short recycled nylon fibers results in a drop in the compressive strength of tested mortar by as much as 37%. S.B. Kim et al. (2007) found that recycled polyethylene terephthalate (PET) and poly-propylene fiber–reinforced specimens had a lower compressive strength than the plain specimen by around 1% to 9%, approximately 1% to 10%, respectively.

TABLE 5.2
Typical Mix Proportions and Related Properties of RAC

Mix Symbol (FrC300C345F0S75)	Cylindrical Strength (ksc)	Cement (kg/m³)	Fly ash (kg/m³)	Water (kg/m³)	Sand (kg/m³)	Recycled Aggregate 3/4" (kg/m³)	Recommended Quantity Admixture (Liters/m³)	Fiber (%)
	280	345	0	155	835	1,145	4.10	1.4

Properties	Average	Standard Deviation	Compressive Strength (ksc)	Average	Standard Deviation
Air content (%)[1]	1.2	0.1	7 days	82.68	0.84
Slump (mm)[2]	27.5	15.2	28 days	191.41	1.03
Initial setting time (min)[3]	101	4.2	90 days	249.62	2.88
Final setting time (min)[3]	250	30.2	365 days	290.74	3.62

Mix Symbol (FrC300C275F0S75)	Cylindrical Strength (ksc)	Cement (kg/m³)	Fly Ash (kg/m³)	Water (kg/m³)	Sand (kg/m³)	Recycled Aggregate 3/4" (kg/m³)	Recommended Quantity Admixture (Liters/m³)	Fiber (%)
	280	275	70	150	835	1,145	4.15	1.5

Properties	Average	Standard deviation	Compressive strength (ksc)	Average	Standard deviation
Air content (%)[1]	1.1	0.3	7 days	93.51	13.10

Mix Symbol (FrC320C370F0S75)

Cement (kg/m³)	Fly Ash (kg/m³)	Water (kg/m³)	Sand (kg/m³)	Recycled Aggregate 3/4" (kg/m³)	Recommended Quantity Admixture (Liters/m³)	Fiber (%)
275	70	145	835	1,145	4.4	1.6

Properties

Property	Average	Standard deviation
Slump (mm)[2]	27.0	4.9
Initial setting time (min)[3]	151	16.7
Final setting time (min)[3]	275	15.9
Cylindrical Strength (ksc)	240	

Compressive strength (ksc)	Average	Standard deviation
28 days	216.36	4.73
90 days	254.73	11.62
365 days	305.81	5.58

Mix Symbol (FrC320C295F74S75)

Cement (kg/m³)	Fly Ash (kg/m³)	Water (kg/m³)	Sand (kg/m³)	Recycled Aggregate 3/4" (kg/m³)	Recommended Quantity Admixture (Liters/m³)	Fiber (%)
287	73	155	850	1,115	4.3	1.7

Properties

Property	Average	Standard deviation
Air content (%)[1]	1.8	0.1
Slump (mm)[2]	25.0	8.6
Initial setting time (min)[3]	192	18.7
Final setting time (min)[3]	340	26.3
Cylindrical Strength (ksc)	240	

Compressive strength (ksc)	Average	Standard deviation
7 days	93.33	14.76
28 days	194.97	12.63
90 days	277.54	8.36
365 days	320.26	6.01

(Continued)

TABLE 5.2 (Continued)
Typical Mix Proportions and Related Properties of RAC

Mix Symbol (FrC320C295F74S75)	Cylindrical Strength (ksc)	Cement (kg/m³)	Fly Ash (kg/m³)	Water (kg/m³)	Sand (kg/m³)	Recycled Aggregate 3/4" (kg/m³)	Recommended Quantity Admixture (Liters/m³)	Fiber (%)
Air content (%)[1]	1.9		0.2		7 days		96.98	24.06
Slump (mm)[2]	25.5		25.0		28 days		181.65	19.54
Initial setting time (min)[3]	172		16.8		90 days		258.33	11.63
Final setting time (min)[3]	300		44.2		365 days		289.06	7.21

Mix Symbol (FrC350C395F0S75)	Cylindrical Strength (ksc)	Cement (kg/m³)	Fly Ash (kg/m³)	Water (kg/m³)	Sand (kg/m³)	Recycled Aggregate 3/4" (kg/m³)	Recommended Quantity Admixture (Liters/m³)	Fiber (%)
	240	345	0	150	835	1,145	4.70	1.8
Properties	Average		Standard deviation		Compressive strength (ksc)		Average	Standard deviation
Air content (%)[1]	1.0		0.2		7 days		115.98	22.01
Slump (mm)[2]	26.5		5.6		28 days		246.32	15.89
Initial setting time (min)[3]	122		20.0		90 days		294.56	13.62
Final setting time (min)[3]	320		10.5		365 days		319.66	4.46

Mix Symbol (FrC350C315F80S75)	Cylindrical Strength (ksc)	Cement (kg/m³)	Fly Ash (kg/m³)	Water (kg/m³)	Sand (kg/m³)	Recycled Aggregate 3/4" (kg/m³)	Recommended Quantity Admixture (Liters/m³)	Fiber (%)
Properties	240	360	0	160	850	1,115	4.50	1.9

	Average		Standard deviation	Compressive strength (ksc)			Average	Standard deviation
Air content (%)[1]	1.6		0.3	7 days			112.58	18.74
Slump (mm)[2]	51.0		13.9	28 days			212.56	11.29
Initial setting time (min)[3]	111		14.5	90 days			284.81	3.24
Final setting time (min)[3]	290		8.6	365 days			341.47	5.26

Mix Symbol (FrC380C420F0S75)	Cylindrical Strength (ksc)	Cement (kg/m³)	Fly Ash (kg/m³)	Water (kg/m³)	Sand (kg/m³)	Recycled Aggregate 3/4" (kg/m³)	Recommended Quantity Admixture (Liters/m³)	Fiber (%)
Properties	240	314	56	160	855	1,120	5.00	2.0

	Average		Standard deviation	Compressive strength (ksc)			Average	Standard deviation
Air content (%)[1]	2.0		0.1	7 days			93.59	24.59
Slump (mm)[2]	29.5		22.8	28 days			181.63	14.77
Initial setting time (min)[3]	121		12.9	90 days			212.08	5.88

(Continued)

TABLE 5.2 (Continued)
Typical Mix Proportions and Related Properties of RAC

Mix Symbol (FrC380C420F0S75)	Cylindrical Strength (ksc)	Cement (kg/m³)	Fly Ash (kg/m³)	Water (kg/m³)	Sand (kg/m³)	Recycled Aggregate 3/4" (kg/m³)	Recommended Quantity Admixture (Liters/m³)	Fiber (%)
Final setting time (min)[3]	280		42.5	365 days			294.59	3.47
Mix Symbol (FrC380C335F84S75)	**Cylindrical Strength (ksc)**	**Cement (kg/m³)**	**Fly Ash (kg/m³)**	**Water (kg/m³)**	**Sand (kg/m³)**	**Recycled Aggregate 3/4" (kg/m³)**	**Recommended Quantity Admixture (Liters/m³)**	**Fiber (%)**
Properties	370	445	0	165	744	1,070	4.90	2.1
Properties	**Average**		**Standard deviation**	**Compressive strength (ksc)**			**Average**	**Standard deviation**
Air content (%)[1]	1.7		0.1	7 days			154.25	19.70
Slump (mm)[2]	28.0		11.8	28 days			172.33	14.02
Initial setting time (min)[3]	141		3.5	90 days			244.97	10.71
Final setting time (min)[3]	295		2.2	365 days			272.48	6.11

Mix Symbol (FrC400C445F0S75)	Cylindrical Strength (ksc)	Cement (kg/m³)	Fly Ash (kg/m³)	Water (kg/m³)	Sand (kg/m³)	Recycled Aggregate 3/4" (kg/m³)	Recommended Quantity Admixture (Liters/m³)	Fiber (%)
	320	395	0	165	825	1,077	5.20	2.0

Properties	Average	Standard deviation	Compressive strength (ksc)	Average	Standard deviation
Air content (%)[1]	1.8	0.2	7 days	124.05	21.47
Slump (mm)[2]	28.5	8.4	28 days	211.59	14.89
Initial setting time (min)[3]	171	8.1	90 days	301.08	5.04
Final setting time (min)[3]	255	9.4	365 days	244.86	3.32

Mix Symbol (FrC400C355F90S75)	Cylindrical Strength (ksc)	Cement (kg/m³)	Fly Ash (kg/m³)	Water (kg/m³)	Sand (kg/m³)	Recycled Aggregate 3/4" (kg/m³)	Recommended Quantity Admixture (Liters/m³)	Fiber (%)
	320	295	100	165	825	1,077	5.30	2.1

Properties	Average	Standard deviation	Compressive strength (ksc)	Average	Standard deviation
Air content (%)[1]	1.9	0.3	7 days	111.74	30.14
Slump (mm)[2]	29.0	7.6	28 days	263.66	25.53
Initial setting time (min)[3]	191	17.2	90 days	325.45	14.88
Final setting time (min)[3]	260	14.9	365 days	385.72	4.09

(Continued)

TABLE 5.2 (Continued)
Typical Mix Proportions and Related Properties of RAC

Mix Symbol (FrC420C470F0S75)	Cylindrical Strength (ksc)	Cement (kg/m³)	Fly Ash (kg/m³)	Water (kg/m³)	Sand (kg/m³)	Recycled Aggregate 3/4" (kg/m³)	Recommended Quantity Admixture (Liters/m³)	Fiber (%)
	210	295	0	155	835	1,130	5.60	2.3
Properties	Average		Standard deviation	Compressive strength (ksc)			Average	Standard deviation
Air content (%)[1]	1.2		0.3	7 days			269.21	22.14
Slump (mm)[2]	56.0		24.7	28 days			227.86	14.60
Initial setting time (min)[3]	102		18.9	90 days			309.01	8.36
Final setting time (min)[3]	340		26.3	365 days			320.82	4.87

Mix Symbol (FrC420C375F94S75)	Cylindrical Strength (ksc)	Cement (kg/m³)	Fly Ash (kg/m³)	Water (kg/m³)	Sand (kg/m³)	Recycled Aggregate 3/4" (kg/m³)	Recommended Quantity Admixture (Liters/m³)	Fiber (%)
	280	295	75	150	825	1,130	5.50	2.4
Properties	Average		Standard deviation	Compressive strength (ksc)			Average	Standard deviation
Air content (%)[1]	1.3		0.1	7 days			157.43	3.24
Slump (mm)[2]	56.5		1.8	28 days			214.71	17.48

Properties	Average	Standard deviation	Compressive strength (ksc)	Average	Standard deviation
Initial setting time (min)[3]	112	3.5	90 days	247.06	5.97
Final setting time (min)[3]	345	14.3	365 days	274.89	4.83

Mix Symbol (FrC450C495F0S75)	Cylindrical Strength (ksc)	Cement (kg/m³)	Fly Ash (kg/m³)	Water (kg/m³)	Sand (kg/m³)	Recycled Aggregate 3/4" (kg/m³)	Recommended Quantity Admixture (Liters/m³)	Fiber (%)
	320	286	184	165	820	1,100	5.90	2.5

Properties	Average	Standard deviation	Compressive strength (ksc)	Average	Standard deviation
Air content (%)[1]	1.8	0.1	7 days	104.88	14.76
Slump (mm)[2]	50.5	22.8	28 days	203.49	14.03
Initial setting time (min)[3]	121	12.9	90 days	239.61	2.08
Final setting time (min)[3]	260	14.9	365 days	287.72	6.55

Mix Symbol (FrC450C395F100S75)	Cylindrical Strength (ksc)	Cement (kg/m³)	Fly Ash (kg/m³)	Water (kg/m³)	Sand (kg/m³)	Recycled Aggregate 3/4" (kg/m³)	Recommended Quantity Admixture (Liters/m³)	Fiber (%)
	320	294	190	165	820	1,100	5.80	2.6

Properties	Average	Standard deviation	Compressive strength (ksc)	Average	Standard deviation
Air content (%)[1]	1.9	0.2	7 days	149.29	20.99

(Continued)

TABLE 5.2 (Continued)
Typical Mix Proportions and Related Properties of RAC

Mix Symbol (FrC450C395F100S75)	Cylindrical Strength (ksc)	Cement (kg/m³)	Fly Ash (kg/m³)	Water (kg/m³)	Sand (kg/m³)	Recycled Aggregate 3/4" (kg/m³)	Recommended Quantity Admixture (Liters/m³)	Fiber (%)
Slump (mm)[2]	51.0		13.9	28 days			167.41	18.71
Initial setting time (min)[3]	111		14.5	90 days			242.54	10.02
Final setting time (min)[3]	265		3.1	365 days			282.36	4.10

Mix Symbol (FrC480C520F0S75)	Cylindrical Strength (ksc)	Cement (kg/m³)	Fly Ash (kg/m³)	Water (kg/m³)	Sand (kg/m³)	Recycled Aggregate 3/4" (kg/m³)	Recommended Quantity Admixture (Liters/m³)	Fiber (%)
Properties	320	274	176	165	820	1,100	6.10	2.7

Properties	Average	Standard deviation		Compressive strength (ksc)	Average	Standard deviation
Air content (%)[1]	2.0	0.1		7 days	182.31	10.06
Slump (mm)[2]	51.5	24.5		28 days	259.47	7.14
Initial setting time (min)[3]	131	2.8		90 days	288.06	12.60
Final setting time (min)[3]	270	25.7		365 days	327.41	7.44

Mix Symbol (FrC480C415F104S75)	Cylindrical Strength (ksc)	Cement (kg/m³)	Fly Ash (kg/m³)	Water (kg/m³)	Sand (kg/m³)	Recycled Aggregate 3/4" (kg/m³)	Recommended Quantity Admixture (Liters/m³)	Fiber (%)
	320	350	84	165	820	1,100	6.20	2.9

Properties	Average	Standard deviation	Compressive strength (ksc)	Average	Standard deviation
Air content (%)[1]	2.1	0.2	7 days	202.01	0.98
Slump (mm)[2]	52.0	4.9	28 days	245.96	0.77
Initial setting time (min)[3]	151	16.7	90 days	284.09	0.85
Final setting time (min)[3]	275	15.9	365 days	324.14	0.34

Mix Symbol (FrC300C345F0S75)	Cylindrical Strength (ksc)	Cement (kg/m³)	Fly Ash (kg/m³)	Water (kg/m³)	Sand (kg/m³)	Recycled Aggregate 3/4" (kg/m³)	Recommended Quantity Admixture (Liters/m³)	Fiber (%)
	210	360	0	190	870	1,135	4.00	3.0

Properties	Average	Standard deviation	Compressive strength (ksc)	Average	Standard deviation
Air content (%)[1]	2.2	0.3	7 days	94.14	23.47
Slump (mm)[2]	52.5	11.8	28 days	175.31	10.20
Initial setting time (min)[3]	141	3.5	90 days	248.48	8.87
Final setting time (min)[3]	280	42.5	365 days	292.54	5.25

(Continued)

TABLE 5.2 (Continued)
Typical Mix Proportions and Related Properties of RAC

Mix Symbol (FrC300C275F70S75)	Cylindrical Strength (ksc)	Cement (kg/m³)	Fly Ash (kg/m³)	Water (kg/m³)	Sand (kg/m³)	Recycled Aggregate 3/4" (kg/m³)	Recommended Quantity Admixture (Liters/m³)	Fiber (%)
	210	226	144	165	870	1,135	4.10	3.1

Properties	Average	Standard deviation		Compressive strength (ksc)	Average	Standard deviation
Air content (%)[1]	2.3	0.1		7 days	95.54	11.17
Slump (mm)[2]	53.0	7.6		28 days	176.23	0.88
Initial setting time (min)[3]	191	17.2		90 days	526.89	5.13
Final setting time (min)[3]	285	13.2		365 days	275.34	2.49

Mix Symbol (FrC320C370F0S75)	Cylindrical Strength (ksc)	Cement (kg/m³)	Fly Ash (kg/m³)	Water (kg/m³)	Sand (kg/m³)	Recycled Aggregate 3/4" (kg/m³)	Recommended Quantity Admixture (Liters/m³)	Fiber (%)
	210	234	150	165	870	1,135	4.40	3.2

Properties	Average	Standard deviation		Compressive strength (ksc)	Average	Standard deviation
Air content (%)[1]	1.0	0.1		7 days	85.53	7.90
Slump (mm)[2]	59.5	10.9		28 days	224.65	4.03

(continuation of properties from previous mix)

Property	Average	Standard deviation		Average	Standard deviation
Initial setting time (min)[3]	162	12.1	90 days	260.74	3.01
Final setting time (min)[3]	300	44.2	365 days	302.24	1.25

Mix Symbol (FrC320C295F74S75)

Cement (kg/m³)	Fly Ash (kg/m³)	Water (kg/m³)	Sand (kg/m³)	Recycled Aggregate 3/4" (kg/m³)	Recommended Quantity Admixture (Liters/m³)	Fiber (%)
214	136	165	870	1,135	4.30	3.3

Properties	Cylindrical Strength (ksc)		Compressive strength (ksc)		
	Average	Standard deviation		Average	Standard deviation
	210				
Air content (%)[1]	1.1	0.2	7 days	78.51	19.84
Slump (mm)[2]	59.0	25.0	28 days	225.36	10.65
Initial setting time (min)[3]	172	16.8	90 days	264.47	2.54
Final setting time (min)[3]	305	35.8	365 days	279.25	1.88

Mix Symbol (FrC350C395F0S75)

Cement (kg/m³)	Fly Ash (kg/m³)	Water (kg/m³)	Sand (kg/m³)	Recycled Aggregate 3/4" (kg/m³)	Recommended Quantity Admixture (Liters/m³)	Fiber (%)
270	64	165	870	1,135	4.70	3.5

Properties	Cylindrical Strength (ksc)		Compressive strength (ksc)		
	Average	Standard deviation		Average	Standard deviation
	210				
Air content (%)[1]	3.2	0.3	7 days	164.44	15.47

(Continued)

TABLE 5.2 (Continued)
Typical Mix Proportions and Related Properties of RAC

Mix Symbol (FrC350C395F0S75)

Properties	Average	Standard deviation
Slump (mm)[2]	55.5	20.7
Initial setting time (min)[3]	101	7.9
Final setting time (min)[3]	274	10.8

Compressive strength (ksc)	Average	Standard deviation
28 days	179.87	19.32
90 days	247.47	10.03
365 days	294.03	5.48

Mix Symbol (FrC350C315F80S75)

Cylindrical Strength (ksc)	Cement (kg/m³)	Fly Ash (kg/m³)	Water (kg/m³)	Sand (kg/m³)	Recycled Aggregate 3/4" (kg/m³)	Recommended Quantity Admixture (Liters/m³)	Fiber (%)
240	384	0	190	860	1,120	4.60	3.6

Properties	Average	Standard deviation
Air content (%)[1]	3.0	0.1
Slump (mm)[2]	58.0	22.6
Initial setting time (min)[3]	102	6.8
Final setting time (min)[3]	335	8.7

Compressive strength (ksc)	Average	Standard deviation
7 days	112.21	14.01
28 days	194.88	11.63
90 days	289.55	7.69
365 days	327.41	4.56

Good Practices in RCA Concrete

Mix Symbol (FrC380C420F0S75)	Cylindrical Strength (ksc)	Cement (kg/m³)	Fly Ash (kg/m³)	Water (kg/m³)	Sand (kg/m³)	Recycled Aggregate 3/4" (kg/m³)	Recommended Quantity Admixture (Liters/m³)	Fiber (%)
Properties	240	240	154	165	860	1,120	5.00	3.7

Properties	Average	Standard deviation
Air content (%)[1]	2.8	0.1
Slump (mm)[2]	58.5	23.9
Initial setting time (min)[3]	103	5.7
Final setting time (min)[3]	272	24.8

Compressive strength (ksc)	Average	Standard deviation
7 days	204.35	28.73
28 days	215.87	24.88
90 days	336.84	13.06
365 days	346.43	8.77

Mix Symbol (FrC380C335F84S75)	Cylindrical Strength (ksc)	Cement (kg/m³)	Fly Ash (kg/m³)	Water (kg/m³)	Sand (kg/m³)	Recycled Aggregate 3/4" (kg/m³)	Recommended Quantity Admixture (Liters/m³)	Fiber (%)
Properties	240	250	160	165	860	1,120	4.90	3.8

Properties	Average	Standard deviation
Air content (%)[1]	2.7	0.3
Slump (mm)[2]	56.0	4.2
Initial setting time (min)[3]	110	19.2
Final setting time (min)[3]	287	9.6

Compressive strength (ksc)	Average	Standard deviation
7 days	200.29	24.74
28 days	210.81	23.63
90 days	306.34	14.28
365 days	357.48	7.52

(Continued)

TABLE 5.2 (Continued)
Typical Mix Proportions and Related Properties of RAC

Mix Symbol (FrC400C445F0S75)	Cylindrical Strength (ksc)	Cement (kg/m³)	Fly Ash (kg/m³)	Water (kg/m³)	Sand (kg/m³)	Recycled Aggregate 3/4" (kg/m³)	Recommended Quantity Admixture (Liters/m³)	Fiber (%)
Properties	300	434	0	190	830	1,115	5.20	3.6
Properties	Average		Standard deviation	Compressive strength (ksc)			Average	Standard deviation
Air content (%)[1]	2.4		0.3	7 days			111.41	13.62
Slump (mm)[2]	59.0		10.7	28 days			211.01	16.87
Initial setting time (min)[3]	108		12.4	90 days			305.62	13.02
Final setting time (min)[3]	374		10.9	365 days			351.44	7.14

Mix Symbol (FrC400C355F90S75)	Cylindrical Strength (ksc)	Cement (kg/m³)	Fly Ash (kg/m³)	Water (kg/m³)	Sand (kg/m³)	Recycled Aggregate 3/4" (kg/m³)	Recommended Quantity Admixture (Liters/m³)	Fiber (%)
Properties	300	270	174	165	830	1,115	5.30	3.8
Properties	Average		Standard deviation	Compressive strength (ksc)			Average	Standard deviation
Air content (%)[1]	2.5		0.1	7 days			209.03	24.86
Slump (mm)[2]	57.5		5.8	28 days			229.58	22.41

Properties	Cylindrical Strength (ksc)	Average	Standard deviation	Compressive strength (ksc)	Average	Standard deviation
Initial setting time (min)[3]		109	15.8	90 days	326.96	6.92
Final setting time (min)[3]		273	13.3	365 days	382.15	4.65

Mix Symbol (FrC420C470F0S75)	Cylindrical Strength (ksc)	Cement (kg/m³)	Fly Ash (kg/m³)	Water (kg/m³)	Sand (kg/m³)	Recycled Aggregate 3/4" (kg/m³)	Recommended Quantity Admixture (Liters/m³)	Fiber (%)
	300	280	180	165	830	1,115	5.50	3.7

Properties	Cylindrical Strength (ksc)	Average	Standard deviation	Compressive strength (ksc)	Average	Standard deviation
Air content (%)[1]	300	1.8	0.2	7 days	191.47	24.89
Slump (mm)[2]		58.5	23.7	28 days	233.36	21.55
Initial setting time (min)[3]		209	19.3	90 days	349.51	14.63
Final setting time (min)[3]		311	22.4	365 days	374.82	5.49

Mix Symbol (FrC420C375F94S75)	Cylindrical Strength (ksc)	Cement (kg/m³)	Fly Ash (kg/m³)	Water (kg/m³)	Sand (kg/m³)	Recycled Aggregate 3/4" (kg/m³)	Recommended Quantity Admixture (Liters/m³)	Fiber (%)
	300	258	166	165	830	1,115	5.60	4.0

Properties	Cylindrical Strength (ksc)	Average	Standard deviation	Compressive strength (ksc)	Average	Standard deviation
Air content (%)[1]	300	1.9	0.3	7 days	148.24	14.62

(Continued)

TABLE 5.2 (Continued)
Typical Mix Proportions and Related Properties of RAC

Mix Symbol (FrC420C375F94S75)	Cylindrical Strength (ksc)	Cement (kg/m³)	Fly Ash (kg/m³)	Water (kg/m³)	Sand (kg/m³)	Recycled Aggregate 3/4" (kg/m³)	Recommended Quantity Admixture (Liters/m³)	Fiber (%)
Slump (mm)[2]	60.5		2.9	28 days			282.26	21.76
Initial setting time (min)[3]	224		18.4	90 days			305.06	13.51
Final setting time (min)[3]	268		4.6	365 days			356.61	4.74
Mix Symbol (FrC450C495F0S75)	Cylindrical Strength (ksc)	Cement (kg/m³)	Fly Ash (kg/m³)	Water (kg/m³)	Sand (kg/m³)	Recycled Aggregate 3/4" (kg/m³)	Recommended Quantity Admixture (Liters/m³)	Fiber (%)
	300	330	80	165	830	1,115	5.90	4.0
Properties	Average		Standard deviation	Compressive strength (ksc)			Average	Standard deviation
Air content (%)[1]	2.0		0.1	7 days			217.48	2.88
Slump (mm)[2]	56.5		4.9	28 days			286.26	1.47
Initial setting time (min)[3]	109		5.9	90 days			396.21	4.99
Final setting time (min)[3]	319		33.5	365 days			417.58	5.31

Mix Symbol (FrC450C395F100S75)	Cylindrical Strength (ksc)	Cement (kg/m³)	Fly Ash (kg/m³)	Water (kg/m³)	Sand (kg/m³)	Recycled Aggregate 3/4" (kg/m³)	Recommended Quantity Admixture (Liters/m³)	Fiber (%)
Properties	320	460	0	190	820	1,100	5.80	4.3
	Average		Standard deviation	Compressive strength (ksc)			Average	Standard deviation
Air content (%)[1]	2.2		0.3	7 days			221.36	31.52
Slump (mm)[2]	57.0		17.4	28 days			249.62	14.69
Initial setting time (min)[3]	236		8.7	90 days			364.87	12.33
Final setting time (min)[3]	367		44.7	365 days			412.61	4.96

Mix Symbol (FrC480C520F0S75)	Cylindrical Strength (ksc)	Cement (kg/m³)	Fly Ash (kg/m³)	Water (kg/m³)	Sand (kg/m³)	Recycled Aggregate 3/4" (kg/m³)	Recommended Quantity Admixture (Liters/m³)	Fiber (%)
Properties	320	286	184	165	820	1,100	6.10	4.1
	Average		Standard deviation	Compressive strength (ksc)			Average	Standard deviation
Air content (%)[1]	1.4		0.1	7 days			145.14	12.48
Slump (mm)[2]	64.0		17.2	28 days			269.52	22.36
Initial setting time (min)[3]	132		12.7	90 days			360.03	10.23
Final setting time (min)[3]	305		2.4	365 days			405.63	9.33

(Continued)

TABLE 5.2 (Continued)
Typical Mix Proportions and Related Properties of RAC

Mix Symbol (FrC480C415F104S75)	Cylindrical Strength (ksc)	Cement (kg/m³)	Fly Ash (kg/m³)	Water (kg/m³)	Sand (kg/m³)	Recycled Aggregate 3/4" (kg/m³)	Recommended Quantity Admixture (Liters/m³)	Fiber (%)
	320	294	190	165	820	1,100	6.20	4.2
Properties	Average		Standard deviation	Compressive strength (ksc)			Average	Standard deviation
Air content (%)[1]	1.6		0.2		7 days		160.36	20.01
Slump (mm)[2]	66.0		3.2		28 days		365.14	17.42
Initial setting time (min)[3]	217		10.1		90 days		432.66	8.45
Final setting time (min)[3]	392		33.9		365 days		483.07	3.54

Mix Symbol (FrC300C345F0S75)	Cylindrical Strength (ksc)	Cement (kg/m³)	Fly Ash (kg/m³)	Water (kg/m³)	Sand (kg/m³)	Recycled Aggregate 3/4" (kg/m³)	Recommended Quantity Admixture (Liters/m³)	Fiber (%)
	280	256	164	165	840	1,130	4.00	4.5
Properties	Average		Standard deviation	Compressive strength (ksc)			Average	Standard deviation
Air content (%)[1]	3.5		0.3		7 days		93.25	7.41
Slump (mm)[2]	62.5		18.6		28 days		176.36	6.39

(continued from previous page)

Properties	Value		Compressive strength (ksc)	Average	Standard deviation
Initial setting time (min)[3]	110	5.4	90 days	247.89	5.48
Final setting time (min)[3]	257	5.4	365 days	285.26	4.22

Mix Symbol (FrC300C345F0S75)

Cylindrical Strength (ksc)	Cement (kg/m³)	Fly Ash (kg/m³)	Water (kg/m³)	Sand (kg/m³)	Recycled Aggregate 3/4" (kg/m³)	Recommended Quantity Admixture (Liters/m³)	Fiber (%)
280	264	170	165	840	1,130	3.20	4.6

Properties	Value	Compressive strength (ksc)	Average	Standard deviation
Air content (%)[1]	3.6	7 days	104.14	17.47
Slump (mm)[2]	63.0	28 days	176.25	16.22
Initial setting time (min)[3]	135	90 days	259.04	0.94
Final setting time (min)[3]	324	365 days	285.25	3.58

Mix Symbol (FrC300C275F70S75)

Cylindrical Strength (ksc)	Cement (kg/m³)	Fly Ash (kg/m³)	Water (kg/m³)	Sand (kg/m³)	Recycled Aggregate 3/4" (kg/m³)	Recommended Quantity Admixture (Liters/m³)	Fiber (%)
280	244	156	165	840	1,130	4.10	4.7

Properties	Value	Compressive strength (ksc)	Average	Standard deviation
Air content (%)[1]	3.1	7 days	112.84	33.59

(Continued)

TABLE 5.2 (Continued)
Typical Mix Proportions and Related Properties of RAC

Mix Symbol (FrC300C275F70S75)	Cylindrical Strength (ksc)	Cement (kg/m³)	Fly Ash (kg/m³)	Water (kg/m³)	Sand (kg/m³)	Recycled Aggregate 3/4" (kg/m³)	Recommended Quantity Admixture (Liters/m³)	Fiber (%)
Slump (mm)[2]	68.5		1.8	28 days			183.65	14.84
Initial setting time (min)[3]	104		20.0	90 days			257.48	13.22
Final setting time (min)[3]	278		5.4	365 days			291.25	4.79

Mix Symbol (FrC300C275F70S75)	Cylindrical Strength (ksc)	Cement (kg/m³)	Fly Ash (kg/m³)	Water (kg/m³)	Sand (kg/m³)	Recycled Aggregate 3/4" (kg/m³)	Recommended Quantity Admixture (Liters/m³)	Fiber (%)
Properties	280	310	74	165	840	1,130	3.30	4.9
	Average		Standard deviation	Compressive strength (ksc)			Average	Standard deviation
Air content (%)[1]	1.4		0.3	7 days			102.59	16.78
Slump (mm)[2]	67.0		5.4	28 days			147.23	12.55
Initial setting time (min)[3]	134		15.1	90 days			172.36	4.81
Final setting time (min)[3]	336		16.3	365 days			205.62	5.03

Mix Symbol (FrC300C255F140S75)	Cylindrical Strength (ksc)	Cement (kg/m³)	Fly Ash (kg/m³)	Water (kg/m³)	Sand (kg/m³)	Recycled Aggregate 3/4" (kg/m³)	Recommended Quantity Admixture (Liters/m³)	Fiber (%)
Properties	250	256	164	165	850	1,145	3.90	5.0
	Average	Standard deviation		Compressive strength (ksc)			Average	Standard deviation
Air content (%)[1]	2.3	0.1	7 days				173.25	27.28
Slump (mm)[2]	64.5	18.9	28 days				216.96	22.03
Initial setting time (min)[3]	153	11.2	90 days				279.14	13.91
Final setting time (min)[3]	370	44.1	365 days				320.25	7.47

Mix Symbol (FrC320C370F0S75)	Cylindrical Strength (ksc)	Cement (kg/m³)	Fly Ash (kg/m³)	Water (kg/m³)	Sand (kg/m³)	Recycled Aggregate 3/4" (kg/m³)	Recommended Quantity Admixture (Liters/m³)	Fiber (%)
Properties	250	234	150	165	850	1,145	4.10	5.1
	Average	Standard deviation		Compressive strength (ksc)			Average	Standard deviation
Air content (%)[1]	2.2	0.2	7 days				91.01	22.05
Slump (mm)[2]	65.0	24.4	28 days				235.23	14.87
Initial setting time (min)[3]	173	6.6	90 days				254.36	3.59
Final setting time (min)[3]	395	7.4	365 days				301.55	0.87

(Continued)

TABLE 5.2 (Continued)

Typical Mix Proportions and Related Properties of RAC

Mix Symbol (FrC320C370F0S75)	Cylindrical Strength (ksc)	Cement (kg/m³)	Fly Ash (kg/m³)	Water (kg/m³)	Sand (kg/m³)	Recycled Aggregate 3/4" (kg/m³)	Recommended Quantity Admixture (Liters/m³)	Fiber (%)
Properties	250	300	72	165	850	1,145	3.60	5.2

Properties	Average	Standard deviation	Compressive strength (ksc)	Average	Standard deviation
Air content (%)[1]	2.4	0.3	7 days	109.37	7.14
Slump (mm)[2]	62.5	4.2	28 days	231.54	12.55
Initial setting time (min)[3]	113	2.4	90 days	268.64	6.21
Final setting time (min)[3]	360	37.4	365 days	289.87	6.29

Mix Symbol (FrC320C295F74S75)	Cylindrical Strength (ksc)	Cement (kg/m³)	Fly Ash (kg/m³)	Water (kg/m³)	Sand (kg/m³)	Recycled Aggregate 3/4" (kg/m³)	Recommended Quantity Admixture (Liters/m³)	Fiber (%)
Properties	280	410	0	190	840	1,130	4.20	5.3

Properties	Average	Standard deviation	Compressive strength (ksc)	Average	Standard deviation
Air content (%)[1]	2.5	0.1	7 days	104.14	13.02
Slump (mm)[2]	65.5	7.4	28 days	175.69	4.16

Initial setting time (min)[3]	183		15.0	90 days		246.32		8.55
Final setting time (min)[3]	350		49.8	365 days		290.47		0.74
Mix Symbol (FrC320C295F74S75)	**Cylindrical Strength (ksc)**	**Cement (kg/m^3)**	**Fly Ash (kg/m^3)**	**Water (kg/m^3)**	**Sand (kg/m^3)**	**Recycled Aggregate 3/4″ (kg/m^3)**	**Recommended Quantity Admixture (Liters/m^3)**	**Fiber (%)**
Properties	240	228	146	165	860	1,120	3.60	5.0
	Average		Standard deviation	Compressive strength (ksc)		Average		Standard deviation
Air content (%)[1]	2.1		0.3	7 days		113.69		14.52
Slump (mm)[2]	50.0		19.6	28 days		142.11		10.02
Initial setting time (min)[3]	243		15.1	90 days		181.63		3.99
Final setting time (min)[3]	354		20.1	365 days		206.97		2.41
Mix Symbol (FrC320C271F148S75)	**Cylindrical Strength (ksc)**	**Cement (kg/m^3)**	**Fly Ash (kg/m^3)**	**Water (kg/m^3)**	**Sand (kg/m^3)**	**Recycled Aggregate 3/4″ (kg/m^3)**	**Recommended Quantity Admixture (Liters/m^3)**	**Fiber (%)**
Properties	240	290	70	165	860	1,120	4.00	5.2
	Average		Standard deviation	Compressive strength (ksc)		Average		Standard deviation
Air content (%)[1]	2.2		0.1	7 days		101.59		16.58
Slump (mm)[2]	50.5		20.7	28 days		173.26		18.63

(Continued)

TABLE 5.2 (Continued)
Typical Mix Proportions and Related Properties of RAC

Mix Symbol (FrC320C271F148S75)	Cylindrical Strength (ksc)	Cement (kg/m³)	Fly Ash (kg/m³)	Water (kg/m³)	Sand (kg/m³)	Recycled Aggregate 3/4" (kg/m³)	Recommended Quantity Admixture (Liters/m³)	Fiber (%)
Initial setting time (min)[3]	246		14.1	90 days			254.47	10.21
Final setting time (min)[3]	358		31.1	365 days			301.35	5.36
Mix Symbol (FrC350C395F0S75)	**Cylindrical Strength (ksc)**	**Cement (kg/m³)**	**Fly Ash (kg/m³)**	**Water (kg/m³)**	**Sand (kg/m³)**	**Recycled Aggregate 3/4" (kg/m³)**	**Recommended Quantity Admixture (Liters/m³)**	**Fiber (%)**
	250	294	0	190	850	1,145	4.50	5.3
Properties	**Average**		**Standard deviation**	**Compressive strength (ksc)**		**Average**		**Standard deviation**
Air content (%)[1]	2.3		0.2	7 days			127.52	4.56
Slump (mm)[2]	51.0		15.2	28 days			240.47	7.44
Initial setting time (min)[3]	241		13.6	90 days			257.05	9.58
Final setting time (min)[3]	342		35.5	365 days			291.91	6.27

Mix Symbol (FrC350C395F0S75)	Cylindrical Strength (ksc)	Cement (kg/m³)	Fly Ash (kg/m³)	Water (kg/m³)	Sand (kg/m³)	Recycled Aggregate 3/4" (kg/m³)	Recommended Quantity Admixture (Liters/m³)	Fiber (%)
Properties	250	246	158	165	850	1,145	3.90	5.5
	Average		Standard deviation	Compressive strength (ksc)			Average	Standard deviation
Air content (%)[1]	2.4		0.3	7 days			134.89	4.98
Slump (mm)[2]	51.5		16.1	28 days			234.14	2.84
Initial setting time (min)[3]	242		13.8	90 days			276.03	1.31
Final setting time (min)[3]	356		23.9	365 days			312.55	0.88

Mix Symbol (FrC350C315F80S75)	Cylindrical Strength (ksc)	Cement (kg/m³)	Fly Ash (kg/m³)	Water (kg/m³)	Sand (kg/m³)	Recycled Aggregate 3/4" (kg/m³)	Recommended Quantity Admixture (Liters/m³)	Fiber (%)
Properties	320	274	176	165	820	1,100	4.70	5.6
	Average		Standard deviation	Compressive strength (ksc)			Average	Standard deviation
Air content (%)[1]	1.3		0.3	7 days			147.96	27.26
Slump (mm)[2]	51.5		17.4	28 days			222.41	21.41
Initial setting time (min)[3]	131		8.9	90 days			268.13	13.59
Final setting time (min)[3]	270		24.5	365 days			285.73	7.62

(Continued)

TABLE 5.2 (Continued)
Typical Mix Proportions and Related Properties of RAC

Mix Symbol (FrC350C315F890S75)	Cylindrical Strength (ksc)	Cement (kg/m³)	Fly Ash (kg/m³)	Water (kg/m³)	Sand (kg/m³)	Recycled Aggregate 3/4" (kg/m³)	Recommended Quantity Admixture (Liters/m³)	Fiber (%)
	320	320	84	165	820	1,100	3.90	5.8

Properties	Average	Standard deviation	Compressive strength (ksc)	Average	Standard deviation
Air content (%)[1]	1.4	0.3	7 days	121.32	30.47
Slump (mm)[2]	54.0	9.4	28 days	221.69	14.95
Initial setting time (min)[3]	161	6.9	90 days	245.99	7.04
Final setting time (min)[3]	285	13.2	365 days	305.87	1.92

Mix Symbol (FrC350C287F158S75)	Cylindrical Strength (ksc)	Cement (kg/m³)	Fly Ash (kg/m³)	Water (kg/m³)	Sand (kg/m³)	Recycled Aggregate 3/4" (kg/m³)	Recommended Quantity Admixture (Liters/m³)	Fiber (%)
	340	484	100	190	810	1,135	4.30	6.0

Properties	Average	Standard deviation	Compressive strength (ksc)	Average	Standard deviation
Air content (%)[1]	3.4	0.3	7 days	121.36	10.87
Slump (mm)[2]	60.5	18.3	28 days	221.55	17.88

	Cylindrical Strength (ksc) / Average	Cement (kg/m³)	Fly Ash (kg/m³) / Standard deviation	Water (kg/m³) / Compressive strength (ksc)	Sand (kg/m³)	Recycled Aggregate 3/4" (kg/m³) / Average	Recommended Quantity Admixture (Liters/m³)	Fiber (%) / Standard deviation
Initial setting time (min)[3]	177		11.3	90 days		244.87		4.39
Final setting time (min)[3]	339		10.9	365 days		297.46		6.41

Mix Symbol (FrC380C420F0S75)	Cylindrical Strength (ksc)	Cement (kg/m³)	Fly Ash (kg/m³)	Water (kg/m³)	Sand (kg/m³)	Recycled Aggregate 3/4" (kg/m³)	Recommended Quantity Admixture (Liters/m³)	Fiber (%)
	350	300	194	165	810	1,135	5.00	6.1
Properties	Average		Standard deviation	Compressive strength (ksc)		Average		Standard deviation
Air content (%)[1]	1.5		0.2	7 days		195.81		24.02
Slump (mm)[2]	55.5		20.3	28 days		216.96		23.68
Initial setting time (min)[3]	181		9.2	90 days		306.74		7.41
Final setting time (min)[3]	265		3.1	365 days		380.15		5.33

Mix Symbol (FrC380C420F0S75)	Cylindrical Strength (ksc)	Cement (kg/m³)	Fly Ash (kg/m³)	Water (kg/m³)	Sand (kg/m³)	Recycled Aggregate 3/4" (kg/m³)	Recommended Quantity Admixture (Liters/m³)	Fiber (%)
	350	310	200	165	810	1,135	4.10	6.0
Properties	Average		Standard deviation	Compressive strength (ksc)		Average		Standard deviation
Air content (%)[1]	3.9		0.2	7 days		186.57		28.47

(Continued)

TABLE 5.2 (Continued)
Typical Mix Proportions and Related Properties of RAC

Mix Symbol (FrC380C420F0S75)	Cylindrical Strength (ksc)	Cement (kg/m³)	Fly Ash (kg/m³)	Water (kg/m³)	Sand (kg/m³)	Recycled Aggregate 3/4" (kg/m³)	Recommended Quantity Admixture (Liters/m³)	Fiber (%)
Slump (mm)[2]	63.0		24.5	28 days			233.69	23.66
Initial setting time (min)[3]	146		2.8	90 days			326.34	14.80
Final setting time (min)[3]	359		25.7	365 days			370.77	5.96
Mix Symbol (FrC380C335F84S75)	**Cylindrical Strength (ksc)**	**Cement (kg/m³)**	**Fly Ash (kg/m³)**	**Water (kg/m³)**	**Sand (kg/m³)**	**Recycled Aggregate 3/4" (kg/m³)**	**Recommended Quantity Admixture (Liters/m³)**	**Fiber (%)**
Properties	350	288	186	165	810	1,135	5.00	6.2
Properties	Average		Standard deviation	Compressive strength (ksc)			Average	Standard deviation
Air content (%)[1]	4.0		0.1	7 days			221.52	29.86
Slump (mm)[2]	73.5		6.8	28 days			357.66	24.11
Initial setting time (min)[3]	207		17.5	90 days			411.03	13.06
Final setting time (min)[3]	349		12.7	365 days			484.36	7.81

Mix Symbol (FrC380C335F84S75)	Cylindrical Strength (ksc)	Cement (kg/m³)	Fly Ash (kg/m³)	Water (kg/m³)	Sand (kg/m³)	Recycled Aggregate 3/4" (kg/m³)	Recommended Quantity Admixture (Liters/m³)	Fiber (%)
Properties	350	370	90	165	810	1,135	4.10	6.3

Properties	Average	Standard deviation	Compressive strength (ksc)	Average	Standard deviation
Air content (%)[1]	2.1	0.2	7 days	254.73	30.21
Slump (mm)[2]	72.5	10.1	28 days	278.59	23.89
Initial setting time (min)[3]	210	15.7	90 days	395.62	2.69
Final setting time (min)[3]	274	33.5	365 days	480.14	1.55

Mix Symbol (FrC380C303F166S75)	Cylindrical Strength (ksc)	Cement (kg/m³)	Fly Ash (kg/m³)	Water (kg/m³)	Sand (kg/m³)	Recycled Aggregate 3/4" (kg/m³)	Recommended Quantity Admixture (Liters/m³)	Fiber (%)
Properties	360	510	0	190	800	1,120	4.50	6.4

Properties	Average	Standard deviation	Compressive strength (ksc)	Average	Standard deviation
Air content (%)[1]	2.3	0.1	7 days	210.48	30.21
Slump (mm)[2]	74.5	15.7	28 days	235.07	23.89
Initial setting time (min)[3]	243	2.2	90 days	342.65	17.86
Final setting time (min)[3]	326	46.9	365 days	369.52	10.26

(Continued)

TABLE 5.2 (Continued)
Typical Mix Proportions and Related Properties of RAC

Mix Symbol (FrC400C445F0S75)	Cylindrical Strength (ksc)	Cement (kg/m³)	Fly Ash (kg/m³)	Water (kg/m³)	Sand (kg/m³)	Recycled Aggregate 3/4" (kg/m³)	Recommended Quantity Admixture (Liters/m³)	Fiber (%)
	380	316	204	165	800	1,120	5.20	6.5
Properties	Average		Standard deviation	Compressive strength (ksc)			Average	Standard deviation
Air content (%)[1]	1.7		0.3	7 days			179.88	28.12
Slump (mm)[2]	73.0		13.7	28 days			304.52	27.74
Initial setting time (min)[3]	216		5.4	90 days			433.34	8.93
Final setting time (min)[3]	339		30.6	365 days			487.68	4.77

Mix Symbol (FrC400C445F0S75)	Cylindrical Strength (ksc)	Cement (kg/m³)	Fly Ash (kg/m³)	Water (kg/m³)	Sand (kg/m³)	Recycled Aggregate 3/4" (kg/m³)	Recommended Quantity Admixture (Liters/m³)	Fiber (%)
	380	324	210	165	800	1,120	4.40	6.4
Properties	Average		Standard deviation	Compressive strength (ksc)			Average	Standard deviation
Air content (%)[1]	2.8		0.1	7 days			231.94	32.74
Slump (mm)[2]	65.0		21.4	28 days			195.46	15.02

(Table continued from previous page — compressive strength and setting-time results for the preceding mix)

Properties		Average	Standard deviation
Compressive strength (ksc)	90 days	399.99	17.77
	365 days	484.64	4.84
Initial setting time (min)[3]		108	5.7
Final setting time (min)[3]		371	16.6

Mix Symbol (FrC400C355F90S75)	Cylindrical Strength (ksc)	Cement (kg/m³)	Fly Ash (kg/m³)	Water (kg/m³)	Sand (kg/m³)	Recycled Aggregate 3/4" (kg/m³)	Recommended Quantity Admixture (Liters/m³)	Fiber (%)
	380	304	196	165	800	1,120	5.30	6.2

Properties		Average	Standard deviation
Compressive strength (ksc)	7 days	210.48	16.54
	28 days	235.07	19.35
	90 days	341.52	30.14
	365 days	374.41	32.66
Air content (%)[1]		2.5	0.1
Slump (mm)[2]		78.5	1.9
Initial setting time (min)[3]		247	9.5
Final setting time (min)[3]		373	36.9

Mix Symbol (FrC400C355F90S75)	Cylindrical Strength (ksc)	Cement (kg/m³)	Fly Ash (kg/m³)	Water (kg/m³)	Sand (kg/m³)	Recycled Aggregate 3/4" (kg/m³)	Recommended Quantity Admixture (Liters/m³)	Fiber (%)
	380	390	94	165	800	1,120	4.10	6.9

Properties		Average	Standard deviation
Compressive strength (ksc)	7 days	218.25	18.74
Air content (%)[1]		3.2	0.2

(Continued)

TABLE 5.2 (Continued)
Typical Mix Proportions and Related Properties of RAC

Mix Symbol (FrC400C355F90S75)	Cylindrical Strength (ksc)	Cement (kg/m^3)	Fly Ash (kg/m^3)	Water (kg/m^3)	Sand (kg/m^3)	Recycled Aggregate 3/4" (kg/m^3)	Recommended Quantity Admixture (Liters/m^3)	Fiber (%)
Slump (mm)[2]	56.5		21.6	28 days			229.04	19.62
Initial setting time (min)[3]	137		14.6	90 days			391.75	14.83
Final setting time (min)[3]	269		28.8	365 days			482.22	10.63
Mix Symbol (FrC400C319F176S75)	Cylindrical Strength (ksc)	Cement (kg/m^3)	Fly Ash (kg/m^3)	Water (kg/m^3)	Sand (kg/m^3)	Recycled Aggregate 3/4" (kg/m^3)	Recommended Quantity Admixture (Liters/m^3)	Fiber (%)
Properties	350	370	90	165	810	1,005	4.90	7.0
	Average		Standard deviation	Compressive strength (ksc)			Average	Standard deviation
Air content (%)[1]	2.6		0.2	7 days			214.85	34.74
Slump (mm)[2]	71.0		8.4	28 days			329.63	16.25
Initial setting time (min)[3]	106		4.8	90 days			423.62	14.24
Final setting time (min)[3]	326		9.8	365 days			442.07	8.46

Mix Symbol (FrC420C470F0S75)	Cylindrical Strength (ksc)	Cement (kg/m³)	Fly Ash (kg/m³)	Water (kg/m³)	Sand (kg/m³)	Recycled Aggregate 3/4" (kg/m³)	Recommended Quantity Admixture (Liters/m³)	Fiber (%)
Properties	380	510	0	190	800	1,035	5.50	7.1
	Average		Standard deviation	Compressive strength (ksc)			Average	Standard deviation
Air content (%)[1]	3.9		0.3		7 days		166.98	6.59
Slump (mm)[2]	74.5		20.8		28 days		241.44	14.15
Initial setting time (min)[3]	104		17.6		90 days		247.84	13.02
Final setting time (min)[3]	378		19.4		365 days		285.55	7.77

Mix Symbol (FrC420C470F0S75)	Cylindrical Strength (ksc)	Cement (kg/m³)	Fly Ash (kg/m³)	Water (kg/m³)	Sand (kg/m³)	Recycled Aggregate 3/4" (kg/m³)	Recommended Quantity Admixture (Liters/m³)	Fiber (%)
Properties	380	316	204	165	800	1,105	4.60	7.2
	Average		Standard deviation	Compressive strength (ksc)			Average	Standard deviation
Air content (%)[1]	2.9		0.2		7 days		164.25	22.67
Slump (mm)[2]	70.5		10.9		28 days		305.67	24.41
Initial setting time (min)[3]	105		15.2		90 days		391.51	12.36
Final setting time (min)[3]	270		41.6		365 days		447.88	7.58

(Continued)

TABLE 5.2 (Continued)
Typical Mix Proportions and Related Properties of RAC

Mix Symbol (FrC420C375F94S75)	Cylindrical Strength (ksc)	Cement (kg/m³)	Fly Ash (kg/m³)	Water (kg/m³)	Sand (kg/m³)	Recycled Aggregate 3/4" (kg/m³)	Recommended Quantity Admixture (Liters/m³)	Fiber (%)
	380	324	210	165	800	1,130	5.10	7.3
Properties	Average			Compressive strength (ksc)			Average	Standard deviation
			Standard deviation					
Air content (%)[1]	3.3		0.1	7 days			147.25	13.89
Slump (mm)[2]	78.5		9.3	28 days			309.32	14.01
Initial setting time (min)[3]	107		3.3	90 days			431.66	10.47
Final setting time (min)[3]	256		38.4	365 days			491.14	6.94

Mix Symbol (FrC420C375F94S75)	Cylindrical Strength (ksc)	Cement (kg/m³)	Fly Ash (kg/m³)	Water (kg/m³)	Sand (kg/m³)	Recycled Aggregate 3/4" (kg/m³)	Recommended Quantity Admixture (Liters/m³)	Fiber (%)
	350	484	0	190	810	1,135	4.65	7.3
Properties	Average			Compressive strength (ksc)			Average	Standard deviation
			Standard deviation					
Air content (%)[1]	2.3		0.2	7 days			153.22	37.49
Slump (mm)[2]	77.0		6.9	28 days			286.25	24.88

(continued from previous page)

	Average	Standard deviation	Compressive strength (ksc)	Average	Standard deviation
Initial setting time (min)[3]	166	1.4	90 days	391.21	13.06
Final setting time (min)[3]	318	14.7	365 days	449.58	7.48

Mix Symbol (FrC420C335F184S75)	Cylindrical Strength (ksc)	Cement (kg/m³)	Fly Ash (kg/m³)	Water (kg/m³)	Sand (kg/m³)	Recycled Aggregate 3/4" (kg/m³)	Recommended Quantity Admixture (Liters/m³)	Fiber (%)
	350	300	194	165	810	1,135	5.20	7.1

Properties	Average	Standard deviation	Compressive strength (ksc)	Average	Standard deviation
Air content (%)[11]	2.4	0.3	7 days	176.24	33.67
Slump (mm)[2]	81.0	12.7	28 days	282.59	24.82
Initial setting time (min)[3]	226	5.4	90 days	391.63	17.06
Final setting time (min)[3]	398	24.6	365 days	481.96	4.48

Mix Symbol (FrC450C495F0S75)	Cylindrical Strength (ksc)	Cement (kg/m³)	Fly Ash (kg/m³)	Water (kg/m³)	Sand (kg/m³)	Recycled Aggregate 3/4" (kg/m³)	Recommended Quantity Admixture (Liters/m³)	Fiber (%)
	350	310	200	165	810	1,135	5.50	7.2

Properties	Average	Standard deviation	Compressive strength (ksc)	Average	Standard deviation
Air content (%)[11]	3.1	0.1	7 days	205.43	24.77

(Continued)

TABLE 5.2 (Continued)
Typical Mix Proportions and Related Properties of RAC

Mix Symbol (FrC450C495F0S75)	Cylindrical Strength (ksc)	Cement (kg/m³)	Fly Ash (kg/m³)	Water (kg/m³)	Sand (kg/m³)	Recycled Aggregate 3/4" (kg/m³)	Recommended Quantity Admixture (Liters/m³)	Fiber (%)
Slump (mm)[2]	80.0		8.3	28 days			266.93	16.32
Initial setting time (min)[3]	176		1.2	90 days			244.89	7.44
Final setting time (min)[3]	298		24.9	365 days			392.61	5.61

Mix Symbol (FrC450C495F0S75)	Cylindrical Strength (ksc)	Cement (kg/m³)	Fly Ash (kg/m³)	Water (kg/m³)	Sand (kg/m³)	Recycled Aggregate 3/4" (kg/m³)	Recommended Quantity Admixture (Liters/m³)	Fiber (%)
	350	288	186	165	810	1,135	4.60	7.4

Properties	Average		Standard deviation	Compressive strength (ksc)			Average	Standard deviation
Air content (%)[1]	3.2		0.3	7 days			202.32	21.96
Slump (mm)[2]	77.5		15.3	28 days			227.81	18.44
Initial setting time (min)[3]	206		7.6	90 days			234.93	4.87
Final setting time (min)[3]	348		5.9	365 days			278.74	1.25

Mix Symbol (FrC450C395F100S75)	Cylindrical Strength (ksc)	Cement (kg/m³)	Fly Ash (kg/m³)	Water (kg/m³)	Sand (kg/m³)	Recycled Aggregate 3/4" (kg/m³)	Recommended Quantity Admixture (Liters/m³)	Fiber (%)
Properties	300	280	180	165	830	1,115	5.50	7.5
Properties	Average		Standard deviation	Compressive strength (ksc)			Average	Standard deviation
Air content (%)[1]	2.7		0.3	7 days			136.25	26.74
Slump (mm)[2]	78.0		16.7	28 days			173.04	19.26
Initial setting time (min)[3]	236		2.6	90 days			247.91	10.07
Final setting time (min)[3]	368		28.4	365 days			302.08	6.92

Mix Symbol (FrC450C395F100S75)	Cylindrical Strength (ksc)	Cement (kg/m³)	Fly Ash (kg/m³)	Water (kg/m³)	Sand (kg/m³)	Recycled Aggregate 3/4" (kg/m³)	Recommended Quantity Admixture (Liters/m³)	Fiber (%)
Properties	300	258	166	165	830	1,115	4.90	7.6
Properties	Average		Standard deviation	Compressive strength (ksc)			Average	Standard deviation
Air content (%)[1]	2.8		0.2	7 days			194.63	30.14
Slump (mm)[2]	79.0		14.6	28 days			247.88	24.03
Initial setting time (min)[3]	156		12.7	90 days			309.66	15.01
Final setting time (min)[3]	388		32.2	365 days			390.54	9.06

(Continued)

TABLE 5.2 (Continued)
Typical Mix Proportions and Related Properties of RAC

Mix Symbol (FrC450C353F192S575)	Cylindrical Strength (ksc)	Cement (kg/m³)	Fly Ash (kg/m³)	Water (kg/m³)	Sand (kg/m³)	Recycled Aggregate 3/4" (kg/m³)	Recommended Quantity Admixture (Liters/m³)	Fiber (%)
	300	330	80	165	830	1,115	5.20	8.0

Properties	Average	Standard deviation	Compressive strength (ksc)	Average	Standard deviation
Air content (%)[1]	2.5	0.3	7 days	152.66	14.85
Slump (mm)[2]	80.5	19.3	28 days	223.07	13.62
Initial setting time (min)[3]	136	14.9	90 days	389.01	7.47
Final setting time (min)[3]	378	16.8	365 days	413.65	3.62

Mix Symbol (FrC480C520F0S75)	Cylindrical Strength (ksc)	Cement (kg/m³)	Fly Ash (kg/m³)	Water (kg/m³)	Sand (kg/m³)	Recycled Aggregate 3/4" (kg/m³)	Recommended Quantity Admixture (Liters/m³)	Fiber (%)
	320	460	60	190	820	1,100	6.00	8.1

Properties	Average	Standard deviation	Compressive strength (ksc)	Average	Standard deviation
Air content (%)[1]	2.6	0.2	7 days	112.52	22.96
Slump (mm)[2]	78.5	18.9	28 days	168.41	18.63

(Continuation of previous mix)

Property	Average	Standard deviation
Compressive strength (ksc) — 90 days	315.17	14.58
Compressive strength (ksc) — 365 days	365.73	4.26
Initial setting time (min)[3]	196	17.7
Final setting time (min)[3]	358	36.6

Mix Symbol (FrC480C520F0S75)

Cylindrical Strength (ksc)	Cement (kg/m³)	Fly Ash (kg/m³)	Water (kg/m³)	Sand (kg/m³)	Recycled Aggregate 3/4" (kg/m³)	Recommended Quantity Admixture (Liters/m³)	Fiber (%)
400	332	202	165	785	1,080	5.10	8.0

Properties	Average	Standard deviation
Compressive strength (ksc) — 7 days	112.58	27.98
Compressive strength (ksc) — 28 days	243.25	23.11
Compressive strength (ksc) — 90 days	282.01	11.08
Compressive strength (ksc) — 365 days	324.77	4.78
Air content (%)[1]	2.9	0.1
Slump (mm)[2]	76.5	8.7
Initial setting time (min)[3]	186	8.7
Final setting time (min)[3]	328	44.8

Mix Symbol (FrC480C415F104S75)

Cylindrical Strength (ksc)	Cement (kg/m³)	Fly Ash (kg/m³)	Water (kg/m³)	Sand (kg/m³)	Recycled Aggregate 3/4" (kg/m³)	Recommended Quantity Admixture (Liters/m³)	Fiber (%)
400	430	104	165	785	1,080	6.00	8.2

Properties	Average	Standard deviation
Compressive strength (ksc) — 7 days	145.28	24.87
Air content (%)[1]	3.0	0.1

(Continued)

TABLE 5.2 (Continued)
Typical Mix Proportions and Related Properties of RAC

Mix Symbol (FrC480C415F104S75)	Cylindrical Strength (ksc)	Cement (kg/m³)	Fly Ash (kg/m³)	Water (kg/m³)	Sand (kg/m³)	Recycled Aggregate 3/4" (kg/m³)	Recommended Quantity Admixture (Liters/m³)	Fiber (%)
Slump (mm)[2]	79.5		18.4	28 days			257.73	20.63
Initial setting time (min)[3]	146		9.2	90 days			285.11	11.08
Final setting time (min)[3]	308		10.4	365 days			332.47	4.76

Mix Symbol (FrC480C415F104S75)	Cylindrical Strength (ksc)	Cement (kg/m³)	Fly Ash (kg/m³)	Water (kg/m³)	Sand (kg/m³)	Recycled Aggregate 3/4" (kg/m³)	Recommended Quantity Admixture (Liters/m³)	Fiber (%)
	400	404	130	165	785	1,080	5.10	8.3
Properties	Average		Standard deviation	Compressive strength (ksc)		Average		Standard deviation
Air content (%)[1]	3.3		0.1	7 days			151.58	21.84
Slump (mm)[2]	73.5		18.1	28 days			265.32	11.59
Initial setting time (min)[3]	185		8.3	90 days			298.36	12.54
Final setting time (min)[3]	357		21.7	365 days			339.89	4.54

Mix Symbol (FrC480C367F202S75)	Cylindrical Strength (ksc)	Cement (kg/m³)	Fly Ash (kg/m³)	Water (kg/m³)	Sand (kg/m³)	Recycled Aggregate 3/4" (kg/m³)	Recommended Quantity Admixture (Liters/m³)	Fiber (%)
Properties	400	376	158	165	785	1,080	5.50	8.5
Properties	Average		Standard deviation	Compressive strength (ksc)			Average	Standard deviation
Air content (%)[1]	3.4		0.2	7 days			102.92	20.14
Slump (mm)[2]	74.5		17.1	28 days			183.45	10.63
Initial setting time (min)[3]	175		1.5	90 days			301.87	12.88
Final setting time (min)[3]	287		42.7	365 days			351.19	5.97

Mix Symbol (FrC400C445F0S75)	Cylindrical Strength (ksc)	Cement (kg/m³)	Fly Ash (kg/m³)	Water (kg/m³)	Sand (kg/m³)	Recycled Aggregate 3/4" (kg/m³)	Recommended Quantity Admixture (Liters/m³)	Fiber (%)
Properties	310	336	108	165	830	1,005	5.50	8.6
Properties	Average		Standard deviation	Compressive strength (ksc)			Average	Standard deviation
Air content (%)[1]	3.7		0.2	7 days			121.51	24.88
Slump (mm)[2]	75.0		15.1	28 days			247.95	14.69
Initial setting time (min)[3]	155		19.8	90 days			452.66	4.26
Final setting time (min)[3]	297		32.6	365 days			503.14	1.63

(Continued)

TABLE 5.2 (Continued)
Typical Mix Proportions and Related Properties of RAC

Mix Symbol (FrC400C355F90S75)	Cylindrical Strength (ksc)	Cement (kg/m³)	Fly Ash (kg/m³)	Water (kg/m³)	Sand (kg/m³)	Recycled Aggregate 3/4" (kg/m³)	Recommended Quantity Admixture (Liters/m³)	Fiber (%)
	310	314	130	165	830	1,100	5.40	8.7
Properties	Average		Standard deviation	Compressive strength (ksc)			Average	Standard deviation
Air content (%)[1]	3.8		0.1	7 days		132.25		24.94
Slump (mm)[2]	72.5		5.9	28 days		251.16		13.58
Initial setting time (min)[3]	105		19.2	90 days		425.47		13.22
Final setting time (min)[3]	307		28.1	365 days		460.18		4.79

Mix Symbol (FrC400C445F0S75)	Cylindrical Strength (ksc)	Cement (kg/m³)	Fly Ash (kg/m³)	Water (kg/m³)	Sand (kg/m³)	Recycled Aggregate 3/4" (kg/m³)	Recommended Quantity Admixture (Liters/m³)	Fiber (%)
	300	434	0	190	830	1,060	5.30	8.5
Properties	Average		Standard deviation	Compressive strength (ksc)			Average	Standard deviation
Air content (%)[1]	1.0		0.2	7 days		180.36		28.47
Slump (mm)[2]	74.0		19.1	28 days		385.14		20.14

Mix Symbol (FrC400C355F90S75)	Cylindrical Strength (ksc)	Cement (kg/m³)	Fly Ash (kg/m³)	Water (kg/m³)	Sand (kg/m³)	Recycled Aggregate 3/4" (kg/m³)	Recommended Quantity Admixture (Liters/m³)	Fiber (%)
Initial setting time (min)[3]	115		9.8		90 days		398.36	12.54
Final setting time (min)[3]	317		12.9		365 days		439.89	4.54
	300	270	174	165	830	1,000	5.20	8.1
Properties	Average		Standard deviation		Compressive strength (ksc)		Average	Standard deviation
Air content (%)[1]	1.1		0.3		7 days		178.98	33.59
Slump (mm)[2]	75.5		13.1		28 days		293.65	14.84
Initial setting time (min)[3]	165		15.7		90 days		301.87	12.88
Final setting time (min)[3]	257		35.3		365 days		351.19	5.97

Mix Symbol (FrC400C355F90S75)	Cylindrical Strength (ksc)	Cement (kg/m³)	Fly Ash (kg/m³)	Water (kg/m³)	Sand (kg/m³)	Recycled Aggregate 3/4" (kg/m³)	Recommended Quantity Admixture (Liters/m³)	Fiber (%)
	300	350	84	165	830	1,100	4.40	9.0
Properties	Average		Standard deviation		Compressive strength (ksc)		Average	Standard deviation
Air content (%)[1]	3.9		0.1		7 days		136.96	20.14

(Continued)

TABLE 5.2 (Continued)
Typical Mix Proportions and Related Properties of RAC

Mix Symbol (FrC400C355F90S75)	Cylindrical Strength (ksc)	Cement (kg/m³)	Fly Ash (kg/m³)	Water (kg/m³)	Sand (kg/m³)	Recycled Aggregate 3/4" (kg/m³)	Recommended Quantity Admixture (Liters/m³)	Fiber (%)
Slump (mm)[2]	71.5		14.1		28 days		248.40	10.29
Initial setting time (min)[3]	195		9.9		90 days		278.63	5.09
Final setting time (min)[3]	267		33.5		365 days		300.01	4.53

Mix Symbol (FrC400C355F90S75)	Cylindrical Strength (ksc)	Cement (kg/m³)	Fly Ash (kg/m³)	Water (kg/m³)	Sand (kg/m³)	Recycled Aggregate 3/4" (kg/m³)	Recommended Quantity Admixture (Liters/m³)	Fiber (%)
Properties	300	328	106	165	830	1,090	4.30	8.9
	Average		Standard deviation		Compressive strength (ksc)		Average	Standard deviation
Air content (%)[1]	4.0		0.3		7 days		139.40	7.90
Slump (mm)[2]	72.0		3.6		28 days		251.37	4.03
Initial setting time (min)[3]	125		11.2		90 days		274.68	16.99
Final setting time (min)[3]	327		7.6		365 days		305.14	9.21

Mix Symbol (FrC480C520F0S75)	Cylindrical Strength (ksc)	Cement (kg/m³)	Fly Ash (kg/m³)	Water (kg/m³)	Sand (kg/m³)	Recycled Aggregate 3/4" (kg/m³)	Recommended Quantity Admixture (Liters/m³)	Fiber (%)
Properties	300	308	126	165	830	1,120	8.20	8.7
Properties	Average		Standard deviation	Compressive strength (ksc)			Average	Standard deviation
Air content (%)[1]	3.5		0.3	7 days			136.89	17.61
Slump (mm)[2]	73.0		12.1	28 days			288.43	12.88
Initial setting time (min)[3]	135		2.4	90 days			334.15	3.01
Final setting time (min)[3]	347		25.0	365 days			369.70	1.25

Mix Symbol (FrC420C470F0S75)	Cylindrical Strength (ksc)	Cement (kg/m³)	Fly Ash (kg/m³)	Water (kg/m³)	Sand (kg/m³)	Recycled Aggregate 3/4" (kg/m³)	Recommended Quantity Admixture (Liters/m³)	Fiber (%)
Properties	250	394	0	190	850	1,130	9.10	9.0
Properties	Average		Standard deviation	Compressive strength (ksc)			Average	Standard deviation
Air content (%)[1]	3.6		0.2	7 days			145.94	26.89
Slump (mm)[2]	76.0		16.1	28 days			285.36	14.80
Initial setting time (min)[3]	145		20.0	90 days			324.10	5.41
Final setting time (min)[3]	337		47.6	365 days			374.28	2.44

(Continued)

TABLE 5.2 (Continued)
Typical Mix Proportions and Related Properties of RAC

Mix Symbol (FrC450C395F100S75)	Cylindrical Strength (ksc)	Cement (kg/m³)	Fly Ash (kg/m³)	Water (kg/m³)	Sand (kg/m³)	Recycled Aggregate 3/4" (kg/m³)	Recommended Quantity Admixture (Liters/m³)	Fiber (%)
	380	304	196	165	800	1,120	4.80	8.9
Properties	Average		Standard deviation	Compressive strength (ksc)			Average	Standard deviation
Air content (%)[1]	2.4		0.3	7 days			179.36	22.74
Slump (mm)[2]	57.5		17.3	28 days			194.52	14.59
Initial setting time (min)[3]	120		6.4	90 days			260.19	24.74
Final setting time (min)[3]	267		10.4	365 days			312.87	16.25

Mix Symbol (FrC450C495F05S175)	Cylindrical Strength (ksc)	Cement (kg/m³)	Fly Ash (kg/m³)	Water (kg/m³)	Sand (kg/m³)	Recycled Aggregate 3/4" (kg/m³)	Recommended Quantity Admixture (Liters/m³)	Fiber (%)
	380	290	94	165	800	1,120	5.70	1.3
Properties	Average		Standard deviation	Compressive strength (ksc)			Average	Standard deviation
Air content (%)[1]	2.5		0.2	7 days			178.36	34.14
Slump (mm)[2]	93.5		18.4	28 days			332.14	24.62

	Average	Standard deviation		Compressive strength	Average	Standard deviation
Initial setting time (min)[3]	125	19.8		90 days	412.63	19.88
Final setting time (min)[3]	270	21.4		365 days	486.81	2.01

Mix Symbol (FrC450C395F100S175)	Cylindrical Strength (ksc)	Cement (kg/m³)	Fly Ash (kg/m³)	Water (kg/m³)	Sand (kg/m³)	Recycled Aggregate 3/4" (kg/m³)	Recommended Quantity Admixture (Liters/m³)	Fiber (%)
Properties	310	358	86	165	830	1,115	5.60	1.2

Properties	Average	Standard deviation		Compressive strength (ksc)	Average	Standard deviation
Air content (%)[1]	3.5	0.3		7 days	132.26	7.75
Slump (mm)[2]	130.5	7.3		28 days	253.71	6.44
Initial setting time (min)[3]	180	5.5		90 days	291.11	5.22
Final setting time (min)[3]	300	9.5		365 days	328.69	3.60

RAC reinforced with fibers is a composite material that may either have the fibers arranged in a specific pattern or randomly scattered throughout the cement matrix. The effectiveness of the stress transmission between the matrix and the fibers is going to be a determining factor in its qualities. Following is a condensed discussion of the factors:

- The relative stiffness of the fiber matrix: For there to be an effective transmission of stress, the modulus of elasticity of the matrix has to be much less than that of the fiber. Fibers with a low modulus, such as nylons and polypropylene, are thus unlikely to produce a gain in strength. On the other hand, these fibers aid in the absorption of vast quantities of energy, imparting a higher degree of toughness and resistance to impact. The composite receives its strength and stiffness from the high-modulus fibers, which may be steel, glass, and carbon.
- The number of fibers in volume: The number of fibers that are included in the composite material has a significant impact on its level of strength. An increase in the volume of fibers results in a roughly linear improvement in both the tensile strength and the toughness of the composite. Compared to using a lower amount of fiber, using a more significant percentage of fiber in RAC and mortar is more likely to result in segregation and harshness. In addition, the strength of the interfacial connection between the matrix and the fiber is a significant factor in determining how well stress is transferred from the matrix to the fiber. In order to increase the tensile strength of the composite, a strong bond is necessary.
- Dimensional distribution of the fiber: The aspect ratio of the fiber is yet another key component that plays a role in determining the characteristics and behavior of the composite material. According to reports, an increase in the aspect ratio has been shown to linearly enhance the final RAC for aspect ratios up to 75.
- The arrangement of the fibers: The bars used in conventional reinforcement are orientated in the intended direction, but the fibers used in fiber reinforcement are oriented in a random pattern. This is one of the key distinctions between conventional reinforcement and fiber reinforcement. Mortar specimens reinforced with 0.5% volume of fibers were put through a series of tests to see the impact of randomization. Fibers were aligned in the direction of the load in one group of specimens; in another set, fibers were aligned in the direction perpendicular to that of the load; and in the third set, fibers were randomly dispersed. It was discovered that the fibers oriented in parallel to the load being applied gave more tensile strength and toughness than the fibers randomly dispersed or perpendicular to one another.
- The workability of RAC and its ability to be compacted: The incorporation of steel fiber results in a significant reduction in workability. Consolidation of a fresh mix is negatively impacted as a result of this circumstance. Even after an extended period of external vibration, the RAC does not become more compact. Another issue that arises as a result of poor workability is a distribution of fibers that are not uniform. The length and diameter of the fiber

both have a role in determining the volume of the fiber at which this condition is attained. In most cases, the mix's workability and compaction standard may be enhanced by increasing the ratio of water to cement in the mixture or by using some additive that reduces the amount of water in the mixture.

- The dimensions of the coarse aggregate particles: In order to prevent a perceptible weakening of the composite, the largest size of the coarse aggregate should not be allowed to exceed 10 mm in diameter. In a sense, fibers also perform the function of an aggregate. Although they have a straightforward geometry, their effect on the characteristics of newly mixed RAC is somewhat complicated. The orientation and distribution of the fibers, and hence the characteristics of the composite, are determined by the inter-particle friction that exists between the fibers themselves as well as between the fibers and the aggregates. Additives that reduce friction and increase the mix's cohesion are two types of admixtures that can improve the mix considerably.

Fiber-Reinforced Recycled Aggregate Concrete Materials

- Type 1 hydraulic Portland cements conforming to ASTM C150 (ASTM C150/C150M–17 Standard specifications for Portland cement) were used throughout concrete mixtures. Their chemical compositions and physical properties are shown in Table 5.3.

TABLE 5.3
Chemical Compositions and Physical Properties of Type 1 Portland Cements and FA

Chemical Compositions (% by Mass)	SiO_2	Al_2O_3	Fe_2O_3	CaO	MgO	Na_2O	SO_3	Free CaO
Type 1 Portland Cement	20.84	5.22	3.20	66.28	1.24	0.10	2.41	0.99
FA	42.10	21.80	11.22	13.56	2.41	2.90	1.88	1.44

Physical Properties	Type 1 Portland Cement	FA
Loss on Ignition (%)	0.96	2.33
Moisture Content (%)	0.19	1.50
Blaine Surface Area (cm^2/g)		
Fineness (Particle Size, % Retained)	3,200	2,850
– ≥ 75 µm	0.50	0.56
– 75 µm	5.25	8.25
– 45 µm	3.60	4.76
– ≤ 36 µm	90.62	86.43
Fineness (Retained) on 45 Micron		
(No. 325)	5.75	4.90
Water Requirement (%)	100	97
Bulk Density (kg / l)	1.03	0.51
Specific Gravity	3.15	2.13

- The ASTM C618 (ASTM C618–15 Standard specification for coal fly ash and raw or calcined natural pozzolan for use in concrete) classifies the PFA as low calcium (Type F).
- Tap water with a pH 7.0 conforming to ASTM C1602 (ASTM C1602/C1602M–12 Standard specification for mixing water used in the production of hydraulic cement concrete).
- River sand with gradation conforming to the ASTM C33 (ASTM C33/C33M–1601 standard specification for concrete aggregates).
- Crushed lime recycled aggregate rock with gradation conforming to the ASTM C33 (ASTM C33/C33M–1601 standard specification for RAC aggregates).
- The chemical admixture used superplasticizer that conforms to the ASTM C494 (ASTM C494/C494M–16 Standard specification for chemical admixtures for concrete); that is, the superplasticizer had a recommended dosage rate of cementitious materials (per 100 of a kilogram of cementitious materials).

5.2.2 HIGH-STRENGTH RAC

The use of high-strength RAC, abbreviated as HSC, has attracted the attention of civil and structural engineers in recent years. The life cycle cost-performance ratio that this relatively new construction material provides, in addition to its outstanding engineering properties, such as higher compressive and tensile strengths, higher stiffness, and better durability when compared to traditional normal-strength RAC, can help to partially explain the expanding commercial use of this relatively new construction material (NSC). The development of high-strength RAC has been slow and steady for many years, even though it is often thought of as relatively new material (Tables 5.4 and 5.5).

The following is a discussion of the relevance of each constituent in the production of high-strength RAC:

- The ratio of water to the binder, sometimes known as w/b, and the cement content: HSC typically consists of one or two mineral additives that partially replace cement in the construction process. As a result, the phrase water/cement ratio (w/c ratio), which was previously used with normal-strength RAC, has been replaced by the term water/binder ratio (w/b ratio), in which the binder is the whole weight of the cementitious components (cement plus additives). Approximately 0.36 is the minimal value for the w/b ratio required for the complete hydration of cement pastes.
- Mineral admixtures: (b.1) The production of silicon and ferrosilicon alloys generates a by-product known as silica fume, which takes the form of highly reactive glass. This by-product is created throughout the manufacturing process. The presence of silica fume boosts the performance of the superplasticizer, which in turn lowers the w/b ratio necessary to get the desired degree of workability in the material. In high-performance RAC, silica fume typically ranges from 3 to 10%. According to Bernard and

TABLE 5.4

Typical Mix Proportions and Related Properties of RAC

Mix Symbol (HSC300C345F0S175)	Cylindrical Strength (ksc)	Cement (kg/m³)	Fly Ash (kg/m³)	Water (kg/m³)	Sand (kg/m³)	Recycled Aggregate 3/4" (kg/m³)	Recommended Quantity Admixture (Liters/m³)	Fiber (%)
	280	275	70	150	835	1,145	4.00	–
Properties	Average		Standard deviation	Compressive strength (ksc)		Average		Standard deviation
Air content (%)[1]	3.7		0.3	7 days		119.51		21.58
Slump (mm)[2]	94.5		14.2	28 days		206.47		13.02
Initial setting time (min)[3]	216		14.6	90 days		288.48		8.67
Final setting time (min)[3]	279		41.7	365 days		362.47		5.33
Mix Symbol (HSC300C275F0S175)	Cylindrical Strength (ksc)	Cement (kg/m³)	Fly Ash (kg/m³)	Water (kg/m³)	Sand (kg/m³)	Recycled Aggregate 3/4" (kg/m³)	Recommended Quantity Admixture (Liters/m³)	Fiber (%)
	240	287	73	155	850	1,115	4.20	–
Properties	Average		Standard deviation	Compressive strength (ksc)		Average		Standard deviation
Air content (%)[1]	3.8		0.1	7 days		107.98		15.47

(Continued)

TABLE 5.4 (Continued)
Typical Mix Proportions and Related Properties of RAC

Mix Symbol (HSC300C275F70S175)	Cylindrical Strength (ksc)	Cement (kg/m³)	Fly Ash (kg/m³)	Water (kg/m³)	Sand (kg/m³)	Recycled Aggregate 3/4" (kg/m³)	Recommended Quantity Admixture (Liters/m³)	Fiber (%)
Slump (mm)[2]	95.0		5.7	28 days			252.67	3.65
Initial setting time (min)[3]	239		1.4	90 days			289.36	10.07
Final setting time (min)[3]	355		14.3	365 days			325.48	6.04

Mix Symbol (HSC320C370F0S175)	Cylindrical Strength (ksc)	Cement (kg/m³)	Fly Ash (kg/m³)	Water (kg/m³)	Sand (kg/m³)	Recycled Aggregate 3/4" (kg/m³)	Recommended Quantity Admixture (Liters/m³)	Fiber (%)
	280	370	0	150	820	1,125	4.50	–
Properties	Average		Standard deviation	Compressive strength (ksc)			Average	Standard deviation
Air content (%)[1]	1.0		0.1	7 days			101.89	21.13
Slump (mm)[2]	92.5		22.4	28 days			197.26	17.59
Initial setting time (min)[3]	102		14.2	90 days			269.58	9.31
Final setting time (min)[3]	396		19.4	365 days			321.54	4.55

Mix Symbol (HSC320C295F74S175)	Cylindrical Strength (ksc)	Cement (kg/m³)	Fly Ash (kg/m³)	Water (kg/m³)	Sand (kg/m³)	Recycled Aggregate 3/4" (kg/m³)	Recommended Quantity Admixture (Liters/m³)	Fiber (%)
	300	290	70	125	850	1,000	4.50	–
Properties	Average		Standard deviation	Compressive strength (ksc)			Average	Standard deviation
Air content (%)[1]	1.1		0.2	7 days			180.14	16.22
Slump (mm)[2]	93.0		2.3	28 days			236.25	7.36
Initial setting time (min)[3]	112		3.2	90 days			273.48	2.48
Final setting time (min)[3]	262		1.4	365 days			321.89	0.87

Mix Symbol (HSC350C395F0S175)	Cylindrical Strength (ksc)	Cement (kg/m³)	Fly Ash (kg/m³)	Water (kg/m³)	Sand (kg/m³)	Recycled Aggregate 3/4" (kg/m³)	Recommended Quantity Admixture (Liters/m³)	Fiber (%)
	280	314	56	150	825	1,130	4.50	–
Properties	Average		Standard deviation	Compressive strength (ksc)			Average	Standard deviation
Air content (%)[1]	1.2		0.1	7 days			184.44	15.47
Slump (mm)[2]	84.5		21.5	28 days			199.87	19.32

(Continued)

TABLE 5.4 (Continued)
Typical Mix Proportions and Related Properties of RAC

Mix Symbol (HSC350C395F0S175)	Cylindrical Strength (ksc)	Cement (kg/m³)	Fly Ash (kg/m³)	Water (kg/m³)	Sand (kg/m³)	Recycled Aggregate 3/4" (kg/m³)	Recommended Quantity Admixture (Liters/m³)	Fiber (%)
Initial setting time (min)[3]	118		9.8	90 days			267.47	10.03
Final setting time (min)[3]	366		23.9	365 days			314.97	5.48

Mix Symbol (HSC350C315F80S175)	Cylindrical Strength (ksc)	Cement (kg/m³)	Fly Ash (kg/m³)	Water (kg/m³)	Sand (kg/m³)	Recycled Aggregate 3/4" (kg/m³)	Recommended Quantity Admixture (Liters/m³)	Fiber (%)
	280	345	30	155	835	1,145	4.70	—

Properties	Average		Standard deviation	Compressive strength (ksc)			Average	Standard deviation
Air content (%)[1]	1.3		0.2	7 days			132.21	14.01
Slump (mm)[2]	85.0		14.7	28 days			214.83	11.63
Initial setting time (min)[3]	244		2.3	90 days			309.55	7.69
Final setting time (min)[3]	276		33.5	365 days			347.81	4.56

Mix Symbol (HSC380C420F0S175)	Cylindrical Strength (ksc)	Cement (kg/m³)	Fly Ash (kg/m³)	Water (kg/m³)	Sand (kg/m³)	Recycled Aggregate 3/4" (kg/m³)	Recommended Quantity Admixture (Liters/m³)	Fiber (%)
Properties	280	370	0	150	825	1,130	4.90	–
	Average		Standard deviation	Compressive strength (ksc)			Average	Standard deviation
Air content (%)[1]	3.1		0.2		7 days		203.31	13.54
Slump (mm)[2]	80.5		4.8		28 days		217.89	10.05
Initial setting time (min)[3]	205		19.4		90 days		299.47	7.52
Final setting time (min)[3]	391		15.7		365 days		348.07	6.41

Mix Symbol (HSC380C335F84S175)	Cylindrical Strength (ksc)	Cement (kg/m³)	Fly Ash (kg/m³)	Water (kg/m³)	Sand (kg/m³)	Recycled Aggregate 3/4" (kg/m³)	Recommended Quantity Admixture (Liters/m³)	Fiber (%)
Properties	280	395	0	160	830	1,065	5.00	–
	Average		Standard deviation	Compressive strength (ksc)			Average	Standard deviation
Air content (%)[1]	2.7		0.3		7 days		204.88	24.89
Slump (mm)[2]	81.5		5.1		28 days		223.69	10.36

(Continued)

TABLE 5.4 (Continued)
Typical Mix Proportions and Related Properties of RAC

Mix Symbol (HSC380C335F84S175)	Cylindrical Strength (ksc)	Cement (kg/m³)	Fly Ash (kg/m³)	Water (kg/m³)	Sand (kg/m³)	Recycled Aggregate 3/4" (kg/m³)	Recommended Quantity Admixture (Liters/m³)	Fiber (%)
Initial setting time (min)[3]	243		8.2	90 days			321.56	3.59
Final setting time (min)[3]	256		44.2	365 days			352.76	4.88

Mix Symbol (HSC400C445F0S175)	Cylindrical Strength (ksc)	Cement (kg/m³)	Fly Ash (kg/m³)	Water (kg/m³)	Sand (kg/m³)	Recycled Aggregate 3/4" (kg/m³)	Recommended Quantity Admixture (Liters/m³)	Fiber (%)
Properties	380	294	190	155	800	1,120	5.10	–

Properties	Average	Standard deviation		Compressive strength (ksc)			Average	Standard deviation
Air content (%)[1]	2.9	0.2		7 days			224.35	24.86
Slump (mm)[2]	79.0	3.2		28 days			235.87	22.41
Initial setting time (min)[3]	217	16.4		90 days			356.84	6.92
Final setting time (min)[3]	289	45.8		365 days			366.43	4.65

Good Practices in RCA Concrete

Mix Symbol (HSC400C355F90S175)	Cylindrical Strength (ksc)	Cement (kg/m³)	Fly Ash (kg/m³)	Water (kg/m³)	Sand (kg/m³)	Recycled Aggregate 3/4" (kg/m³)	Recommended Quantity Admixture (Liters/m³)	Fiber (%)
Properties	380	274	176	155	800	1,120	5.30	–
	Average		Standard deviation	Compressive strength (ksc)			Average	Standard deviation
Air content (%)[1]	3.0		0.3	7 days			220.29	24.74
Slump (mm)[2]	83.5		9.2	28 days			230.81	23.63
Initial setting time (min)[3]	116		18.4	90 days			326.34	14.28
Final setting time (min)[3]	318		7.0	365 days			377.48	7.52

Mix Symbol (HSC420C470F0S175)	Cylindrical Strength (ksc)	Cement (kg/m³)	Fly Ash (kg/m³)	Water (kg/m³)	Sand (kg/m³)	Recycled Aggregate 3/4" (kg/m³)	Recommended Quantity Admixture (Liters/m³)	Fiber (%)
Properties	380	350	84	155	800	1,120	5.50	–
	Average		Standard deviation	Compressive strength (ksc)			Average	Standard deviation
Air content (%)[1]	2.0		0.2	7 days			181.01	22.24
Slump (mm)[2]	88.5		17.9	28 days			274.43	14.98

(Continued)

TABLE 5.4 (Continued)
Typical Mix Proportions and Related Properties of RAC

Mix Symbol (HSC420C470F0S175)	Cylindrical Strength (ksc)	Cement (kg/m³)	Fly Ash (kg/m³)	Water (kg/m³)	Sand (kg/m³)	Recycled Aggregate 3/4" (kg/m³)	Recommended Quantity Admixture (Liters/m³)	Fiber (%)
Initial setting time (min)[3]	190		5.5	90 days			307.97	3.57
Final setting time (min)[3]	343		19.3	365 days			376.78	1.09
Mix Symbol (HSC420C375F94S175)	Cylindrical Strength (ksc)	Cement (kg/m³)	Fly Ash (kg/m³)	Water (kg/m³)	Sand (kg/m³)	Recycled Aggregate 3/4" (kg/m³)	Recommended Quantity Admixture (Liters/m³)	Fiber (%)
	380	350	84	155	800	1,120	5.60	–
Properties	Average		Standard deviation	Compressive strength (ksc)			Average	Standard deviation
Air content (%)[1]	1.2		0.1	7 days			137.33	15.25
Slump (mm)[2]	94.0		13.1	28 days			230.29	4.96
Initial setting time (min)[3]	228		5.5	90 days			358.63	1.07
Final setting time (min)[3]	308		10.4	365 days			395.14	0.98

Mix Symbol (HSC450C495F0S175)	Cylindrical Strength (ksc)	Cement (kg/m³)	Fly Ash (kg/m³)	Water (kg/m³)	Sand (kg/m³)	Recycled Aggregate 3/4" (kg/m³)	Recommended Quantity Admixture (Liters/m³)	Fiber (%)
Properties	350	330	80	155	810	1,135	5.50	–
	Average		Standard deviation	Compressive strength (ksc)			Average	Standard deviation
Air content (%)[1]	1.3		0.3	7 days			138.27	22.18
Slump (mm)[2]	94.5		11.5	28 days			327.34	9.23
Initial setting time (min)[3]	224		3.0	90 days			399.44	15.74
Final setting time (min)[3]	365		29.5	365 days			405.78	6.44

Mix Symbol (HSC450C395F100S175)	Cylindrical Strength (ksc)	Cement (kg/m³)	Fly Ash (kg/m³)	Water (kg/m³)	Sand (kg/m³)	Recycled Aggregate 3/4" (kg/m³)	Recommended Quantity Admixture (Liters/m³)	Fiber (%)
Properties	350	330	80	155	810	1,135	5.30	–
	Average		Standard deviation	Compressive strength (ksc)			Average	Standard deviation
Air content (%)[1]	1.8		0.2	7 days			272.36	22.84
Slump (mm)[2]	80.0		14.6	28 days			274.57	11.39

(Continued)

TABLE 5.4 (Continued)
Typical Mix Proportions and Related Properties of RAC

Mix Symbol (HSC450C395F100S175)	Cylindrical Strength (ksc)	Cement (kg/m³)	Fly Ash (kg/m³)	Water (kg/m³)	Sand (kg/m³)	Recycled Aggregate 3/4″ (kg/m³)	Recommended Quantity Admixture (Liters/m³)	Fiber (%)
Initial setting time (min)[3]	176		1.2	90 days			430.14	1.29
Final setting time (min)[3]	364		28.0	365 days			447.53	1.03

Mix Symbol (HSC480C520F0S175)	Cylindrical Strength (ksc)	Cement (kg/m³)	Fly Ash (kg/m³)	Water (kg/m³)	Sand (kg/m³)	Recycled Aggregate 3/4″ (kg/m³)	Recommended Quantity Admixture (Liters/m³)	Fiber (%)
	380	460	0	165	800	1,120	6.00	–

Properties	Average		Standard deviation	Compressive strength (ksc)			Average	Standard deviation
Air content (%)[1]	1.5		0.2	7 days			276.11	5.48
Slump (mm)[2]	90.0		8.8	28 days			296.84	3.99
Initial setting time (min)[3]	146		9.2	90 days			456.28	2.73
Final setting time (min)[3]	362		24.9	365 days			508.69	1.04

Mix Symbol (HSC480C415F104S175)	Cylindrical Strength (ksc)	Cement (kg/m³)	Fly Ash (kg/m³)	Water (kg/m³)	Sand (kg/m³)	Recycled Aggregate 3/4" (kg/m³)	Recommended Quantity Admixture (Liters/m³)	Fiber (%)
	380	286	184	155	800	1,120	6.10	–
Properties	Average		Standard deviation	Compressive strength (ksc)			Average	Standard deviation
Air content (%)[1]	1.6		0.3	7 days			269.88	24.74
Slump (mm)[2]	81.0		12.7	28 days			391.66	21.63
Initial setting time (min)[3]	226		5.4	90 days			467.23	14.39
Final setting time (min)[3]	328		34.8	365 days			584.81	2.88

Mix Symbol (HSC300C345F0S175)	Cylindrical Strength (ksc)	Cement (kg/m³)	Fly Ash (kg/m³)	Water (kg/m³)	Sand (kg/m³)	Recycled Aggregate 3/4" (kg/m³)	Recommended Quantity Admixture (Liters/m³)	Fiber (%)
	300	360	0	130	900	1,115	4.20	–
Properties	Average		Standard deviation	Compressive strength (ksc)			Average	Standard deviation
Air content (%)[1]	1.1		0.3	7 days			282.41	23.21
Slump (mm)[2]	90.5		18.3	28 days			402.43	19.76

(Continued)

TABLE 5.4 (Continued)
Typical Mix Proportions and Related Properties of RAC

Mix Symbol (HSC300C345F0S175)	Cylindrical Strength (ksc)	Cement (kg/m³)	Fly Ash (kg/m³)	Water (kg/m³)	Sand (kg/m³)	Recycled Aggregate 3/4" (kg/m³)	Recommended Quantity Admixture (Liters/m³)	Fiber (%)
Initial setting time (min)[3]	206		7.6	90 days			536.89	8.62
Final setting time (min)[3]	358		36.6	365 days			622.63	6.85

Mix Symbol (HSC300C275F70S175)	Cylindrical Strength (ksc)	Cement (kg/m³)	Fly Ash (kg/m³)	Water (kg/m³)	Sand (kg/m³)	Recycled Aggregate 3/4" (kg/m³)	Recommended Quantity Admixture (Liters/m³)	Fiber (%)
	300	290	70	130	900	1,115	4.20	–

Properties	Average	Standard deviation	Compressive strength (ksc)	Average	Standard deviation
Air content (%)[1]	1.9	0.1	7 days	167.94	14.32
Slump (mm)[2]	85.0	11.5	28 days	341.87	9.67
Initial setting time (min)[3]	196	13.2	90 days	387.61	5.41
Final setting time (min)[3]	348	42.4	365 days	417.18	3.99

Mix Symbol (HSC320C370F0S175)	Cylindrical Strength (ksc)	Cement (kg/m³)	Fly Ash (kg/m³)	Water (kg/m³)	Sand (kg/m³)	Recycled Aggregate 3/4" (kg/m³)	Recommended Quantity Admixture (Liters/m³)	Fiber (%)
	320	385	0	130	890	1,105	4.50	–
Properties	Average		Standard deviation	Compressive strength (ksc)			Average	Standard deviation
Air content (%)[1]	2.1		0.1	7 days			290.55	7.61
Slump (mm)[2]	98.5		14.2	28 days			380.02	5.11
Initial setting time (min)[3]	199		2.8	90 days			193.98	3.06
Final setting time (min)[3]	255		6.7	365 days			569.57	1.87

Mix Symbol (HSC320C295F74S175)	Cylindrical Strength (ksc)	Cement (kg/m³)	Fly Ash (kg/m³)	Water (kg/m³)	Sand (kg/m³)	Recycled Aggregate 3/4" (kg/m³)	Recommended Quantity Admixture (Liters/m³)	Fiber (%)
	320	310	74	130	890	1,100	4.50	–
Properties	Average		Standard deviation	Compressive strength (ksc)			Average	Standard deviation
Air content (%)[1]	2.2		0.3	7 days			191.32	14.98
Slump (mm)[2]	97.5		7.3	28 days			276.54	8.31

(Continued)

TABLE 5.4 (Continued)
Typical Mix Proportions and Related Properties of RAC

Mix Symbol (HSC320C295F74S175)	Cylindrical Strength (ksc)	Cement (kg/m³)	Fly Ash (kg/m³)	Water (kg/m³)	Sand (kg/m³)	Recycled Aggregate 3/4" (kg/m³)	Recommended Quantity Admixture (Liters/m³)	Fiber (%)
Initial setting time (min)[3]	193		6.8	90 days			317.28	4.68
Final setting time (min)[3]	340		32.7	365 days			354.54	2.22

Mix Symbol (HSC350C395F0S175)	Cylindrical Strength (ksc)	Cement (kg/m³)	Fly Ash (kg/m³)	Water (kg/m³)	Sand (kg/m³)	Recycled Aggregate 3/4" (kg/m³)	Recommended Quantity Admixture (Liters/m³)	Fiber (%)
	320	256	0	120	820	1,100	4.80	–
Properties	Average		Standard deviation	Compressive strength (ksc)			Average	Standard deviation
Air content (%)[1]	3.7		0.1	7 days			286.62	6.54
Slump (mm)[2]	104.5		15.8	28 days			353.83	3.57
Initial setting time (min)[3]	246		10.1	90 days			458.84	2.25
Final setting time (min)[3]	299		21.2	365 days			521.03	1.98

Mix Symbol (HSC350C315F80S175)	Cylindrical Strength (ksc)	Cement (kg/m³)	Fly Ash (kg/m³)	Water (kg/m³)	Sand (kg/m³)	Recycled Aggregate 3/4" (kg/m³)	Recommended Quantity Admixture (Liters/m³)	Fiber (%)
Properties	320	264	75	120	820	1,100	4.80	–

Property	Average	Standard deviation
Air content (%)[1]	3.6	0.2
Slump (mm)[2]	101.5	11.2
Initial setting time (min)[3]	187	5.7
Final setting time (min)[3]	339	10.9

Compressive strength (ksc)	Average	Standard deviation
7 days	178.04	15.87
28 days	362.56	12.30
90 days	476.35	10.91
365 days	551.59	5.12

Mix Symbol (HSC380C420F0S175)	Cylindrical Strength (ksc)	Cement (kg/m³)	Fly Ash (kg/m³)	Water (kg/m³)	Sand (kg/m³)	Recycled Aggregate 3/4" (kg/m³)	Recommended Quantity Admixture (Liters/m³)	Fiber (%)
Properties	320	244	0	120	820	1,100	5.10	–

Property	Average	Standard deviation
Air content (%)[1]	2.7	0.1
Slump (mm)[2]	121.0	9.3

Compressive strength (ksc)	Average	Standard deviation
7 days	270.24	24.56
28 days	321.25	22.41

(Continued)

TABLE 5.4 (Continued)
Typical Mix Proportions and Related Properties of RAC

Mix Symbol (HSC380C420F0S175)	Cylindrical Strength (ksc)	Cement (kg/m³)	Fly Ash (kg/m³)	Water (kg/m³)	Sand (kg/m³)	Recycled Aggregate 3/4" (kg/m³)	Recommended Quantity Admixture (Liters/m³)	Fiber (%)
Initial setting time (min)[3]	107		3.3	90 days			447.98	19.32
Final setting time (min)[3]	273		13.3	365 days			500.14	2.77

Mix Symbol (HSC380C335F84S175)	Cylindrical Strength (ksc)	Cement (kg/m³)	Fly Ash (kg/m³)	Water (kg/m³)	Sand (kg/m³)	Recycled Aggregate 3/4" (kg/m³)	Recommended Quantity Admixture (Liters/m³)	Fiber (%)
Properties	320	310	80	120	820	1,100	5.10	–
	Average		Standard deviation	Compressive strength (ksc)			Average	Standard deviation
Air content (%)[1]	3.0		0.2	7 days			286.14	36.14
Slump (mm)[2]	120.5		23.7	28 days			380.59	24.88
Initial setting time (min)[3]	209		19.3	90 days			518.36	20.95
Final setting time (min)[3]	319		33.5	365 days			536.42	10.05

Mix Symbol (HSC400C445F0S175)	Cylindrical Strength (ksc)	Cement (kg/m³)	Fly Ash (kg/m³)	Water (kg/m³)	Sand (kg/m³)	Recycled Aggregate 3/4" (kg/m³)	Recommended Quantity Admixture (Liters/m³)	Fiber (%)
	350	434	0	165	810	1,100	5.10	–

Properties	Average	Standard deviation	Compressive strength (ksc)		Average	Standard deviation
Air content (%)[1]	3.1	0.3		7 days	241.88	9.47
Slump (mm)[2]	96.5	10.9		28 days	289.86	8.23
Initial setting time (min)[3]	216	7.5		90 days	392.04	10.51
Final setting time (min)[3]	269	15.8		365 days	447.75	2.24

Mix Symbol (HSC400C355F90S175)	Cylindrical Strength (ksc)	Cement (kg/m³)	Fly Ash (kg/m³)	Water (kg/m³)	Sand (kg/m³)	Recycled Aggregate 3/4" (kg/m³)	Recommended Quantity Admixture (Liters/m³)	Fiber (%)
	350	270	174	155	810	1,100	5.20	–

Properties	Average	Standard deviation	Compressive strength (ksc)		Average	Standard deviation
Air content (%)[1]	1.2	0.2		7 days	249.81	3.42
Slump (mm)[2]	90.5	24.3		28 days	314.44	2.77

(Continued)

TABLE 5.4 (Continued)
Typical Mix Proportions and Related Properties of RAC

Mix Symbol (HSC400C355F90S175)	Cylindrical Strength (ksc)	Cement (kg/m³)	Fly Ash (kg/m³)	Water (kg/m³)	Sand (kg/m³)	Recycled Aggregate 3/4" (kg/m³)	Recommended Quantity Admixture (Liters/m³)	Fiber (%)
Initial setting time (min)[3]	248		12.8	90 days			432.93	1.08
Final setting time (min)[3]	355		23.9	365 days			458.36	0.98

Mix Symbol (HSC420C470F0S175)	Cylindrical Strength (ksc)	Cement (kg/m³)	Fly Ash (kg/m³)	Water (kg/m³)	Sand (kg/m³)	Recycled Aggregate 3/4" (kg/m³)	Recommended Quantity Admixture (Liters/m³)	Fiber (%)
Properties	350	280	180	155	810	1,100	5.30	–

Properties	Average		Standard deviation	Compressive strength (ksc)			Average	Standard deviation
Air content (%)[1]	1.7		0.1	7 days			205.87	15.74
Slump (mm)[2]	91.5		15.2	28 days			375.55	20.31
Initial setting time (min)[3]	241		13.6	90 days			574.39	14.93
Final setting time (min)[3]	344		34.4	365 days			630.47	7.44

Mix Symbol (HSC420C375F94S175)	Cylindrical Strength (ksc)	Cement (kg/m³)	Fly Ash (kg/m³)	Water (kg/m³)	Sand (kg/m³)	Recycled Aggregate 3/4" (kg/m³)	Recommended Quantity Admixture (Liters/m³)	Fiber (%)
	350	258	166	155	810	1,100	5.50	–
Properties	Average		Standard deviation	Compressive strength (ksc)			Average	Standard deviation
Air content (%)[1]	3.4		0.3	7 days			290.01	9.84
Slump (mm)[2]	84.0		5.5	28 days			500.44	6.47
Initial setting time (min)[3]	197		2.4	90 days			525.71	5.63
Final setting time (min)[3]	329		3.6	365 days			610.63	3.74

Mix Symbol (HSC450C495F0S175)	Cylindrical Strength (ksc)	Cement (kg/m³)	Fly Ash (kg/m³)	Water (kg/m³)	Sand (kg/m³)	Recycled Aggregate 3/4" (kg/m³)	Recommended Quantity Admixture (Liters/m³)	Fiber (%)
	300	484	0	165	830	1,115	5.90	–
Properties	Average		Standard deviation	Compressive strength (ksc)			Average	Standard deviation
Air content (%)[1]	4.0		0.1	7 days			296.88	7.63
Slump (mm)[2]	87.5		21.7	28 days			469.63	1.10

(Continued)

TABLE 5.4 (Continued)
Typical Mix Proportions and Related Properties of RAC

Mix Symbol (HSC450C495F0S175)	Cylindrical Strength (ksc)	Cement (kg/m³)	Fly Ash (kg/m³)	Water (kg/m³)	Sand (kg/m³)	Recycled Aggregate 3/4″ (kg/m³)	Recommended Quantity Admixture (Liters/m³)	Fiber (%)
Initial setting time (min)[3]	167		6.4		90 days		493.63	4.75
Final setting time (min)[3]	359		23.9		365 days		580.31	0.66
Mix Symbol (HSC450C395F100S175)	**Cylindrical Strength (ksc)**	**Cement (kg/m³)**	**Fly Ash (kg/m³)**	**Water (kg/m³)**	**Sand (kg/m³)**	**Recycled Aggregate 3/4″ (kg/m³)**	**Recommended Quantity Admixture (Liters/m³)**	**Fiber (%)**
	300	240	154	155	830	1,115	5.50	–
Properties	**Average**		**Standard deviation**		**Compressive strength (ksc)**		**Average**	**Standard deviation**
Air content (%)[1]	1.8		0.3		7 days		211.88	22.78
Slump (mm)[2]	92.0		14.3		28 days		291.37	14.99
Initial setting time (min)[3]	245		13.2		90 days		489.54	6.57
Final setting time (min)[3]	353		24.2		365 days		431.67	2.37

Mix Symbol (HSC480C520F0S175)	Cylindrical Strength (ksc)	Cement (kg/m³)	Fly Ash (kg/m³)	Water (kg/m³)	Sand (kg/m³)	Recycled Aggregate 3/4" (kg/m³)	Recommended Quantity Admixture (Liters/m³)	Fiber (%)
	300	290	70	155	830	1,115	6.00	–
Properties	Average		Standard deviation	Compressive strength (ksc)			Average	Standard deviation
Air content (%)[1]	1.0		0.2	7 days			248.33	12.10
Slump (mm)[2]	86.5		22.1	28 days			445.84	28.48
Initial setting time (min)[3]	249		10.7	90 days			502.45	18.04
Final setting time (min)[3]	356		20.1	365 days			580.09	5.49
Mix Symbol (HSC480C415F104S175)	Cylindrical Strength (ksc)	Cement (kg/m³)	Fly Ash (kg/m³)	Water (kg/m³)	Sand (kg/m³)	Recycled Aggregate 3/4" (kg/m³)	Recommended Quantity Admixture (Liters/m³)	Fiber (%)
	320	410	0	165	820	1,100	6.10	–
Properties	Average		Standard deviation	Compressive strength (ksc)			Average	Standard deviation
Air content (%)[1]	1.3		0.3	7 days			307.52	43.22
Slump (mm)[2]	88.0		17.4	28 days			413.89	24.93

(Continued)

TABLE 5.4 (Continued)
Typical Mix Proportions and Related Properties of RAC

Mix Symbol (HSC480C415F104S175)	Cylindrical Strength (ksc)	Cement (kg/m³)	Fly Ash (kg/m³)	Water (kg/m³)	Sand (kg/m³)	Recycled Aggregate 3/4" (kg/m³)	Recommended Quantity Admixture (Liters/m³)	Fiber (%)
Initial setting time (min)[3]	243		8.9	90 days			509.07	18.45
Final setting time (min)[3]	350		32.2	365 days			553.94	3.94
Mix Symbol (HSC310C375F0S175)	**Cylindrical Strength (ksc)**	**Cement (kg/m³)**	**Fly Ash (kg/m³)**	**Water (kg/m³)**	**Sand (kg/m³)**	**Recycled Aggregate 3/4" (kg/m³)**	**Recommended Quantity Admixture (Liters/m³)**	**Fiber (%)**
Properties	280	214	136	155	840	1,110	5.00	–
	Average		Standard deviation	Compressive strength (ksc)			Average	Standard deviation
Air content (%)[1]	1.5		0.2	7 days			163.92	18.05
Slump (mm)[2]	97.0		19.6	28 days			361.64	11.76
Initial setting time (min)[3]	242		15.1	90 days			434.56	10.52
Final setting time (min)[3]	349		33.3	365 days			510.08	8.56

Mix Symbol (HSC310C295F74S175)	Cylindrical Strength (ksc)	Cement (kg/m³)	Fly Ash (kg/m³)	Water (kg/m³)	Sand (kg/m³)	Recycled Aggregate 3/4" (kg/m³)	Recommended Quantity Admixture (Liters/m³)	Fiber (%)
	280	270	64	155	840	1,110	4.90	–
Properties	Average		Standard deviation	Compressive strength (ksc)			Average	Standard deviation
Air content (%)[1]	3.1		0.3	7 days			168.97	5.94
Slump (mm)[2]	123.0		14.9	28 days			378.12	7.59
Initial setting time (min)[3]	245		3.3	90 days			454.08	12.47
Final setting time (min)[3]	268		4.6	365 days			506.49	8.03

Mix Symbol (HSC500C545F0S175)	Cylindrical Strength (ksc)	Cement (kg/m³)	Fly Ash (kg/m³)	Water (kg/m³)	Sand (kg/m³)	Recycled Aggregate 3/4" (kg/m³)	Recommended Quantity Admixture (Liters/m³)	Fiber (%)
	450	510	0	130	840	1,110	6.50	–
Properties	Average		Standard deviation	Compressive strength (ksc)			Average	Standard deviation
Air content (%)[1]	2.1		0.3	7 days			267.98	36.49
Slump (mm)[2]	102.0		13.2	28 days			449.62	27.99

(Continued)

TABLE 5.4 (Continued)
Typical Mix Proportions and Related Properties of RAC

Mix Symbol (HSC500C545F0S175)	Cylindrical Strength (ksc)	Cement (kg/m³)	Fly Ash (kg/m³)	Water (kg/m³)	Sand (kg/m³)	Recycled Aggregate 3/4" (kg/m³)	Recommended Quantity Admixture (Liters/m³)	Fiber (%)
Initial setting time (min)[3]	132		0.5		90 days		467.14	7.60
Final setting time (min)[3]	318		23.7		365 days		576.26	4.45
Mix Symbol (HSC500C435F110S175)	**Cylindrical Strength (ksc)**	**Cement (kg/m³)**	**Fly Ash (kg/m³)**	**Water (kg/m³)**	**Sand (kg/m³)**	**Recycled Aggregate 3/4" (kg/m³)**	**Recommended Quantity Admixture (Liters/m³)**	**Fiber (%)**
	450	410	100	130	840	1,110	6.30	–
Properties	Average		Standard deviation		Compressive strength (ksc)		Average	Standard deviation
Air content (%)[1]	1.4		0.1		7 days		279.15	25.44
Slump (mm)[2]	93.0		18.5		28 days		423.68	13.98
Initial setting time (min)[3]	247		13.9		90 days		497.06	5.02
Final setting time (min)[3]	354		24.5		365 days		547.23	1.84

Mix Symbol (HSC500C435F110S175) Properties	Cylindrical Strength (ksc)	Cement (kg/m³)	Fly Ash (kg/m³)	Water (kg/m³)	Sand (kg/m³)	Recycled Aggregate 3/4" (kg/m³)	Recommended Quantity Admixture (Liters/m³)	Fiber (%)
	350	410	0	130	880	1,135	5.10	–
	Average		Standard deviation	Compressive strength (ksc)			Average	Standard deviation
Air content (%)[1]	1.7		0.1	7 days			328.25	11.05
Slump (mm)[2]	84.5		18.9	28 days			436.16	24.98
Initial setting time (min)[3]	119		5.7	90 days			494.98	7.27
Final setting time (min)[3]	317		9.4	365 days			542.74	3.55

Mix Symbol (HSC425C327F142S175) Properties	Cylindrical Strength (ksc)	Cement (kg/m³)	Fly Ash (kg/m³)	Water (kg/m³)	Sand (kg/m³)	Recycled Aggregate 3/4" (kg/m³)	Recommended Quantity Admixture (Liters/m³)	Fiber (%)
	350	330	80	130	880	1,135	5.50	–
	Average		Standard deviation	Compressive strength (ksc)			Average	Standard deviation
Air content (%)[1]	1.6		0.3	7 days			298.75	38.52
Slump (mm)[2]	95.5		24.7	28 days			426.25	9.42

(Continued)

TABLE 5.4 (Continued)
Typical Mix Proportions and Related Properties of RAC

Mix Symbol (HSC425C327F142S175)	Cylindrical Strength (ksc)	Cement (kg/m³)	Fly Ash (kg/m³)	Water (kg/m³)	Sand (kg/m³)	Recycled Aggregate 3/4" (kg/m³)	Recommended Quantity Admixture (Liters/m³)	Fiber (%)
Initial setting time (min)[3]	205		12.7	90 days			489.63	14.51
Final setting time (min)[3]	395		44.1	365 days			553.77	10.76

Mix Symbol (HSC500C545F0S80)	Cylindrical Strength (ksc)	Cement (kg/m³)	Fly Ash (kg/m³)	Water (kg/m³)	Sand (kg/m³)	Recycled Aggregate 3/4" (kg/m³)	Recommended Quantity Admixture (Liters/m³)	Fiber (%)
	350	330	80	130	880	1,135	6.30	–

Properties	Average	Standard deviation	Compressive strength (ksc)	Average	Standard deviation
Air content (%)[1]	1.2	0.1	7 days	300.14	6.34
Slump (mm)[2]	75.0	11.7	28 days	319.58	4.12
Initial setting time (min)[3]	117	7.1	90 days	478.13	3.52
Final setting time (min)[3]	344	32.4	365 days	532.55	2.77

Mix Symbol (HSC500C435F110S80)	Cylindrical Strength (ksc)	Cement (kg/m³)	Fly Ash (kg/m³)	Water (kg/m³)	Sand (kg/m³)	Recycled Aggregate 3/4" (kg/m³)	Recommended Quantity Admixture (Liters/m³)	Fiber (%)
Properties	300	410	0	130	880	1,135	6.40	–
	Average		Standard deviation	Compressive strength (ksc)			Average	Standard deviation
Air content (%)[1]	2.7		0.1	7 days			328.28	3.89
Slump (mm)[2]	62.0		5.5	28 days			372.64	2.11
Initial setting time (min)[3]	163		0.9	90 days			536.89	7.98
Final setting time (min)[3]	334		45.2	365 days			602.08	5.33

Mix Symbol (HSC500C435F110S80)	Cylindrical Strength (ksc)	Cement (kg/m³)	Fly Ash (kg/m³)	Water (kg/m³)	Sand (kg/m³)	Recycled Aggregate 3/4" (kg/m³)	Recommended Quantity Admixture (Liters/m³)	Fiber (%)
Properties	425	342	142	130	850	1,135	5.10	–
	Average		Standard deviation	Compressive strength (ksc)			Average	Standard deviation
Air content (%)[1]	2.1		0.3	7 days			305.88	3.12
Slump (mm)[2]	56.0		4.2	28 days			349.55	27.16

(Continued)

TABLE 5.4 (Continued)
Typical Mix Proportions and Related Properties of RAC

Mix Symbol (HSC500C435F110S80)	Cylindrical Strength (ksc)	Cement (kg/m³)	Fly Ash (kg/m³)	Water (kg/m³)	Sand (kg/m³)	Recycled Aggregate 3/4" (kg/m³)	Recommended Quantity Admixture (Liters/m³)	Fiber (%)
Initial setting time (min)[3]	110		19.2	90 days			540.11	6.52
Final setting time (min)[3]	256		38.4	365 days			621.07	1.68

Mix Symbol (HSC400C445F0S60)	Cylindrical Strength (ksc)	Cement (kg/m³)	Fly Ash (kg/m³)	Water (kg/m³)	Sand (kg/m³)	Recycled Aggregate 3/4" (kg/m³)	Recommended Quantity Admixture (Liters/m³)	Fiber (%)
Properties	300	250	160	155	830	1,115	5.30	–

Properties	Average	Standard deviation	Standard deviation	Compressive strength (ksc)			Average	Standard deviation
Air content (%)[1]	2.0		0.2	7 days			219.56	31.96
Slump (mm)[2]	55.5		20.7	28 days			390.24	30.11
Initial setting time (min)[3]	101		7.9	90 days			442.76	9.28
Final setting time (min)[3]	374		10.9	365 days			513.27	6.16

Mix Symbol (HSC400C355F90S60)	Cylindrical Strength (ksc)	Cement (kg/m³)	Fly Ash (kg/m³)	Water (kg/m³)	Sand (kg/m³)	Recycled Aggregate 3/4" (kg/m³)	Recommended Quantity Admixture (Liters/m³)	Fiber (%)
Properties	300	228	146	155	830	1,115	5.20	–

Properties	Average	Standard deviation	Compressive strength (ksc)	Average	Standard deviation
Air content (%)[1]	1.1	0.3	7 days	273.28	41.58
Slump (mm)[2]	27.0	4.9	28 days	357.58	6.67
Initial setting time (min)[3]	151	16.7	90 days	471.35	14.95
Final setting time (min)[3]	275	15.9	365 days	536.71	9.78

Mix Symbol (HSC400C355F90S60)	Cylindrical Strength (ksc)	Cement (kg/m³)	Fly Ash (kg/m³)	Water (kg/m³)	Sand (kg/m³)	Recycled Aggregate 3/4" (kg/m³)	Recommended Quantity Admixture (Liters/m³)	Fiber (%)
Properties	390	330	80	120	800	1,105	4.40	–

Properties	Average	Standard deviation	Compressive strength (ksc)	Average	Standard deviation
Air content (%)[1]	2.5	0.2	7 days	243.63	10.36
Slump (mm)[2]	59.0	10.7	28 days	386.24	18.36

(Continued)

TABLE 5.4 (Continued)
Typical Mix Proportions and Related Properties of RAC

Mix Symbol (HSC400C355F90S60)	Cylindrical Strength (ksc)	Cement (kg/m³)	Fly Ash (kg/m³)	Water (kg/m³)	Sand (kg/m³)	Recycled Aggregate 3/4" (kg/m³)	Recommended Quantity Admixture (Liters/m³)	Fiber (%)
Initial setting time (min)[3]		108	12.4		90 days		449.77	16.60
Final setting time (min)[3]		272	24.8		365 days		510.76	9.29

Mix Symbol (HSC350C395F0S60)	Cylindrical Strength (ksc)	Cement (kg/m³)	Fly Ash (kg/m³)	Water (kg/m³)	Sand (kg/m³)	Recycled Aggregate 3/4" (kg/m³)	Recommended Quantity Admixture (Liters/m³)	Fiber (%)
	380	274	176	155	800	1,120	4.50	–

Properties	Average	Standard deviation		Compressive strength (ksc)		Average	Standard deviation
Air content (%)[1]	1.8	0.2		7 days		233.94	31.63
Slump (mm)[2]	28.5	8.4		28 days		455.22	21.88
Initial setting time (min)[3]	171	8.1		90 days		535.31	27.31
Final setting time (min)[3]	255	9.4		365 days		616.18	12.28

Mix Symbol (HSC350C315F80S60)	Cylindrical Strength (ksc)	Cement (kg/m³)	Fly Ash (kg/m³)	Water (kg/m³)	Sand (kg/m³)	Recycled Aggregate 3/4″ (kg/m³)	Recommended Quantity Admixture (Liters/m³)	Fiber (%)
	380	350	84	155	800	1,120	4.30	–

Properties	Average		Standard deviation	Compressive strength (ksc)			Average	Standard deviation
Air content (%)[1]	1.2		0.1	7 days			201.95	28.74
Slump (mm)[2]	27.5		15.2	28 days			288.47	14.77
Initial setting time (min)[3]	101		4.2	90 days			327.01	9.41
Final setting time (min)[3]	250		30.2	365 days			407.48	4.28

Mix Symbol (HSC350C315F80S60)	Cylindrical Strength (ksc)	Cement (kg/m³)	Fly Ash (kg/m³)	Water (kg/m³)	Sand (kg/m³)	Recycled Aggregate 3/4″ (kg/m³)	Recommended Quantity Admixture (Liters/m³)	Fiber (%)
	380	286	184	155	800	1,120	3.90	–

Properties	Average		Standard deviation	Compressive strength (ksc)			Average	Standard deviation
Air content (%)[1]	1.3		0.2	7 days			277.49	4.05
Slump (mm)[2]	51.5		24.5	28 days			297.73	1.47

(Continued)

TABLE 5.4 (Continued)
Typical Mix Proportions and Related Properties of RAC

Mix Symbol (HSC350C315F80S60)	Cylindrical Strength (ksc)	Cement (kg/m³)	Fly Ash (kg/m³)	Water (kg/m³)	Sand (kg/m³)	Recycled Aggregate 3/4" (kg/m³)	Recommended Quantity Admixture (Liters/m³)	Fiber (%)
Initial setting time (min)[3]	131		2.8	90 days			466.32	2.21
Final setting time (min)[3]	270		25.7	365 days			518.16	2.58
Mix Symbol (HSC300C395F0S100)	Cylindrical Strength (ksc)	Cement (kg/m³)	Fly Ash (kg/m³)	Water (kg/m³)	Sand (kg/m³)	Recycled Aggregate 3/4" (kg/m³)	Recommended Quantity Admixture (Liters/m³)	Fiber (%)
	380	294	190	155	800	1,120	2.10	–
Properties	Average	Standard deviation		Compressive strength (ksc)			Average	Standard deviation
Air content (%)[1]	2.2	0.1		7 days			151.11	12.44
Slump (mm)[2]	50.5	20.7		28 days			217.65	11.47
Initial setting time (min)[3]	246	14.1		90 days			327.75	2.59
Final setting time (min)[3]	358	31.1		365 days			377.94	0.78

Mix Symbol (HSC300C395F0S100)	Cylindrical Strength (ksc)	Cement (kg/m³)	Fly Ash (kg/m³)	Water (kg/m³)	Sand (kg/m³)	Recycled Aggregate 3/4" (kg/m³)	Recommended Quantity Admixture (Liters/m³)	Fiber (%)
	280	360	0	165	840	1,130	2.30	–
Properties	Average		Standard deviation	Compressive strength (ksc)			Average	Standard deviation
Air content (%)[1]	3.1		0.2	7 days			222.89	29.14
Slump (mm)[2]	76.0		7.1	28 days			267.05	20.81
Initial setting time (min)[3]	226		9.2	90 days			414.51	13.57
Final setting time (min)[3]	309		14.8	365 days			457.14	5.22

Mix Symbol (HSC300C275F120S100)	Cylindrical Strength (ksc)	Cement (kg/m³)	Fly Ash (kg/m³)	Water (kg/m³)	Sand (kg/m³)	Recycled Aggregate 3/4" (kg/m³)	Recommended Quantity Admixture (Liters/m³)	Fiber (%)
	250	260	62	155	850	1,145	2.30	–
Properties	Average		Standard deviation	Compressive strength (ksc)			Average	Standard deviation
Air content (%)[1]	3.2		0.3	7 days			157.59	21.87
Slump (mm)[2]	88.5		14.8	28 days			329.58	3.41

(Continued)

TABLE 5.4 (Continued)
Typical Mix Proportions and Related Properties of RAC

Mix Symbol (HSC300C275F120S100)	Cylindrical Strength (ksc)	Cement (kg/m³)	Fly Ash (kg/m³)	Water (kg/m³)	Sand (kg/m³)	Recommended Quantity Admixture (Liters/m³)	Fiber (%)
Initial setting time (min)[3]	236		6.7	90 days		467.25	2.98
Final setting time (min)[3]	319		19.5	365 days		493.47	1.22

Mix Symbol (HSC300C275F120S100)	Cylindrical Strength (ksc)	Cement (kg/m³)	Fly Ash (kg/m³)	Water (kg/m³)	Sand (kg/m³)	Recycled Aggregate 3/4″ (kg/m³)	Recommended Quantity Admixture (Liters/m³)	Fiber (%)
	280	226	144	155	840	1,130	2.30	–
Properties	Average		Standard deviation		Compressive strength (ksc)		Average	Standard deviation
Air content (%)[1]	1.0		0.2		7 days		217.98	28.12
Slump (mm)[2]	110.5		4.4		28 days		271.05	19.53
Initial setting time (min)[3]	143		3.6		90 days		4,243.47	12.55
Final setting time (min)[3]	281		1.2		365 days		440.21	2.12

Mix Symbol (HSC300C295F100S100)	Cylindrical Strength (ksc)	Cement (kg/m³)	Fly Ash (kg/m³)	Water (kg/m³)	Sand (kg/m³)	Recycled Aggregate 3/4" (kg/m³)	Recommended Quantity Admixture (Liters/m³)	Fiber (%)
Properties	280	234	150	155	840	1,130	2.30	—
	Average		Standard deviation	Compressive strength (ksc)			Average	Standard deviation
Air content (%)[1]	1.1		0.3	7 days			178.74	44.50
Slump (mm)[2]	105.0		5.9	28 days			322.56	28.57
Initial setting time (min)[3]	234		1.4	90 days			509.64	9.17
Final setting time (min)[3]	399		2.4	365 days			556.59	5.12

Mix Symbol (HSC300C295F100S100)	Cylindrical Strength (ksc)	Cement (kg/m³)	Fly Ash (kg/m³)	Water (kg/m³)	Sand (kg/m³)	Recycled Aggregate 3/4" (kg/m³)	Recommended Quantity Admixture (Liters/m³)	Fiber (%)
Properties	250	226	144	155	840	1,130	2.30	—
	Average		Standard deviation	Compressive strength (ksc)			Average	Standard deviation
Air content (%)[1]	3.1		0.2	7 days			235.22	35.62
Slump (mm)[2]	81.5		23.1	28 days			327.92	9.64

(Continued)

TABLE 5.4 (Continued)
Typical Mix Proportions and Related Properties of RAC

Mix Symbol (HSC300C295F100S100)	Cylindrical Strength (ksc)	Cement (kg/m³)	Fly Ash (kg/m³)	Water (kg/m³)	Sand (kg/m³)	Recycled Aggregate 3/4" (kg/m³)	Recommended Quantity Admixture (Liters/m³)	Fiber (%)
Initial setting time (min)[3]	241		6.5	90 days			442.76	4.88
Final setting time (min)[3]	281		1.7	365 days			510.24	3.92
Mix Symbol (HSC300C315F80S100)	**Cylindrical Strength (ksc)**	**Cement (kg/m³)**	**Fly Ash (kg/m³)**	**Water (kg/m³)**	**Sand (kg/m³)**	**Recycled Aggregate 3/4" (kg/m³)**	**Recommended Quantity Admixture (Liters/m³)**	**Fiber (%)**
	280	214	136	155	840	1,130	2.30	
Properties	Average	Standard deviation		Compressive strength (ksc)			Average	Standard deviation
Air content (%)[1]	2.6	0.2		7 days			175.65	17.52
Slump (mm)[2]	89.0	21.5		28 days			346.51	19.81
Initial setting time (min)[3]	229	7.1		90 days			365.56	15.75
Final setting time (min)[3]	271	1.8		365 days			424.43	9.45

Mix Symbol (HSC300C315F80S100)	Cylindrical Strength (ksc)	Cement (kg/m³)	Fly Ash (kg/m³)	Water (kg/m³)	Sand (kg/m³)	Recycled Aggregate 3/4" (kg/m³)	Recommended Quantity Admixture (Liters/m³)	Fiber (%)
	280	270	64	155	840	1,130	2.30	–
Properties	Average		Standard deviation	Compressive strength (ksc)			Average	Standard deviation
Air content (%)[1]	2.4		0.3		7 days		224.28	37.67
Slump (mm)[2]	78.5		12.2		28 days		432.55	16.12
Initial setting time (min)[3]	177		7.5		90 days		431.75	11.47
Final setting time (min)[3]	258		47.7		365 days		483.26	9.91

Mix Symbol (HSC400C445F0S125)	Cylindrical Strength (ksc)	Cement (kg/m³)	Fly Ash (kg/m³)	Water (kg/m³)	Sand (kg/m³)	Recycled Aggregate 3/4" (kg/m³)	Recommended Quantity Admixture (Liters/m³)	Fiber (%)
	300	384	0	165	830	1,115	3.00	–
Properties	Average		Standard deviation	Compressive strength (ksc)			Average	Standard deviation
Air content (%)[1]	3.0		0.1		7 days		273.65	40.05
Slump (mm)[2]	85.0		6.1		28 days		434.41	9.99

(Continued)

TABLE 5.4 (Continued)
Typical Mix Proportions and Related Properties of RAC

Mix Symbol (HSC400C445F0S125)	Cylindrical Strength (ksc)	Cement (kg/m³)	Fly Ash (kg/m³)	Water (kg/m³)	Sand (kg/m³)	Recycled Aggregate 3/4" (kg/m³)	Recommended Quantity Admixture (Liters/m³)	Fiber (%)
Initial setting time (min)[3]	128		17.7		90 days		516.42	8.55
Final setting time (min)[3]	332		18.8		365 days		530.08	4.45
Mix Symbol (HSC400C355F90S125)	**Cylindrical Strength (ksc)**	**Cement (kg/m³)**	**Fly Ash (kg/m³)**	**Water (kg/m³)**	**Sand (kg/m³)**	**Recycled Aggregate 3/4" (kg/m³)**	**Recommended Quantity Admixture (Liters/m³)**	**Fiber (%)**
	350	330	80	155	810	1,135	3.10	–
Properties	**Average**		**Standard deviation**		**Compressive strength (ksc)**		**Average**	**Standard deviation**
Air content (%)[1]	1.7		0.1		7 days		252.53	32.48
Slump (mm)[2]	97.5		13.2		28 days		378.76	12.54
Initial setting time (min)[3]	235		5.7		90 days		439.51	13.12
Final setting time (min)[3]	306		23.7		365 days		476.87	4.64

Mix Symbol (HSC400C355F90S125)	Cylindrical Strength (ksc)	Cement (kg/m³)	Fly Ash (kg/m³)	Water (kg/m³)	Sand (kg/m³)	Recycled Aggregate 3/4" (kg/m³)	Recommended Quantity Admixture (Liters/m³)	Fiber (%)
	380	460	0	165	800	1,120	3.10	–
Properties	Average		Standard deviation	Compressive strength (ksc)			Average	Standard deviation
Air content (%)[1]	1.6		0.3	7 days			264.34	29.31
Slump (mm)[2]	87.0		14.1	28 days			394.37	12.05
Initial setting time (min)[3]	228		4.2	90 days			426.71	3.25
Final setting time (min)[3]	325		45.7	365 days			517.54	4.66

Mix Symbol (HSC450C495F0S125)	Cylindrical Strength (ksc)	Cement (kg/m³)	Fly Ash (kg/m³)	Water (kg/m³)	Sand (kg/m³)	Recycled Aggregate 3/4" (kg/m³)	Recommended Quantity Admixture (Liters/m³)	Fiber (%)
	350	334	0	165	810	1,135	4.00	–
Properties	Average		Standard deviation	Compressive strength (ksc)			Average	Standard deviation
Air content (%)[1]	2.2		0.1	7 days			252.64	13.34
Slump (mm)[2]	74.0		18.3	28 days			469.74	25.92

(Continued)

TABLE 5.4 (Continued)
Typical Mix Proportions and Related Properties of RAC

Mix Symbol (HSC450C495F0S125)	Cylindrical Strength (ksc)	Cement (kg/m³)	Fly Ash (kg/m³)	Water (kg/m³)	Sand (kg/m³)	Recycled Aggregate 3/4" (kg/m³)	Recommended Quantity Admixture (Liters/m³)	Fiber (%)
Initial setting time (min)[3]	185		18.5	90 days			525.27	22.56
Final setting time (min)[3]	400		17.8	365 days			620.81	13.29

Mix Symbol (HSC450C495F0S125)	Cylindrical Strength (ksc)	Cement (kg/m³)	Fly Ash (kg/m³)	Water (kg/m³)	Sand (kg/m³)	Recycled Aggregate 3/4" (kg/m³)	Recommended Quantity Admixture (Liters/m³)	Fiber (%)
	350	270	174	155	810	1,135	4.00	
Properties	Average	Standard deviation		Compressive strength (ksc)			Average	Standard deviation
Air content (%)[1]	1.9		0.3	7 days			214.14	24.11
Slump (mm)[2]	81.0		5.5	28 days			308.64	20.74
Initial setting time (min)[3]	136		8.2	90 days			354.87	9.45
Final setting time (min)[3]	288		29.7	365 days			369.14	4.86

Mix Symbol (HSC450C395F100S125)	Cylindrical Strength (ksc)	Cement (kg/m³)	Fly Ash (kg/m³)	Water (kg/m³)	Sand (kg/m³)	Recycled Aggregate 3/4" (kg/m³)	Recommended Quantity Admixture (Liters/m³)	Fiber (%)
	320	256	164	155	820	1,100	4.00	–
Properties	Average		Standard deviation		Compressive strength (ksc)		Average	Standard deviation
Air content (%)[1]	2.6		0.2		7 days		264.46	9.25
Slump (mm)[2]	97.5		3.5		28 days		412.98	14.52
Initial setting time (min)[3]	184		7.1		90 days		489.17	3.67
Final setting time (min)[3]	295		33.9		365 days		562.57	2.11

Mix Symbol (HSC450C395F100S125)	Cylindrical Strength (ksc)	Cement (kg/m³)	Fly Ash (kg/m³)	Water (kg/m³)	Sand (kg/m³)	Recycled Aggregate 3/4" (kg/m³)	Recommended Quantity Admixture (Liters/m³)	Fiber (%)
	320	264	170	155	820	1,100	4.00	–
Properties	Average		Standard deviation		Compressive strength (ksc)		Average	Standard deviation
Air content (%)[1]	3.0		0.1		7 days		298.32	7.51
Slump (mm)[2]	88.0		7.9		28 days		307.29	2.96

(Continued)

TABLE 5.4 (Continued)
Typical Mix Proportions and Related Properties of RAC

Mix Symbol (HSC450C395F100S125)	Cylindrical Strength (ksc)	Cement (kg/m³)	Fly Ash (kg/m³)	Water (kg/m³)	Sand (kg/m³)	Recycled Aggregate 3/4" (kg/m³)	Recommended Quantity Admixture (Liters/m³)	Fiber (%)
Initial setting time (min)[3]	125		15.4	90 days			473.61	0.73
Final setting time (min)[3]	345		1.1	365 days			547.37	0.54
Mix Symbol (HSC450C495F0S150)	**Cylindrical Strength (ksc)**	**Cement (kg/m³)**	**Fly Ash (kg/m³)**	**Water (kg/m³)**	**Sand (kg/m³)**	**Recycled Aggregate 3/4" (kg/m³)**	**Recommended Quantity Admixture (Liters/m³)**	**Fiber (%)**
Properties	320	244	156	155	820	1,100	5.00	–
Properties	Average	Standard deviation			Compressive strength (ksc)		Average	Standard deviation
Air content (%)[1]	3.2	0.3		7 days			203.67	11.27
Slump (mm)[2]	98.5	23.1		28 days			284.83	9.03
Initial setting time (min)[3]	139	12.2		90 days			321.41	8.44
Final setting time (min)[3]	289	3.5		365 days			376.54	5.72

Mix Symbol (HSC450C495F0S150)	Cylindrical Strength (ksc)	Cement (kg/m³)	Fly Ash (kg/m³)	Water (kg/m³)	Sand (kg/m³)	Recycled Aggregate 3/4" (kg/m³)	Recommended Quantity Admixture (Liters/m³)	Fiber (%)
Properties	320	310	74	155	820	1,100	5.00	–
	Average		Standard deviation	Compressive strength (ksc)			Average	Standard deviation
Air content (%)[1]	2.2		0.1	7 days			261.13	11.08
Slump (mm)[2]	95.0		11.3	28 days			425.69	3.41
Initial setting time (min)[3]	183		9.9	90 days			470.98	5.58
Final setting time (min)[3]	394		33.4	365 days			509.67	9.82

Mix Symbol (HSC450C395F100S150)	Cylindrical Strength (ksc)	Cement (kg/m³)	Fly Ash (kg/m³)	Water (kg/m³)	Sand (kg/m³)	Recycled Aggregate 3/4" (kg/m³)	Recommended Quantity Admixture (Liters/m³)	Fiber (%)
Properties	350	280	180	155	810	1,135	5.00	–
	Average		Standard deviation	Compressive strength (ksc)			Average	Standard deviation
Air content (%)[1]	3.4		0.3	7 days			181.56	6.90
Slump (mm)[2]	97.0		7.9	28 days			282.54	7.39

(Continued)

TABLE 5.4 (Continued)
Typical Mix Proportions and Related Properties of RAC

Mix Symbol (HSC450C395F100S150)	Cylindrical Strength (ksc)	Cement (kg/m³)	Fly Ash (kg/m³)	Water (kg/m³)	Sand (kg/m³)	Recycled Aggregate 3/4" (kg/m³)	Recommended Quantity Admixture (Liters/m³)	Fiber (%)
Initial setting time (min)[3]	135	16.5		90 days			315.63	14.23
Final setting time (min)[3]	287	37.7		365 days			381.52	5.74
Mix Symbol (HSC450C395F100S150)	**Cylindrical Strength (ksc)**	**Cement (kg/m³)**	**Fly Ash (kg/m³)**	**Water (kg/m³)**	**Sand (kg/m³)**	**Recycled Aggregate 3/4" (kg/m³)**	**Recommended Quantity Admixture (Liters/m³)**	**Fiber (%)**
Properties	350	256	166	155	810	1,135	5.00	–
Properties	Average	Standard deviation		Compressive strength (ksc)			Average	Standard deviation
Air content (%)[1]	3.6	0.2		7 days			213.87	13.75
Slump (mm)[2]	95.5	22.8		28 days			288.55	9.14
Initial setting time (min)[3]	132	12.8		90 days			435.74	5.23
Final setting time (min)[3]	398	7.8		365 days			558.98	1.46

Mix Symbol (HSC450C395F100S150)	Cylindrical Strength (ksc)	Cement (kg/m³)	Fly Ash (kg/m³)	Water (kg/m³)	Sand (kg/m³)	Recycled Aggregate 3/4" (kg/m³)	Recommended Quantity Admixture (Liters/m³)	Fiber (%)
	380	294	190	155	800	1,120	4.00	–
Properties	Average		Standard deviation	Compressive strength (ksc)			Average	Standard deviation
Air content (%)[1]	3.3		0.2	7 days			260.03	3.74
Slump (mm)[2]	98.5		7.9	28 days			318.38	2.98
Initial setting time (min)[3]	214		5.2	90 days			476.02	1.56
Final setting time (min)[3]	359		4.1	365 days			504.11	0.98

Mix Symbol (HSC450C395F100S150)	Cylindrical Strength (ksc)	Cement (kg/m³)	Fly Ash (kg/m³)	Water (kg/m³)	Sand (kg/m³)	Recycled Aggregate 3/4" (kg/m³)	Recommended Quantity Admixture (Liters/m³)	Fiber (%)
	380	274	176	155	800	1,120	4.00	–
Properties	Average		Standard deviation	Compressive strength (ksc)			Average	Standard deviation
Air content (%)[1]	3.5		0.1	7 days			335.41	17.52
Slump (mm)[2]	73.5		9.4	28 days			369.17	24.36

(Continued)

TABLE 5.4 (Continued)
Typical Mix Proportions and Related Properties of RAC

Mix Symbol (HSC450C395F100S150)	Cylindrical Strength (ksc)	Cement (kg/m³)	Fly Ash (kg/m³)	Water (kg/m³)	Sand (kg/m³)	Recycled Aggregate 3/4" (kg/m³)	Recommended Quantity Admixture (Liters/m³)	Fiber (%)
Initial setting time (min)[3]		242	4.7	90 days			550.96	13.52
Final setting time (min)[3]		381	1.5	365 days			611.24	7.96

Mix Symbol (HSC450C495F0S150)	Cylindrical Strength (ksc)	Cement (kg/m³)	Fly Ash (kg/m³)	Water (kg/m³)	Sand (kg/m³)	Recycled Aggregate 3/4" (kg/m³)	Recommended Quantity Admixture (Liters/m³)	Fiber (%)
	350	258	166	155	810	1,135	6.20	–

Properties	Average	Standard deviation		Compressive strength (ksc)	Average	Standard deviation
Air content (%)[1]	3.7	0.3		7 days	304.72	22.27
Slump (mm)[2]	74.0	8.4		28 days	331.25	14.98
Initial setting time (min)[3]	198	3.9		90 days	470.04	3.15
Final setting time (min)[3]	276	1.9		365 days	278.55	2.01

Mix Symbol (HSC450C495F0S150)	Cylindrical Strength (ksc)	Cement (kg/m³)	Fly Ash (kg/m³)	Water (kg/m³)	Sand (kg/m³)	Recycled Aggregate 3/4" (kg/m³)	Recommended Quantity Admixture (Liters/m³)	Fiber (%)
	350	330	80	155	810	1,135	6.20	—
Properties	Average		Standard deviation	Compressive strength (ksc)		Average		Standard deviation
Air content (%)[1]	3.9		0.3	7 days		341.15		42.88
Slump (mm)[2]	84.0		11.1	28 days		502.87		36.98
Initial setting time (min)[3]	187		17.4	90 days		578.41		23.02
Final setting time (min)[3]	319		2.4	365 days		625.22		11.47

Mix Symbol (HSC450C469F26S150)	Cylindrical Strength (ksc)	Cement (kg/m³)	Fly Ash (kg/m³)	Water (kg/m³)	Sand (kg/m³)	Recycled Aggregate 3/4" (kg/m³)	Recommended Quantity Admixture (Liters/m³)	Fiber (%)
	380	460	0	165	800	1,120	6.20	—
Properties	Average		Standard deviation	Compressive strength (ksc)		Average		Standard deviation
Air content (%)[1]	1.3		0.2	7 days		204.35		24.86
Slump (mm)[2]	68.0		21.5	28 days		215.87		22.41

(Continued)

TABLE 5.4 (Continued)
Typical Mix Proportions and Related Properties of RAC

Mix Symbol (HSC450C469F26S150)	Cylindrical Strength (ksc)	Cement (kg/m³)	Fly Ash (kg/m³)	Water (kg/m³)	Sand (kg/m³)	Recycled Aggregate 3/4" (kg/m³)	Recommended Quantity Admixture (Liters/m³)	Fiber (%)
Initial setting time (min)[3]	117		14.2	90 days			305.06	13.51
Final setting time (min)[3]	271		38.9	365 days			356.61	4.74

Mix Symbol (HSC450C469F26S150)	Cylindrical Strength (ksc)	Cement (kg/m³)	Fly Ash (kg/m³)	Water (kg/m³)	Sand (kg/m³)	Recycled Aggregate 3/4" (kg/m³)	Recommended Quantity Admixture (Liters/m³)	Fiber (%)
	380	186	184	155	800	1,120	6.20	–

Properties	Average		Standard deviation	Compressive strength (ksc)			Average	Standard deviation
Air content (%)[1]	1.1		0.2	7 days			221.52	26.48
Slump (mm)[2]	72.5		14.1	28 days			357.56	21.66
Initial setting time (min)[3]	145		19.8	90 days			411.69	13.23
Final setting time (min)[3]	347		7.6	365 days			484.36	7.04

Mix Symbol (HSC450C445F50S150)	Cylindrical Strength (ksc)	Cement (kg/m³)	Fly Ash (kg/m³)	Water (kg/m³)	Sand (kg/m³)	Recycled Aggregate 3/4" (kg/m³)	Recommended Quantity Admixture (Liters/m³)	Fiber (%)
Properties	350	279	30	155	820	1,120	6.00	—

Properties	Average	Standard deviation	Compressive strength (ksc)	Average	Standard deviation
Air content (%)[1]	1.2	0.1	7 days	200.29	13.62
Slump (mm)[2]	80.5	7.2	28 days	210.81	16.87
Initial setting time (min)[3]	152	11.4	90 days	306.34	7.11
Final setting time (min)[3]	339	26.9	365 days	357.48	2.68

Mix Symbol (HSC450C445F50S150)	Cylindrical Strength (ksc)	Cement (kg/m³)	Fly Ash (kg/m³)	Water (kg/m³)	Sand (kg/m³)	Recycled Aggregate 3/4" (kg/m³)	Recommended Quantity Admixture (Liters/m³)	Fiber (%)
Properties	350	259	35	155	820	1,120	6.00	—

Properties	Average	Standard deviation	Compressive strength (ksc)	Average	Standard deviation
Air content (%)[1]	1.6	0.2	7 days	195.81	24.02
Slump (mm)[2]	73.0	5.9	28 days	216.96	23.68

(Continued)

TABLE 5.4 (Continued)
Typical Mix Proportions and Related Properties of RAC

Mix Symbol (HSC450C445F50S150)	Cylindrical Strength (ksc)	Cement (kg/m³)	Fly Ash (kg/m³)	Water (kg/m³)	Sand (kg/m³)	Recycled Aggregate 3/4" (kg/m³)	Recommended Quantity Admixture (Liters/m³)	Fiber (%)
Initial setting time (min)[3]	115		20.0		90 days		305.14	13.02
Final setting time (min)[3]	297		25.0		365 days		371.51	7.58
Mix Symbol (HSC450C419F76S150)	Cylindrical Strength (ksc)	Cement (kg/m³)	Fly Ash (kg/m³)	Water (kg/m³)	Sand (kg/m³)	Recycled Aggregate 3/4" (kg/m³)	Recommended Quantity Admixture (Liters/m³)	Fiber (%)
Properties	350	335	45	155	820	1,120	6.00	–
	Average		Standard deviation		Compressive strength (ksc)		Average	Standard deviation
Air content (%)[1]	1.0		0.2		7 days		209.03	24.89
Slump (mm)[2]	78.5		5.6		28 days		224.63	21.55
Initial setting time (min)[3]	239		0.2		90 days		329.27	14.63
Final setting time (min)[3]	318		23.7		365 days		382.39	5.49

Mix Symbol (HSC450C419F76S150)	Cylindrical Strength (ksc)	Cement (kg/m³)	Fly Ash (kg/m³)	Water (kg/m³)	Sand (kg/m³)	Recycled Aggregate 3/4" (kg/m³)	Recommended Quantity Admixture (Liters/m³)	Fiber (%)
	380	469	50	165	810	1,105	6.00	–
Properties	Average		Standard deviation	Compressive strength (ksc)			Average	Standard deviation
Air content (%)[1]	1.7		0.3	7 days			310.49	7.84
Slump (mm)[2]	75.5		12.1	28 days			344.68	10.32
Initial setting time (min)[3]	125		9.8	90 days			467.27	1.69
Final setting time (min)[3]	327		12.9	365 days			582.41	2.34

Mix Symbol (HSC450C395F100S150)	Cylindrical Strength (ksc)	Cement (kg/m³)	Fly Ash (kg/m³)	Water (kg/m³)	Sand (kg/m³)	Recycled Aggregate 3/4" (kg/m³)	Recommended Quantity Admixture (Liters/m³)	Fiber (%)
	320	243	90	155	830	1,085	6.00	–
Properties	Average		Standard deviation	Compressive strength (ksc)			Average	Standard deviation
Air content (%)[1]	3.3		0.2	7 days			217.48	2.88
Slump (mm)[2]	87.5		21.7	28 days			286.27	0.47

(Continued)

TABLE 5.4 (Continued)
Typical Mix Proportions and Related Properties of RAC

Mix Symbol (HSC450C395F100S150)	Cylindrical Strength (ksc)	Cement (kg/m³)	Fly Ash (kg/m³)	Water (kg/m³)	Sand (kg/m³)	Recycled Aggregate 3/4" (kg/m³)	Recommended Quantity Admixture (Liters/m³)	Fiber (%)
Initial setting time (min)[3]	167		6.4	90 days			364.87	13.06
Final setting time (min)[3]	329		3.6	365 days			412.61	8.77
Mix Symbol (HSC450C395F100S150)	**Cylindrical Strength (ksc)**	**Cement (kg/m³)**	**Fly Ash (kg/m³)**	**Water (kg/m³)**	**Sand (kg/m³)**	**Recycled Aggregate 3/4" (kg/m³)**	**Recommended Quantity Admixture (Liters/m³)**	**Fiber (%)**
Properties	320	315	80	155	830	1,085	6.00	–
	Average		Standard deviation	Compressive strength (ksc)			Average	Standard deviation
Air content (%)[1]	1.3		0.1	7 days			257.55	12.91
Slump (mm)[2]	73.5		18.1	28 days			321.28	11.28
Initial setting time (min)[3]	185		8.3	90 days			427.63	9.34
Final setting time (min)[3]	257		35.3	365 days			512.81	7.22

Mix Symbol (HSC450C495F0S150)	Cylindrical Strength (ksc)	Cement (kg/m³)	Fly Ash (kg/m³)	Water (kg/m³)	Sand (kg/m³)	Recycled Aggregate 3/4" (kg/m³)	Recommended Quantity Admixture (Liters/m³)	Fiber (%)
Properties	350	445	10	165	820	1,120	6.00	–
	Average		Standard deviation	Compressive strength (ksc)			Average	Standard deviation
Air content (%)[1]	3.2		0.3	7 days			221.36	28.73
Slump (mm)[2]	88.5		14.8	28 days			249.62	24.88
Initial setting time (min)[3]	236		6.7	90 days			396.21	4.99
Final setting time (min)[3]	319		19.5	365 days			417.58	5.31

Mix Symbol (HSC450C495F0S150)	Cylindrical Strength (ksc)	Cement (kg/m³)	Fly Ash (kg/m³)	Water (kg/m³)	Sand (kg/m³)	Recycled Aggregate 3/4" (kg/m³)	Recommended Quantity Admixture (Liters/m³)	Fiber (%)
Properties	350	271	10	155	820	1,120	6.00	–
	Average		Standard deviation	Compressive strength (ksc)			Average	Standard deviation
Air content (%)[1]	1.5		0.1	7 days			261.94	38.54
Slump (mm)[2]	72.0		3.6	28 days			338.57	22.65

(Continued)

TABLE 5.4 (Continued)
Typical Mix Proportions and Related Properties of RAC

Mix Symbol (HSC450C495F0S150)	Cylindrical Strength (ksc)	Cement (kg/m³)	Fly Ash (kg/m³)	Water (kg/m³)	Sand (kg/m³)	Recycled Aggregate 3/4" (kg/m³)	Recommended Quantity Admixture (Liters/m³)	Fiber (%)
Initial setting time (min)[3]		105	19.2		90 days		437.14	17.74
Final setting time (min)[3]		287	42.7		365 days		492.36	3.41
Mix Symbol (HSC450C469F26S150)	Cylindrical Strength -(ksc)	Cement (kg/m³)	Fly Ash (kg/m³)	Water (kg/m³)	Sand (kg/m³)	Recycled Aggregate 3/4" (kg/m³)	Recommended Quantity Admixture (Liters/m³)	Fiber (%)
Properties	380	285	194	155	810	1,105	6.00	–
	Average		Standard deviation		Compressive strength (ksc)		Average	Standard deviation
Air content (%)[1]	1.6		0.3		7 days		203.31	13.54
Slump (mm)[2]	84.5		20.7		28 days		217.89	10.05
Initial setting time (min)[3]	246		14.1		90 days		297.84	3.59
Final setting time (min)[3]	352		25.4		365 days		340.88	4.88

Mix Symbol (HSC450C469F26S150)	Cylindrical Strength (ksc)	Cement (kg/m³)	Fly Ash (kg/m³)	Water (kg/m³)	Sand (kg/m³)	Recycled Aggregate 3/4" (kg/m³)	Recommended Quantity Admixture (Liters/m³)	Fiber (%)
Properties	380	295	200	155	810	1,105	6.00	–
	Average		Standard deviation		Compressive strength (ksc)		Average	Standard deviation
Air content (%)[1]	1.9		0.1		7 days		224.35	24.74
Slump (mm)[2]	60.5		12.4		28 days		235.87	23.63
Initial setting time (min)[3]	240		10.4		90 days		358.39	13.51
Final setting time (min)[3]	342		35.5		365 days		377.64	4.74

Mix Symbol (HSC450C445F50S150)	Cylindrical Strength (ksc)	Cement (kg/m³)	Fly Ash (kg/m³)	Water (kg/m³)	Sand (kg/m³)	Recycled Aggregate 3/4" (kg/m³)	Recommended Quantity Admixture (Liters/m³)	Fiber (%)
Properties	380	273	186	155	810	1,105	6.00	–
	Average		Standard deviation		Compressive strength (ksc)		Average	Standard deviation
Air content (%)[1]	3.5		0.1		7 days		220.29	19.86
Slump (mm)[2]	96.5		10.9		28 days		230.81	14.33

(Continued)

TABLE 5.4 (Continued)
Typical Mix Proportions and Related Properties of RAC

Mix Symbol (HSC450C445F50S150)	Cylindrical Strength (ksc)	Cement (kg/m³)	Fly Ash (kg/m³)	Water (kg/m³)	Sand (kg/m³)	Recycled Aggregate 3/4" (kg/m³)	Recommended Quantity Admixture (Liters/m³)	Fiber (%)
Initial setting time (min)[3]		216	7.5		90 days		325.62	13.02
Final setting time (min)[3]		279	13.6		365 days		371.44	7.14

Mix Symbol (HSC450C445F50S150)	Cylindrical Strength (ksc)	Cement (kg/m³)	Fly Ash (kg/m³)	Water (kg/m³)	Sand (kg/m³)	Recycled Aggregate 3/4" (kg/m³)	Recommended Quantity Admixture (Liters/m³)	Fiber (%)
	380	355	90	155	810	1,105	6.00	–

Properties	Average	Standard deviation		Compressive strength (ksc)	Average	Standard deviation
Air content (%)[1]	1.1	0.3		7 days	225.43	28.73
Slump (mm)[2]	69.5	17.8		28 days	286.93	24.88
Initial setting time (min)[3]	224	0.9		90 days	416.21	4.99
Final setting time (min)[3]	387	30.6		365 days	437.58	5.31

Mix Symbol (HSC450C419F76S150)	Cylindrical Strength (ksc)	Cement (kg/m³)	Fly Ash (kg/m³)	Water (kg/m³)	Sand (kg/m³)	Recycled Aggregate 3/4" (kg/m³)	Recommended Quantity Admixture (Liters/m³)	Fiber (%)
Properties	400	301	204	155	810	1,105	6.00	–
	Average		Standard deviation	Compressive strength (ksc)			Average	Standard deviation
Air content (%)[1]	3.8		0.1	7 days			229.03	30.01
Slump (mm)[2]	85.0		11.2	28 days			249.58	14.55
Initial setting time (min)[3]	187		5.7	90 days			349.20	8.74
Final setting time (min)[3]	299		21.2	365 days			402.39	2.63

Mix Symbol (HSC450C419F76S150)	Cylindrical Strength (ksc)	Cement (kg/m³)	Fly Ash (kg/m³)	Water (kg/m³)	Sand (kg/m³)	Recycled Aggregate 3/4" (kg/m³)	Recommended Quantity Admixture (Liters/m³)	Fiber (%)
Properties	400	395	100	155	810	1,105	6.00	–
	Average		Standard deviation	Compressive strength (ksc)			Average	Standard deviation
Air content (%)[1]	3.1		0.2	7 days			237.48	2.88
Slump (mm)[2]	76.0		7.1	28 days			306.26	0.47

(Continued)

TABLE 5.4 (Continued)
Typical Mix Proportions and Related Properties of RAC

Mix Symbol (HSC450C419F76S150)	Cylindrical Strength (ksc)	Cement (kg/m³)	Fly Ash (kg/m³)	Water (kg/m³)	Sand (kg/m³)	Recycled Aggregate 3/4" (kg/m³)	Recommended Quantity Admixture (Liters/m³)	Fiber (%)
Initial setting time (min)[3]	226		9.2		90 days		409.83	13.06
Final setting time (min)[3]	309		14.8		365 days		433.65	8.77
Mix Symbol (HSC450C395F100S150)	Cylindrical Strength (ksc)	Cement (kg/m³)	Fly Ash (kg/m³)	Water (kg/m³)	Sand (kg/m³)	Recycled Aggregate 3/4" (kg/m³)	Recommended Quantity Admixture (Liters/m³)	Fiber (%)
	400	371	124	155	810	1,105	6.00	–
Properties	Average		Standard deviation	Compressive strength (ksc)			Average	Standard deviation
Air content (%)[1]	3.7		0.3		7 days		213.67	14.82
Slump (mm)[2]	84.0		5.5		28 days		253.63	23.02
Initial setting time (min)[3]	197		2.4		90 days		369.59	14.63
Final setting time (min)[3]	269		15.8		365 days		394.82	5.49

Mix Symbol (HSC450C395F100S150)	Cylindrical Strength (ksc)	Cement (kg/m³)	Fly Ash (kg/m³)	Water (kg/m³)	Sand (kg/m³)	Recycled Aggregate 3/4" (kg/m³)	Recommended Quantity Admixture (Liters/m³)	Fiber (%)
	400	345	150	155	810	1,105	6.00	–
Properties	Average		Standard deviation	Compressive strength (ksc)			Average	Standard deviation
Air content (%)[1]	2.9		0.3	7 days			245.36	12.20
Slump (mm)[2]	93.0		18.5	28 days			269.62	11.01
Initial setting time (min)[3]	247		13.9	90 days			383.87	13.59
Final setting time (min)[3]	344		34.4	365 days			432.61	8.47

Mix Symbol (HSC450C469F0S150)	Cylindrical Strength (ksc)	Cement (kg/m³)	Fly Ash (kg/m³)	Water (kg/m³)	Sand (kg/m³)	Recycled Aggregate 3/4" (kg/m³)	Recommended Quantity Admixture (Liters/m³)	Fiber (%)
	320	370	0	150	815	1,115	6.10	–
Properties	Average		Standard deviation	Compressive strength (ksc)			Average	Standard deviation
Air content (%)[1]	3.6		0.2	7 days			211.47	24.89
Slump (mm)[2]	83.5		15.8	28 days			253.36	21.55

(Continued)

TABLE 5.4 (Continued)
Typical Mix Proportions and Related Properties of RAC

Mix Symbol (HSC450C469F0S150)	Cylindrical Strength (ksc)	Cement (kg/m³)	Fly Ash (kg/m³)	Water (kg/m³)	Sand (kg/m³)	Recycled Aggregate 3/4" (kg/m³)	Recommended Quantity Admixture (Liters/m³)	Fiber (%)
Initial setting time (min)[3]	246		10.1	90 days			348.47	7.47
Final setting time (min)[3]	289		9.8	365 days			389.22	4.58

Mix Symbol (HSC450C469F0S150)	Cylindrical Strength (ksc)	Cement (kg/m³)	Fly Ash (kg/m³)	Water (kg/m³)	Sand (kg/m³)	Recycled Aggregate 3/4" (kg/m³)	Recommended Quantity Admixture (Liters/m³)	Fiber (%)
Properties	350	420	0	150	805	1,150	6.10	–
	Average		Standard deviation	Compressive strength (ksc)		Average	Average	Standard deviation
Air content (%)[1]	3.0		0.1	7 days			241.52	29.86
Slump (mm)[2]	87.5		23.2	28 days			377.56	24.11
Initial setting time (min)[3]	244		9.7	90 days			415.26	13.23
Final setting time (min)[3]	349		33.3	365 days			479.62	7.04

Mix Symbol (HSC450C375F94S150)	Cylindrical Strength (ksc)	Cement (kg/m³)	Fly Ash (kg/m³)	Water (kg/m³)	Sand (kg/m³)	Recycled Aggregate 3/4" (kg/m³)	Recommended Quantity Admixture (Liters/m³)	Fiber (%)
	380	445	0	150	795	1,135	6.10	–
Properties	Average		Standard deviation	Compressive strength (ksc)			Average	Standard deviation
Air content (%)[1]	1.1		0.3	7 days			298.15	8.44
Slump (mm)[2]	85.5		17.8	28 days			374.26	2.54
Initial setting time (min)[3]	162		4.6	90 days			463.29	13.52
Final setting time (min)[3]	268		19.1	365 days			514.95	3.14

Mix Symbol (HSC450C375F94S150)	Cylindrical Strength (ksc)	Cement (kg/m³)	Fly Ash (kg/m³)	Water (kg/m³)	Sand (kg/m³)	Recycled Aggregate 3/4" (kg/m³)	Recommended Quantity Admixture (Liters/m³)	Fiber (%)
	400	470	0	150	790	1,140	6.10	–
Properties	Average		Standard deviation	Compressive strength (ksc)			Average	Standard deviation
Air content (%)[1]	2.7		0.1	7 days			207.14	4.62
Slump (mm)[2]	105.5		7.7	28 days			279.71	3.85

(Continued)

TABLE 5.4 (Continued)
Typical Mix Proportions and Related Properties of RAC

Mix Symbol (HSC450C375F94S150)	Cylindrical Strength (ksc)	Cement (kg/m³)	Fly Ash (kg/m³)	Water (kg/m³)	Sand (kg/m³)	Recycled Aggregate 3/4" (kg/m³)	Recommended Quantity Admixture (Liters/m³)	Fiber (%)
Initial setting time (min)[3]	163		12.9		90 days		330.08	5.33
Final setting time (min)[3]	270		15.2		365 days		397.61	1.07

Mix Symbol (HSC450C469F0S125)	Cylindrical Strength (ksc)	Cement (kg/m³)	Fly Ash (kg/m³)	Water (kg/m³)	Sand (kg/m³)	Recycled Aggregate 3/4" (kg/m³)	Recommended Quantity Admixture (Liters/m³)	Fiber (%)
Properties	380	355	90	145	795	1,135	4.50	–

Properties	Average	Standard deviation
Air content (%)[1]	1.2	0.1
Slump (mm)[2]	86.0	18.9
Initial setting time (min)[3]	172	3.7
Final setting time (min)[3]	269	20.0

Compressive strength (ksc)	Average	Standard deviation
7 days	214.56	24.10
28 days	296.25	20.36
90 days	356.24	4.77
365 days	373.64	5.63

Mix Symbol (HSC450C469F0S125)	Cylindrical Strength (ksc)	Cement (kg/m³)	Fly Ash (kg/m³)	Water (kg/m³)	Sand (kg/m³)	Recycled Aggregate 3/4" (kg/m³)	Recommended Quantity Admixture (Liters/m³)	Fiber (%)
	400	375	95	145	790	1,140	4.50	–
Properties	Average		Standard deviation	Compressive strength (ksc)			Average	Standard deviation
Air content (%)[1]	2.8		0.2	7 days			236.51	31.88
Slump (mm)[2]	98.0		8.8	28 days			249.32	24.09
Initial setting time (min)[3]	173		3.5	90 days			344.59	13.46
Final setting time (min)[3]	271		17.9	365 days			390.33	7.62

Mix Symbol (HSC450C375F945125)	Cylindrical Strength (ksc)	Cement (kg/m³)	Fly Ash (kg/m³)	Water (kg/m³)	Sand (kg/m³)	Recycled Aggregate 3/4" (kg/m³)	Recommended Quantity Admixture (Liters/m³)	Fiber (%)
	350	335	85	145	805	1,150	4.50	–
Properties	Average		Standard deviation	Compressive strength (ksc)			Average	Standard deviation
Air content (%)[1]	1.5		0.2	7 days			204.69	24.84
Slump (mm)[2]	90.0		11.2	28 days			236.54	20.13

(Continued)

TABLE 5.4 (Continued)
Typical Mix Proportions and Related Properties of RAC

Mix Symbol (HSC450C375F94S125)	Cylindrical Strength (ksc)	Cement (kg/m³)	Fly Ash (kg/m³)	Water (kg/m³)	Sand (kg/m³)	Recycled Aggregate 3/4″ (kg/m³)	Recommended Quantity Admixture (Liters/m³)	Fiber (%)
Initial setting time (min)[3]	102		10.0	90 days			326.10	5.78
Final setting time (min)[3]	266		12.3	365 days			379.36	6.89

Mix Symbol (HSC450C375F94S125)	Cylindrical Strength (ksc)	Cement (kg/m³)	Fly Ash (kg/m³)	Water (kg/m³)	Sand (kg/m³)	Recycled Aggregate 3/4″ (kg/m³)	Recommended Quantity Admixture (Liters/m³)	Fiber (%)
Properties	320	295	75	145	815	1,115	4.50	–
	Average		Standard deviation	Compressive strength (ksc)			Average	Standard deviation
Air content (%)[1]	2.9		0.3	7 days			204.81	27.26
Slump (mm)[2]	98.5		5.5	28 days			273.14	21.41
Initial setting time (min)[3]	143		8.3	90 days			326.96	13.59
Final setting time (min)[3]	272		20.5	365 days			342.40	7.62

Mix Symbol (HSC500C545F0S175)	Cylindrical Strength (ksc)	Cement (kg/m³)	Fly Ash (kg/m³)	Water (kg/m³)	Sand (kg/m³)	Recycled Aggregate 3/4″ (kg/m³)	Recommended Quantity Admixture (Liters/m³)	Fiber (%)
	400	351	119	145	790	1,140	8.50	–
Properties	Average		Standard deviation	Compressive strength (ksc)			Average	Standard deviation
Air content (%)[1]	2.0		0.2	7 days			207.61	38.66
Slump (mm)[2]	88.5		20.5	28 days			327.14	29.58
Initial setting time (min)[3]	192		1.9	90 days			428.84	18.58
Final setting time (min)[3]	265		16.4	365 days			499.44	11.02

Mix Symbol (HSC500C435F110S175)	Cylindrical Strength (ksc)	Cement (kg/m³)	Fly Ash (kg/m³)	Water (kg/m³)	Sand (kg/m³)	Recycled Aggregate 3/4″ (kg/m³)	Recommended Quantity Admixture (Liters/m³)	Fiber (%)
	350	293	127	165	736	1,059	8.50	–
Properties	Average		Standard deviation	Compressive strength (ksc)			Average	Standard deviation
Air content (%)[1]	3.7		0.1	7 days			254.26	0.94
Slump (mm)[2]	123.5		16.6	28 days			359.24	0.83

(*Continued*)

TABLE 5.4 (Continued)
Typical Mix Proportions and Related Properties of RAC

Mix Symbol (HSC500C435F110S175)	Cylindrical Strength (ksc)	Cement (kg/m³)	Fly Ash (kg/m³)	Water (kg/m³)	Sand (kg/m³)	Recycled Aggregate 3/4″ (kg/m³)	Recommended Quantity Admixture (Liters/m³)	Fiber (%)
Initial setting time (min)[3]		154	8.2	90 days			380.36	1.58
Final setting time (min)[3]		286	33.3	365 days			489.24	5.44
Mix Symbol (HSC500C545F0S175)	**Cylindrical Strength (ksc)**	**Cement (kg/m³)**	**Fly Ash (kg/m³)**	**Water (kg/m³)**	**Sand (kg/m³)**	**Recycled Aggregate 3/4″ (kg/m³)**	**Recommended Quantity Admixture (Liters/m³)**	**Fiber (%)**
	380	310	135	165	727	1,045	8.50	–
Properties		Average	Standard deviation	Compressive strength (ksc)			Average	Standard deviation
Air content (%)[1]		2.4	0.3	7 days			257.25	27.30
Slump (mm)[2]		96.5	4.4	28 days			348.95	22.45
Initial setting time (min)[3]		133	15.7	90 days			384.55	16.45
Final setting time (min)[3]		277	24.4	365 days			464.18	6.65

Mix Symbol (HSC500C435F110S175)	Cylindrical Strength (ksc)	Cement (kg/m³)	Fly Ash (kg/m³)	Water (kg/m³)	Sand (kg/m³)	Recycled Aggregate 3/4" (kg/m³)	Recommended Quantity Admixture (Liters/m³)	Fiber (%)
	320	275	120	165	764	1,054	8.50	–
Properties	Average		Standard deviation	Compressive strength (ksc)			Average	Standard deviation
Air content (%)[1]	3.6		0.2	7 days			234.85	34.74
Slump (mm)[2]	111.5		12.2	28 days			349.84	16.25
Initial setting time (min)[3]	114		17.7	90 days			426.87	10.62
Final setting time (min)[3]	289		47.9	365 days			510.03	11.04

Mix Symbol (HSC350C395F0S100)	Cylindrical Strength (ksc)	Cement (kg/m³)	Fly Ash (kg/m³)	Water (kg/m³)	Sand (kg/m³)	Recycled Aggregate 3/4" (kg/m³)	Recommended Quantity Admixture (Liters/m³)	Fiber (%)
	280	283	96	150	825	1,130	3.90	–
Properties	Average		Standard deviation	Compressive strength (ksc)			Average	Standard deviation
Air content (%)[1]	3.9		0.2	7 days			134.89	14.76
Slump (mm)[2]	58.0		6.5	28 days			234.14	12.63

(*Continued*)

TABLE 5.4 (Continued)
Typical Mix Proportions and Related Properties of RAC

Mix Symbol (HSC350C395F0S100)	Cylindrical Strength (ksc)	Cement (kg/m³)	Fly Ash (kg/m³)	Water (kg/m³)	Sand (kg/m³)	Recycled Aggregate 3/4" (kg/m³)	Recommended Quantity Admixture (Liters/m³)	Fiber (%)
Initial setting time (min)[3]	142		19.8		90 days		276.03	8.36
Final setting time (min)[3]	330		19.4		365 days		312.58	6.01

Mix Symbol (HSC350C315F80S100)	Cylindrical Strength (ksc)	Cement (kg/m³)	Fly Ash (kg/m³)	Water (kg/m³)	Sand (kg/m³)	Recycled Aggregate 3/4" (kg/m³)	Recommended Quantity Admixture (Liters/m³)	Fiber (%)
	280	263	116	150	825	1,130	3.90	–
Properties	Average		Standard deviation		Compressive strength (ksc)		Average	Standard deviation
Air content (%)[1]	1.4		0.3		7 days		183.31	13.54
Slump (mm)[2]	60.5		12.7		28 days		197.89	10.05
Initial setting time (min)[3]	182		9.5		90 days		277.84	3.69
Final setting time (min)[3]	310		29.7		365 days		320.14	4.58

Mix Symbol (HSC350C395F0S100)	Cylindrical Strength (ksc)	Cement (kg/m³)	Fly Ash (kg/m³)	Water (kg/m³)	Sand (kg/m³)	Recycled Aggregate 3/4" (kg/m³)	Recommended Quantity Admixture (Liters/m³)	Fiber (%)
Properties	280	303	76	150	825	1,130	3.90	–
Properties	Average		Standard deviation	Compressive strength (ksc)			Average	Standard deviation
Air content (%)[1]	1.5		0.3	7 days			147.96	20.18
Slump (mm)[2]	58.5		19.3	28 days			222.41	10.06
Initial setting time (min)[3]	152		17.4	90 days			268.13	7.58
Final setting time (min)[3]	335		25.4	365 days			285.73	4.21
Mix Symbol (HSC350C315F80S100)	Cylindrical Strength (ksc)	Cement (kg/m³)	Fly Ash (kg/m³)	Water (kg/m³)	Sand (kg/m³)	Recycled Aggregate 3/4" (kg/m³)	Recommended Quantity Admixture (Liters/m³)	Fiber (%)
Properties	280	225	157	150	825	1,130	3.90	
Properties	Average		Standard deviation	Compressive strength (ksc)			Average	Standard deviation
Air content (%)[1]	2.8		0.1	7 days			184.88	24.89
Slump (mm)[2]	80.5		4.8	28 days			203.69	10.36

(Continued)

TABLE 5.4 (Continued)
Typical Mix Proportions and Related Properties of RAC

Mix Symbol (HSC350C315F80S100)	Cylindrical Strength (ksc)	Cement (kg/m³)	Fly Ash (kg/m³)	Water (kg/m³)	Sand (kg/m³)	Recycled Aggregate 3/4" (kg/m³)	Recommended Quantity Admixture (Liters/m³)	Fiber (%)
Initial setting time (min)[3]	205		19.4		90 days		286.14	10.59
Final setting time (min)[3]	391		15.7		365 days		321.77	7.87

Mix Symbol (HSC240C295F0S100)	Cylindrical Strength (ksc)	Cement (kg/m³)	Fly Ash (kg/m³)	Water (kg/m³)	Sand (kg/m³)	Recycled Aggregate 3/4" (kg/m³)	Recommended Quantity Admixture (Liters/m³)	Fiber (%)
	210	320	10	160	855	1,105	2.90	–

Properties	Average	Standard deviation
Air content (%)[1]	3.4	0.2
Slump (mm)[2]	81.5	5.1
Initial setting time (min)[3]	243	8.2
Final setting time (min)[3]	373	36.9

Compressive strength (ksc)	Average	Standard deviation
7 days	127.52	4.98
28 days	240.47	2.84
90 days	257.05	1.31
365 days	291.91	0.88

Mix Symbol (HSC240C235F60S100)	Cylindrical Strength (ksc)	Cement (kg/m³)	Fly Ash (kg/m³)	Water (kg/m³)	Sand (kg/m³)	Recycled Aggregate 3/4" (kg/m³)	Recommended Quantity Admixture (Liters/m³)	Fiber (%)
	210	395	10	165	780	1,086	2.90	—
Properties	Average		Standard deviation	Compressive strength (ksc)			Average	Standard deviation
Air content (%)[1]	3.9		0.2	7 days			148.24	32.93
Slump (mm)[2]	93.0		2.3	28 days			282.26	24.74
Initial setting time (min)[3]	112		3.2	90 days			336.84	10.33
Final setting time (min)[3]	324		2.4	365 days			346.43	5.47

Mix Symbol (HSC240C295F0S100)	Cylindrical Strength (ksc)	Cement (kg/m³)	Fly Ash (kg/m³)	Water (kg/m³)	Sand (kg/m³)	Recycled Aggregate 3/4" (kg/m³)	Recommended Quantity Admixture (Liters/m³)	Fiber (%)
	210	235	50	150	865	1,050	2.90	—
Properties	Average		Standard deviation	Compressive strength (ksc)			Average	Standard deviation
Air content (%)[1]	3.1		0.1	7 days			181.93	14.62
Slump (mm)[2]	83.5		9.2	28 days			243.25	21.76

(Continued)

TABLE 5.4 (Continued)
Typical Mix Proportions and Related Properties of RAC

Mix Symbol (HSC240C295F0S100)	Cylindrical Strength (ksc)	Cement (kg/m³)	Fly Ash (kg/m³)	Water (kg/m³)	Sand (kg/m³)	Recycled Aggregate 3/4" (kg/m³)	Recommended Quantity Admixture (Liters/m³)	Fiber (%)
Initial setting time (min)[3]	116		18.4		90 days		311.47	6.92
Final setting time (min)[3]	318		7.0		365 days		347.61	4.65
Mix Symbol (HSC240C235F60S100)	**Cylindrical Strength (ksc)**	**Cement (kg/m³)**	**Fly Ash (kg/m³)**	**Water (kg/m³)**	**Sand (kg/m³)**	**Recycled Aggregate 3/4" (kg/m³)**	**Recommended Quantity Admixture (Liters/m³)**	**Fiber (%)**
	210	255	50	165	806	1,000	2.90	–
Properties	Average		Standard deviation		Compressive strength (ksc)		Average	Standard deviation
Air content (%)[1]	1.4		0.3		7 days		191.47	22.14
Slump (mm)[2]	95.0		5.7		28 days		229.58	10.65
Initial setting time (min)[3]	239		1.4		90 days		326.96	6.26
Final setting time (min)[3]	355		14.3		365 days		382.15	4.87

Mix Symbol (HSC700C545F0S175)	Cylindrical Strength (ksc)	Cement (kg/m³)	Fly Ash (kg/m³)	Water (kg/m³)	Sand (kg/m³)	Recycled Aggregate 3/4" (kg/m³)	Recommended Quantity Admixture (Liters/m³)	Fiber (%)
	350	314	106	145	805	1,150	9.50	—
Properties	Average		Standard deviation	Compressive strength (ksc)		Average		Standard deviation
Air content (%)[1]	3.8		0.1	7 days		201.74		21.78
Slump (mm)[2]	92.5		22.4	28 days		253.06		13.32
Initial setting time (min)[3]	102		14.2	90 days		291.44		7.41
Final setting time (min)[3]	257		5.4	365 days		336.32		2.24

Mix Symbol (HSC700C435F110S175)	Cylindrical Strength (ksc)	Cement (kg/m³)	Fly Ash (kg/m³)	Water (kg/m³)	Sand (kg/m³)	Recycled Aggregate 3/4" (kg/m³)	Recommended Quantity Admixture (Liters/m³)	Fiber (%)
	380	333	112	145	795	1,135	9.50	—
Properties	Average		Standard deviation	Compressive strength (ksc)		Average		Standard deviation
Air content (%)[1]	1.3		0.2	7 days		288.88		7.41
Slump (mm)[2]	94.5		14.2	28 days		387.12		6.39

(*Continued*)

TABLE 5.4 (Continued)
Typical Mix Proportions and Related Properties of RAC

Mix Symbol (HSC700C435F110S175)	Cylindrical Strength (ksc)	Cement (kg/m³)	Fly Ash (kg/m³)	Water (kg/m³)	Sand (kg/m³)	Recycled Aggregate 3/4" (kg/m³)	Recommended Quantity Admixture (Liters/m³)	Fiber (%)
Initial setting time (min)[3]	216		14.6	90 days			407.93	5.48
Final setting time (min)[3]	279		41.7	365 days			519.28	4.22

Mix Symbol (HSC700C545F0S175)	Cylindrical Strength (ksc)	Cement (kg/m³)	Fly Ash (kg/m³)	Water (kg/m³)	Sand (kg/m³)	Recycled Aggregate 3/4" (kg/m³)	Recommended Quantity Admixture (Liters/m³)	Fiber (%)
Properties	400	351	119	145	790	1,140	9.50	–

Properties	Average	Standard deviation		Compressive strength (ksc)			Average	Standard deviation
Air content (%)[1]	1.0	0.1		7 days			315.77	43.89
Slump (mm)[2]	87.5	4.8		28 days			510.31	35.41
Initial setting time (min)[3]	148	12.8		90 days			563.88	22.63
Final setting time (min)[3]	398	11.0		365 days			656.24	10.74

Mix Symbol (HSC700C435F110S175)	Cylindrical Strength (ksc)	Cement (kg/m³)	Fly Ash (kg/m³)	Water (kg/m³)	Sand (kg/m³)	Recycled Aggregate 3/4" (kg/m³)	Recommended Quantity Admixture (Liters/m³)	Fiber (%)
	500	240	105	145	825	1,130	9.50	–
Properties	Average	Standard deviation	Compressive strength (ksc)				Average	Standard deviation
Air content (%)[1]	2.1	0.1	7 days				310.15	28.91
Slump (mm)[2]	82.0	11.6	28 days				440.88	30.61
Initial setting time (min)[3]	197	11.2	90 days				501.47	14.32
Final setting time (min)[3]	359	28.0	365 days				509.63	6.01

Mix Symbol (HSC240C321F0S175)	Cylindrical Strength (ksc)	Cement (kg/m³)	Fly Ash (kg/m³)	Water (kg/m³)	Sand (kg/m³)	Recycled Aggregate 3/4" (kg/m³)	Recommended Quantity Admixture (Liters/m³)	Fiber (%)
	190	183	62	160	885	1,180	3.50	–
Properties	Average	Standard deviation	Compressive strength (ksc)				Average	Standard deviation
Air content (%)[1]	1.1	0.2	7 days				167.62	39.41
Slump (mm)[2]	88.0	6.0	28 days				411.84	22.63

(Continued)

TABLE 5.4 (Continued)
Typical Mix Proportions and Related Properties of RAC

Mix Symbol (HSC240C321F0S175)	Cylindrical Strength (ksc)	Cement (kg/m³)	Fly Ash (kg/m³)	Water (kg/m³)	Sand (kg/m³)	Recycled Aggregate 3/4" (kg/m³)	Recommended Quantity Admixture (Liters/m³)	Fiber (%)
Initial setting time (min)[3]	208		16.6		90 days		436.85	14.97
Final setting time (min)[3]	321		3.4		365 days		534.77	2.99

Mix Symbol (HSC240C255F64S175)	Cylindrical Strength (ksc)	Cement (kg/m³)	Fly Ash (kg/m³)	Water (kg/m³)	Sand (kg/m³)	Recycled Aggregate 3/4" (kg/m³)	Recommended Quantity Admixture (Liters/m³)	Fiber (%)
	180	201	69	160	875	1,160	3.70	–
Properties	Average		Standard deviation	Compressive strength (ksc)			Average	Standard deviation
Air content (%)[1]	2.5		0.3		7 days		158.27	22.18
Slump (mm)[2]	86.0		7.2		28 days		247.34	9.23
Initial setting time (min)[3]	217		12.3		90 days		279.44	15.74
Final setting time (min)[3]	339		29.1		365 days		425.78	6.44

Mix Symbol (HSC240C255F64S175)	Cylindrical Strength (ksc)	Cement (kg/m³)	Fly Ash (kg/m³)	Water (kg/m³)	Sand (kg/m³)	Recycled Aggregate 3/4" (kg/m³)	Recommended Quantity Admixture (Liters/m³)	Fiber (%)
	210	220	75	160	865	1,150	3.20	–
Properties	Average		Standard deviation	Compressive strength (ksc)			Average	Standard deviation
Air content (%)[1]	1.2		0.1	7 days			201.74	20.18
Slump (mm)[2]	86.5		9.4	28 days			253.06	10.06
Initial setting time (min)[3]	138		9.0	90 days			291.44	7.58
Final setting time (min)[3]	298		46.2	365 days			336.32	4.21

Mix Symbol (HSC350C395F0S175)	Cylindrical Strength (ksc)	Cement (kg/m³)	Fly Ash (kg/m³)	Water (kg/m³)	Sand (kg/m³)	Recycled Aggregate 3/4" (kg/m³)	Recommended Quantity Admixture (Liters/m³)	Fiber (%)
	240	239	81	160	855	1,135	4.50	–
Properties	Average		Standard deviation	Compressive strength (ksc)			Average	Standard deviation
Air content (%)[1]	2.4		0.2	7 days			288.88	32.93
Slump (mm)[2]	85.5		6.1	28 days			387.12	24.74

(Continued)

TABLE 5.4 (Continued)
Typical Mix Proportions and Related Properties of RAC

Mix Symbol (HSC350C395F0S175)	Cylindrical Strength (ksc)	Cement (kg/m³)	Fly Ash (kg/m³)	Water (kg/m³)	Sand (kg/m³)	Recycled Aggregate 3/4" (kg/m³)	Recommended Quantity Admixture (Liters/m³)	Fiber (%)
Initial setting time (min)[3]		227	17.8			90 days	407.93	10.33
Final setting time (min)[3]		379	27.9			365 days	519.28	5.47
Mix Symbol (HSC700C545F0S175)	**Cylindrical Strength (ksc)**	**Cement (kg/m³)**	**Fly Ash (kg/m³)**	**Water (kg/m³)**	**Sand (kg/m³)**	**Recycled Aggregate 3/4" (kg/m³)**	**Recommended Quantity Admixture (Liters/m³)**	**Fiber (%)**
	400	370	90	130	860	1,105	9.00	–
Properties	Average	Standard deviation				Compressive strength (ksc)	Average	Standard deviation
Air content (%)[1]	1.3	0.1				7 days	327.50	22.41
Slump (mm)[2]	87.0	5.1				28 days	462.85	10.55
Initial setting time (min)[3]	198	17.4				90 days	511.47	4.98
Final setting time (min)[3]	355	10.7				365 days	594.21	1.30

Mix Symbol (HSC700C545F0S175)	Cylindrical Strength (ksc)	Cement (kg/m³)	Fly Ash (kg/m³)	Water (kg/m³)	Sand (kg/m³)	Recycled Aggregate 3/4" (kg/m³)	Recommended Quantity Admixture (Liters/m³)	Fiber (%)
	425	302	158	130	860	1,102	9.00	–

Properties	Average	Standard deviation	Compressive strength (ksc)	Average	Standard deviation
Air content (%)[1]	2.6	0.1	7 days	363.14	36.51
Slump (mm)[2]	82.5	3.7	28 days	517.62	25.44
Initial setting time (min)[3]	187	15.6	90 days	594.91	23.96
Final setting time (min)[3]	349	47.6	365 days	610.11	10.30

Mix Symbol (HSC425C327F142S175)	Cylindrical Strength (ksc)	Cement (kg/m³)	Fly Ash (kg/m³)	Water (kg/m³)	Sand (kg/m³)	Recycled Aggregate 3/4" (kg/m³)	Recommended Quantity Admixture (Liters/m³)	Fiber (%)
	350	314	106	160	805	1,150	5.00	

Properties	Average	Standard deviation	Compressive strength (ksc)	Average	Standard deviation
Air content (%)[1]	1.4	0.2	7 days	302.41	5.48
Slump (mm)[2]	93.5	10.9	28 days	422.43	3.99

(Continued)

TABLE 5.4 (Continued)
Typical Mix Proportions and Related Properties of RAC

Mix Symbol (HSC425C327F142S175)	Cylindrical Strength (ksc)	Cement (kg/m³)	Fly Ash (kg/m³)	Water (kg/m³)	Sand (kg/m³)	Recycled Aggregate 3/4" (kg/m³)	Recommended Quantity Admixture (Liters/m³)	Fiber (%)
Initial setting time (min)[3]	108		10.2	90 days			556.89	2.73
Final setting time (min)[3]	379		25.7	365 days			602.63	1.04

Mix Symbol (HSC425C327F142S175)	Cylindrical Strength (ksc)	Cement (kg/m³)	Fly Ash (kg/m³)	Water (kg/m³)	Sand (kg/m³)	Recycled Aggregate 3/4" (kg/m³)	Recommended Quantity Admixture (Liters/m³)	Fiber (%)
	380	333	112	160	795	1,135	5.00	–

Properties	Average		Standard deviation	Compressive strength (ksc)			Average	Standard deviation
Air content (%)[1]	2.9		0.2	7 days			306.14	36.14
Slump (mm)[2]	85.0		8.3	28 days			400.59	24.88
Initial setting time (min)[3]	237		13.4	90 days			538.51	20.95
Final setting time (min)[3]	369		48.7	365 days			556.42	10.05

Mix Symbol (HSC425C327F142S175)	Cylindrical Strength (ksc)	Cement (kg/m³)	Fly Ash (kg/m³)	Water (kg/m³)	Sand (kg/m³)	Recycled Aggregate 3/4" (kg/m³)	Recommended Quantity Admixture (Liters/m³)	Fiber (%)
Properties	250	246	84	160	845	1,160	5.50	–

Properties	Average	Standard deviation	Compressive strength (ksc)	Average	Standard deviation
Air content (%)[1]	3.0	0.2	7 days	204.88	22.11
Slump (mm)[2]	120.5	23.7	28 days	223.69	14.55
Initial setting time (min)[3]	209	19.3	90 days	321.56	3.69
Final setting time (min)[3]	319	33.5	365 days	352.76	4.58

Mix Symbol (HSC400C355F90S175)	Cylindrical Strength (ksc)	Cement (kg/m³)	Fly Ash (kg/m³)	Water (kg/m³)	Sand (kg/m³)	Recycled Aggregate 3/4" (kg/m³)	Recommended Quantity Admixture (Liters/m³)	Fiber (%)
Properties	280	258	87	160	835	1,145	5.30	–

Properties	Average	Standard deviation	Compressive strength (ksc)	Average	Standard deviation
Air content (%)[1]	1.7	0.1	7 days	215.81	22.02
Slump (mm)[2]	120.0	1.5	28 days	236.96	14.77

(Continued)

TABLE 5.4 (Continued)
Typical Mix Proportions and Related Properties of RAC

Mix Symbol (HSC400C355F90S175)	Cylindrical Strength (ksc)	Cement (kg/m³)	Fly Ash (kg/m³)	Water (kg/m³)	Sand (kg/m³)	Recycled Aggregate 3/4" (kg/m³)	Recommended Quantity Admixture (Liters/m³)	Fiber (%)
Initial setting time (min)[3]	140		12.7	90 days			319.66	13.02
Final setting time (min)[3]	381		4.8	365 days			373.28	7.58

Mix Symbol (HSC400C355F90S175)	Cylindrical Strength (ksc)	Cement (kg/m³)	Fly Ash (kg/m³)	Water (kg/m³)	Sand (kg/m³)	Recycled Aggregate 3/4" (kg/m³)	Recommended Quantity Admixture (Liters/m³)	Fiber (%)
Properties	300	276	94	160	825	1,130	4.00	–
	Average		Standard deviation	Compressive strength (ksc)			Average	Standard deviation
Air content (%)[1]	3.1		0.3	7 days			186.98	22.48
Slump (mm)[2]	123.0		14.9	28 days			261.44	21.06
Initial setting time (min)[3]	245		3.3	90 days			297.78	7.56
Final setting time (min)[3]	268		4.6	365 days			342.59	4.22

Mix Symbol (HSC425C287F158S175)	Cylindrical Strength (ksc)	Cement (kg/m³)	Fly Ash (kg/m³)	Water (kg/m³)	Sand (kg/m³)	Recycled Aggregate 3/4" (kg/m³)	Recommended Quantity Admixture (Liters/m³)	Fiber (%)
Properties	320	295	100	160	815	1,115	5.00	—

Properties	Average	Standard deviation	Compressive strength (ksc)	Average	Standard deviation
Air content (%)[1]	1.0	0.2	7 days	196.24	33.67
Slump (mm)[2]	110.5	4.4	28 days	302.59	24.82
Initial setting time (min)[3]	143	3.6	90 days	432.19	17.44
Final setting time (min)[3]	281	1.2	365 days	500.84	6.89

Mix Symbol (HSC425C287F158S175)	Cylindrical Strength (ksc)	Cement (kg/m³)	Fly Ash (kg/m³)	Water (kg/m³)	Sand (kg/m³)	Recycled Aggregate 3/4" (kg/m³)	Recommended Quantity Admixture (Liters/m³)	Fiber (%)
Properties	350	275	70	150	835	1,100	5.30	—

Properties	Average	Standard deviation	Compressive strength (ksc)	Average	Standard deviation
Air content (%)[1]	2.0	0.2	7 days	206.57	24.16
Slump (mm)[2]	106.5	6.5	28 days	253.69	20.03

(Continued)

TABLE 5.4 (Continued)
Typical Mix Proportions and Related Properties of RAC

Mix Symbol (HSC425C287F158S175)	Cylindrical Strength (ksc)	Cement (kg/m³)	Fly Ash (kg/m³)	Water (kg/m³)	Sand (kg/m³)	Recycled Aggregate 3/4" (kg/m³)	Recommended Quantity Admixture (Liters/m³)	Fiber (%)
Initial setting time (min)[3]	180		19.2	90 days			351.58	14.02
Final setting time (min)[3]	336		29.1	365 days			389.42	5.87

Mix Symbol (HSC425C303F166S175)	Cylindrical Strength (ksc)	Cement (kg/m³)	Fly Ash (kg/m³)	Water (kg/m³)	Sand (kg/m³)	Recycled Aggregate 3/4" (kg/m³)	Recommended Quantity Admixture (Liters/m³)	Fiber (%)
Properties	320	315	80	150	815	1,115	5.30	–
	Average		Standard deviation		Compressive strength (ksc)		Average	Standard deviation
Air content (%)[1]	1.2		0.1		7 days		222.01	27.28
Slump (mm)[2]	101.5		19.4		28 days		265.96	22.03
Initial setting time (min)[3]	244		13.3		90 days		299.14	13.91
Final setting time (min)[3]	304		23.7		365 days		340.25	7.47

Mix Symbol (HSC425C287F158S175)	Cylindrical Strength (ksc)	Cement (kg/m³)	Fly Ash (kg/m³)	Water (kg/m³)	Sand (kg/m³)	Recycled Aggregate 3/4" (kg/m³)	Recommended Quantity Admixture (Liters/m³)	Fiber (%)
	350	335	85	150	805	1,100	5.30	–
Properties	Average		Standard deviation	Compressive strength (ksc)			Average	Standard deviation
Air content (%)[1]	2.0		0.2	7 days			230.48	30.21
Slump (mm)[2]	111.5		20.2	28 days			255.07	23.89
Initial setting time (min)[3]	155		20.0	90 days			361.52	2.69
Final setting time (min)[3]	303		12.5	365 days			394.77	1.55

Mix Symbol (HSC425C287F158S175)	Cylindrical Strength (ksc)	Cement (kg/m³)	Fly Ash (kg/m³)	Water (kg/m³)	Sand (kg/m³)	Recycled Aggregate 3/4" (kg/m³)	Recommended Quantity Admixture (Liters/m³)	Fiber (%)
	380	355	90	150	795	1,090	5.30	–
Properties	Average		Standard deviation	Compressive strength (ksc)			Average	Standard deviation
Air content (%)[1]	1.6		0.3	7 days			251.56	12.87
Slump (mm)[2]	121.0		5.6	28 days			317.48	23.66

(Continued)

TABLE 5.4 (Continued)
Typical Mix Proportions and Related Properties of RAC

Mix Symbol (HSC425C287F158S175)	Cylindrical Strength (ksc)	Cement (kg/m³)	Fly Ash (kg/m³)	Water (kg/m³)	Sand (kg/m³)	Recycled Aggregate 3/4" (kg/m³)	Recommended Quantity Admixture (Liters/m³)	Fiber (%)
Initial setting time (min)[3]	135		15.4		90 days		432.58	10.05
Final setting time (min)[3]	400		3.7		365 days		514.74	5.28
Mix Symbol (HSC425C327F142S175)	Cylindrical Strength (ksc)	Cement (kg/m³)	Fly Ash (kg/m³)	Water (kg/m³)	Sand (kg/m³)	Recycled Aggregate 3/4" (kg/m³)	Recommended Quantity Admixture (Liters/m³)	Fiber (%)
	300	257	112	160	825	1,130	5.00	–
Properties	Average		Standard deviation		Compressive strength (ksc)		Average	Standard deviation
Air content (%)[1]	1.8		0.2		7 days		205.99	0.93
Slump (mm)[2]	110.5		2.9		28 days		263.47	4.87
Initial setting time (min)[3]	145		8.8		90 days		362.48	14.24
Final setting time (min)[3]	396		11.7		365 days		389.52	8.46

Mix Symbol (HSC425C327F1425175)	Cylindrical Strength (ksc)	Cement (kg/m³)	Fly Ash (kg/m³)	Water (kg/m³)	Sand (kg/m³)	Recycled Aggregate 3/4" (kg/m³)	Recommended Quantity Admixture (Liters/m³)	Fiber (%)
	320	275	120	160	815	1,115	5.00	–

Properties	Average	Standard deviation	Compressive strength (ksc)	Average	Standard deviation
Air content (%)[1]	1.3	0.2	7 days	276.91	5.94
Slump (mm)[2]	121.5	24.9	28 days	306.24	3.11
Initial setting time (min)[3]	249	15.5	90 days	420.87	2.05
Final setting time (min)[3]	305	24.6	365 days	504.44	0.89

Mix Symbol (HSC350C255F140S175)	Cylindrical Strength (ksc)	Cement (kg/m³)	Fly Ash (kg/m³)	Water (kg/m³)	Sand (kg/m³)	Recycled Aggregate 3/4" (kg/m³)	Recommended Quantity Admixture (Liters/m³)	Fiber (%)
	350	292	127	160	805	1,150	5.00	–

Properties	Average	Standard deviation	Compressive strength (ksc)	Average	Standard deviation
Air content (%)[1]	1.3	0.2	7 days	235.48	29.88
Slump (mm)[2]	101.5	8.5	28 days	251.66	24.66

(Continued)

TABLE 5.4 (Continued)
Typical Mix Proportions and Related Properties of RAC

Mix Symbol (HSC350C255F140S175)	Cylindrical Strength (ksc)	Cement (kg/m³)	Fly Ash (kg/m³)	Water (kg/m³)	Sand (kg/m³)	Recycled Aggregate 3/4" (kg/m³)	Recommended Quantity Admixture (Liters/m³)	Fiber (%)
Initial setting time (min)[3]	144		14.5		90 days		349.25	12.02
Final setting time (min)[3]	365		21.5		365 days		412.87	4.57
Mix Symbol (HSC350C255F140S175)	**Cylindrical Strength (ksc)**	**Cement (kg/m³)**	**Fly Ash (kg/m³)**	**Water (kg/m³)**	**Sand (kg/m³)**	**Recycled Aggregate 3/4" (kg/m³)**	**Recommended Quantity Admixture (Liters/m³)**	**Fiber (%)**
	380	310	135	160	795	1,135	5.00	–
Properties	Average		Standard deviation	Compressive strength (ksc)			Average	Standard deviation
Air content (%)[1]	1.6		0.2		7 days		261.88	22.77
Slump (mm)[2]	103.0		12.9		28 days		348.26	13.85
Initial setting time (min)[3]	148		18.9		90 days		388.01	10.36
Final setting time (min)[3]	370		30.1		365 days		426.58	2.80

TABLE 5.5

Chemical Compositions and Physical Properties of Type 1 Portland Cements and FA

Chemical Compositions (% by Mass)	SiO_2	Al_2O_3	Fe_2O_3	CaO	MgO	Na_2O	SO_3	Free CaO
Type 1 Portland Cement	20.84	5.22	3.20	66.28	1.24	0.10	2.41	0.99
FA	42.10	21.80	11.22	13.56	2.41	2.90	1.88	1.44

Physical Properties	Type 1 Portland Cement	FA
Loss on Ignition (%)	0.96	2.33
Moisture Content (%)	0.19	1.50
Blaine Surface Area (cm^2/g)		
Fineness (Particle Size, % Retained)	3,200	2,850
– ≥ 75 μm	0.50	0.56
– 75 μm	5.25	8.25
– 45 μm	3.60	4.76
– ≤ 36 μm	90.62	86.43
Fineness (Retained) on 45 Micron (No. 325)	5.75	4.90
Water Requirement (%)	100	97
Bulk Density (kg/l)	1.03	0.51
Specific Gravity	3.15	2.13

Ziaria, the effects of silica fume on compressive strength are more significant than those of a reduction in the w/b ratio. According to the estimates, the best value for silica fume is 6%, while the best value for w/b is 0.35.

- Superplasticizers: Superplasticizers are absolutely necessary to make high-strength RAC that is good to work with. There are fundamentally three primary categories of superplasticizers, which are as follows: lignosulfonate-based, melamine sulfonate, and naphthalene sulfonate. In most cases, a mixture of the abovementioned kinds is used to create high-strength RAC. The level of workability that must be achieved determines the quantity of superplasticizer added to a mixture.

High-Strength Recycled Aggregate Concrete Materials

- Type 1 hydraulic Portland cements conforming to ASTM C150 (ASTM C150/C150M–17 Standard specifications for Portland cement) were used throughout concrete mixtures. Their chemical compositions and physical properties are shown in Table 5.5.
- The ASTM C618 (ASTM C618–15 Standard specification for coal fly ash and raw or calcined natural pozzolan for use in concrete) classifies the PFA as low calcium (Type F).

- Tap water with a pH 7.0 conforming to ASTM C1602 (ASTM C1602/C1602M–12 Standard specification for mixing water used in the production of hydraulic cement concrete).
- River sand with gradation conforming to the ASTM C33 (ASTM C33/C33M–1601 standard specification for concrete aggregates).
- Crushed lime recycled aggregate rock with gradation conforming to the ASTM C33 (ASTM C33/C33M–1601 standard specification for RAC aggregates).
- The chemical admixture used superplasticizer that conforms to the ASTM C494 (ASTM C494/C494M–16 Standard specification for chemical admixtures for concrete); that is, the superplasticizer had a recommended dosage rate of cementitious materials (per 100 of a kilogram of cementitious materials).

5.2.3 Marine RAC

Marine RAC is a type of RAC that is capable of enduring marine environments. Marine environments include the natural and biological resources that make up any coastal, sea, seabed, or subsoil ecosystem. This includes the living and nonliving components and the ecological patterns and processes that occur therein. Marine RAC can be defined as a type of RAC that can withstand marine environments. Marine RAC must withstand some of the most severe conditions that can be found in an engineering setting. It is often used in key applications where the material's service life and structural dependability are essential considerations. Not only is marine RAC vulnerable to the corrosive effects of saltwater, but it is also exposed to the ongoing wave loadings and the abrasive impact of bed and suspended loads. Various loading circumstances might trigger potential degradation processes, which could lead to several distinct outcomes. Because loadings, processes, and effects may all possibly interact with one another in several dimensions, complete knowledge is essential before proper design and construction procedures can be implemented for any specific project.

It was evident that the permeability of RAC is the essential factor in determining the long-time durability of RAC based on the performance of RAC in a marine environment as well as a review of the primary causes of RAC deterioration. This conclusion was reached after looking at both of these factors. Therefore, with any new construction, it is not only essential to select materials and proportions for the RAC mixture that are most likely to produce a low-permeability product on curing, but it is also necessary to maintain the watertightness of the structure for as long as possible through the intended service life of the structure. This can be accomplished by keeping the water-tightness of the structure intact for as long as possible through the intended service life of the structure. In a nutshell, while constructing with RAC, it is essential to give careful consideration to each of the selections of ingredients for creating RAC and the amounts of the mix, best practices in RAC placing, and precautions should be taken during service to prevent pre-existing RAC microcracks from becoming larger and more widespread.

Mostly, there is no such thing as a standard solution for marine RAC constructions. This is primarily due to the severe climate and the long service life demand. It is necessary to carefully consider a wide range of issues to develop a robust, site-specific solution. These issues include functional design requirements, environmental conditions, the availability and selection of materials, low-risk construction methodologies, future maintenance approaches, and health and safety constraints.

Regarding the standards governing the sector, it is noteworthy that the American RAC Institute does not have any specialized codes for typical coastal marine projects. The ACI Manual of RAC Practice includes guidelines for general structures (such as buildings and bridges, parking garages, silos and bins, chimneys and cooling towers, and nuclear and sanitary structures) used for the design of conventional marine structures. These guidelines are adhered to. A suggested practice for the design of offshore structures, a state-of-the-art study on Arctic offshore structures, and another report on barge-like structures have all been produced by ACI Committee 357. This committee is currently drafting an ACI Guide for Design of Concrete Marine Structures as part of creating the guide.

The ideas that underpin the ACI Recommended Practice are adaptable to determining the proportions of RAC mixtures for most marine structures, except those subjected to highly hostile environments. the example that comes up next demonstrates how one may go about doing this.

- Choice of slump. RAC mixes should have a consistency that allows for total homogeneity on mixing and ease of transportation, placing, and consolidation without segregation. This is referred to as the slump. It is important to keep in mind that the slump requirement for superplasticizer RAC for substantially reinforced marine constructions that will be put by pumping is typically between 150 and 200 mm.
- The choice of the maximum aggregate size. Because of the proximity of the reinforcing bars in heavily reinforced structural elements and the importance of maintaining the low permeability of RAC, it is recommended that the maximum aggregate size not exceed 10 mm. RAC mixes with maximum aggregate sizes ranging from 25 to 37 mm may be utilized for either unreinforced or mildly reinforced constructions.
- Determining the quantity of mixing water and the amount of air mixed in. When using well-graded normal aggregates, the amount of mixing water required is determined by the maximum aggregate size, the consistency that is wanted, and the amount of air that is mixed in.
- Selection of water/cement ratio. The ACI Committee 201 recommends a maximum permissible water/cement ratio of 040 for structures exposed to seawater. This recommendation is made even though a higher water/cement ratio may be acceptable when considering the RAC's strength. Assume that the strength of the prescribed 28-day RAC is 30 MPa.
- Selection of admixtures. In addition to the high-range water-reducing admixture (superplasticizer), a high-quality pozzolan should be used to increase workability and minimize permeability in the RAC mixture.

When it comes to a chemical assault on hydrated Portland cement in unreinforced RAC, one would assume that sulfate and magnesium are the detrimental ingredients in saltwater. This is because reactive alkali aggregates are not present, which makes it possible for one to make this prediction. It is important to remember that sulfate attack on groundwater is considered severe when the sulfate ion concentration is higher than 1,500 mg/L. Similarly, Portland cement paste can deteriorate due to cat–ion-exchange reactions when the magnesium ion concentration is higher than, for example, 500 mg/L. Even when a high-C_3A Portland cement has been used, and large amounts of ettringite are present due to sulfate attack on the cement paste, it is a common observation that RAC deterioration is not characterized by expansion. Instead, it mostly takes the form of erosion or loss of the solid constituents from the mass. This is interesting because seawater contains an undesirable high concentration of sulfate. It has been hypothesized that ettringite growth is inhibited when the OH- and Cl- ions have substituted mainly ions. This approach, by the way, is compatible with the concept that the enlargement of ettringite by water adsorption requires the presence of an alkaline environment.

It should be pointed out that according to ACI Building Code 318-83 (Mahmood et al., 2021), sulfate exposure to seawater is categorized as moderate, which allows for the utilization of ASTM Type-II Portland cement (maximum 8% C_3A) with a maximum water/cement ratio of 0.50 in normal-weight RAC. The ACI 318R-21 Building Code Commentary states that types of cement with C_3A levels of up to ten percent may be used; the maximum water-to-cement ratio is further decreased to 0.40. This is something that should be taken into consideration. Moreover, for international specifications/practices, many institutions determine the RAC that can withstand marine environments, including ACI 357-84 (1997) "Guide for Design and Construction of Fixed Offshore Structures," BS 6349-1:2000 "Marine structures. Code of practice for general criteria," "Offshore Standard DNV-OS-C502, Offshore Concrete Structures" (July 2004), USACE EM 1110-2-2000, "Engineering and Design – Standard Practice for Concrete for Civil Works Structures," and RILEM Technical Committee 32-RCA state-of-the-art report "Seawater Attack on Concrete and Precautionary Measures" (1985). Overall, it can be summarized that C3A content: 4/5–10% range, or 10% maximum; water/cement ratio: 0.40–0.45, depending on the severity of exposure (tidal vs. submerged); compressive strength: > 35 MPa (RILEM); and supplementary cementitious materials (SCMs), such as slag, fly ash, and natural pozzolan are recognized as beneficial (Table 5.6).

Marine Recycled Aggregate Concrete Materials

- Type 1 hydraulic Portland cements conforming to ASTM C150 (ASTM C150/C150M–17 Standard specifications for Portland cement) were used throughout concrete mixtures. Their chemical compositions and physical properties are shown in Table 5.7.
- The ASTM C618 (ASTM C618–15 Standard specification for coal fly ash and raw or calcined natural pozzolan for use in concrete) classifies the PFA as low calcium (Type F).
- Tap water with a pH 7.0 conforming to ASTM C1602 (ASTM C1602/C1602M–12 Standard specification for mixing water used in the production of hydraulic cement concrete).

TABLE 5.6
Typical Mix Proportions and Related Properties of RAC

Mix Symbol (MC210C310F0S75)	Cylindrical Strength (ksc)	Cement (kg/m³)	Fly Ash (kg/m³)	Water (kg/m³)	Sand (kg/m³)	Recycled Aggregate 3/4" (kg/m³)	Recommended Quantity Admixture (Liters/m³)	Fiber (%)
Properties	180	314	0	190	850	1,105	1.16	

Properties	Average	Standard deviation
Air content (%)[1]	2.0	0.3
Slump (mm)[2]	49.0	2.9
Initial setting time (min)[3]	95	9.8
Final setting time (min)[3]	250	26.8

Compressive strength (ksc)	Average	Standard deviation
7 days	69.55	11.98
28 days	140.96	14.23
90 days	203.63	6.25
365 days	241.47	3.64

Mix Symbol (MC210C263F47S75)	Cylindrical Strength (ksc)	Cement (kg/m³)	Fly Ash (kg/m³)	Water (kg/m³)	Sand (kg/m³)	Recycled Aggregate 3/4" (kg/m³)	Recommended Quantity Admixture (Liters/m³)	Fiber (%)
Properties	180	225	134	165	850	1,115	1.36	–

Properties	Average	Standard deviation
Air content (%)[1]	2.0	0.1

Compressive strength (ksc)	Average	Standard deviation
7 days	77.45	19.27

(Continued)

TABLE 5.6 (Continued)
Typical Mix Proportions and Related Properties of RAC

Mix Symbol (MC210C263F47S75)	Cylindrical Strength (ksc)	Cement (kg/m³)	Fly Ash (kg/m³)	Water (kg/m³)	Sand (kg/m³)	Recycled Aggregate 3/4" (kg/m³)	Recommended Quantity Admixture (Liters/m³)	Fiber (%)
Slump (mm)[2]	50.0		1.5	28 days			151.32	15.88
Initial setting time (min)[3]	100		5.7	90 days			222.67	9.56
Final setting time (min)[3]	250		3.3	365 days			235.73	4.36
Mix Symbol (MC320C327F83S75)	**Cylindrical Strength (ksc)**	**Cement (kg/m³)**	**Fly Ash (kg/m³)**	**Water (kg/m³)**	**Sand (kg/m³)**	**Recycled Aggregate 3/4" (kg/m³)**	**Recommended Quantity Admixture (Liters/m³)**	**Fiber (%)**
	180	198	126	165	880	1,105	1.36	–
Properties	Average		Standard deviation	Compressive strength (ksc)		Average		Standard deviation
Air content (%)[1]	2.4		0.3	7 days			93.51	13.10
Slump (mm)[2]	51.5		11.4	28 days			216.25	4.73
Initial setting time (min)[3]	100		7.4	90 days			254.73	11.62
Final setting time (min)[3]	250		8.9	365 days			305.81	5.58

Mix Symbol (MC240C247F63S75)	Cylindrical Strength (ksc)	Cement (kg/m³)	Fly Ash (kg/m³)	Water (kg/m³)	Sand (kg/m³)	Recycled Aggregate 3/4" (kg/m³)	Recommended Quantity Admixture (Liters/m³)	Fiber (%)
Properties	180	250	60	165	880	1,105	1.36	–

Properties	Average	Standard deviation	Compressive strength (ksc)	Average	Standard deviation
Air content (%)[1]	2.5	0.3	7 days	70.66	17.41
Slump (mm)[2]	50.0	2.9	28 days	147.14	8.44
Initial setting time (min)[3]	100	8.2	90 days	210.37	5.36
Final setting time (min)[3]	242	19.2	365 days	258.41	2.20

Mix Symbol (MC180C227F58S75)	Cylindrical Strength (ksc)	Cement (kg/m³)	Fly Ash (kg/m³)	Water (kg/m³)	Sand (kg/m³)	Recycled Aggregate 3/4" (kg/m³)	Recommended Quantity Admixture (Liters/m³)	Fiber (%)
Properties	140	210	0	190	850	1,115	1.06	–

Properties	Average	Standard deviation	Compressive strength (ksc)	Average	Standard deviation
Air content (%)[1]	2.7	0.3	7 days	54.65	13.09
Slump (mm)[2]	54.0	4.4	28 days	93.63	11.11
Initial setting time (min)[3]	100	11.2	90 days	144.25	6.68
Final setting time (min)[3]	246	40.2	365 days	168.16	3.99

(Continued)

TABLE 5.6 (Continued)
Typical Mix Proportions and Related Properties of RAC

Mix Symbol (MC240C267F68S75)	Cylindrical Strength (ksc)	Cement (kg/m³)	Fly Ash (kg/m³)	Water (kg/m³)	Sand (kg/m³)	Recycled Aggregate 3/4" (kg/m³)	Recommended Quantity Admixture (Liters/m³)	Fiber (%)
Properties	180	220	140	145	850	1,125	1.16	–
	Average		Standard deviation	Compressive strength (ksc)			Average	Standard deviation
Air content (%)[1]	2.8		0.1	7 days			142.17	10.24
Slump (mm)[2]	49.0		7.9	28 days			216.84	6.99
Initial setting time (min)[3]	100		15.4	90 days			235.63	4.21
Final setting time (min)[3]	248		1.1	365 days			290.41	0.83

Mix Symbol (MC240C267F68S75)	Cylindrical Strength (ksc)	Cement (kg/m³)	Fly Ash (kg/m³)	Water (kg/m³)	Sand (kg/m³)	Recycled Aggregate 3/4" (kg/m³)	Recommended Quantity Admixture (Liters/m³)	Fiber (%)
Properties	210	360	0	190	870	1,135	1.16	–
	Average		Standard deviation	Compressive strength (ksc)			Average	Standard deviation
Air content (%)[1]	2.8		0.1	7 days			74.54	6.48
Slump (mm)[2]	53.0		15.7	28 days			98.93	3.11

(Property)	Cylindrical Strength (ksc)	Cement (kg/m³)	Fly Ash (kg/m³)	Water (kg/m³)	Sand (kg/m³)	Recycled Aggregate 3/4" (kg/m³)	Recommended Quantity Admixture (Liters/m³)	Fiber (%)
Initial setting time (min)[3]	100		18.1		90 days		134.87	4.63
Final setting time (min)[3]	250		4.7		365 days		150.22	2.45
Mix Symbol (MC240C284F51S75)	210	226	144	165	870	1,135	1.16	–
Properties	Average		Standard deviation		Compressive strength (ksc)		Average	Standard deviation
Air content (%)[1]	3.0		0.3		7 days		80.02	16.36
Slump (mm)[2]	51.0		23.1		28 days		161.26	12.41
Initial setting time (min)[3]	101		12.2		90 days		214.69	10.09
Final setting time (min)[3]	248		3.5		365 days		236.17	2.14
Mix Symbol (MC280C275F70S75)	240	384	0	190	860	1,110	1.40	–
Properties	Average		Standard deviation		Compressive strength (ksc)		Average	Standard deviation
Air content (%)[1]	3.0		0.2		7 days		85.06	4.96
Slump (mm)[2]	50.0		3.5		28 days		171.87	3.21

(Continued)

TABLE 5.6 (Continued)
Typical Mix Proportions and Related Properties of RAC

Mix Symbol (MC280C275F70S75)	Cylindrical Strength (ksc)	Cement (kg/m³)	Fly Ash (kg/m³)	Water (kg/m³)	Sand (kg/m³)	Recycled Aggregate 3/4" (kg/m³)	Recommended Quantity Admixture (Liters/m³)	Fiber (%)
Initial setting time (min)[3]	100		7.1		90 days		207.99	1.88
Final setting time (min)[3]	255		33.9		365 days		245.36	0.68
Mix Symbol (MC150C170F75S75)	Cylindrical Strength (ksc)	Cement (kg/m³)	Fly Ash (kg/m³)	Water (kg/m³)	Sand (kg/m³)	Recycled Aggregate 3/4" (kg/m³)	Recommended Quantity Admixture (Liters/m³)	Fiber (%)
	140	185	116	165	890	1,100	1.00	–
Properties	Average		Standard deviation		Compressive strength (ksc)		Average	Standard deviation
Air content (%)[1]	3.1		0.1		7 days		44.11	12.70
Slump (mm)[2]	52.0		9.2		28 days		90.69	11.02
Initial setting time (min)[3]	102		12.3		90 days		120.48	5.21
Final setting time (min)[3]	245		27.4		365 days		152.66	3.02

Mix Symbol	Cylindrical Strength (ksc)	Cement (kg/m³)	Fly Ash (kg/m³)	Water (kg/m³)	Sand (kg/m³)	Recycled Aggregate 3/4" (kg/m³)	Recommended Quantity Admixture (Liters/m³)	Fiber (%)
MC150C170F75S75	140	156	124	165	890	1,100	1.00	–
Properties	Average		Standard deviation	Compressive strength (ksc)		Average		Standard deviation
Air content (%)[1]	1.5		0.3	7 days		34.63		9.36
Slump (mm)[2]	94.0		20.8	28 days		101.46		7.41
Initial setting time (min)[3]	158		7.6	90 days		118.67		5.33
Final setting time (min)[3]	319		2.5	365 days		134.14		2.01
MC240C223F97S75	210	128	90	150	788	1,050	1.30	–
Properties	Average		Standard deviation	Compressive strength (ksc)		Average		Standard deviation
Air content (%)[1]	1.8		0.1	7 days		81.65		6.41
Slump (mm)[2]	55.5		6.7	28 days		118.48		5.33
Initial setting time (min)[3]	118		14.9	90 days		132.41		4.01
Final setting time (min)[3]	364		41.7	365 days		159.58		2.89

(Continued)

TABLE 5.6 (Continued)
Typical Mix Proportions and Related Properties of RAC

Mix Symbol (MC240C223F97S75)	Cylindrical Strength (ksc)	Cement (kg/m³)	Fly Ash (kg/m³)	Water (kg/m³)	Sand (kg/m³)	Recycled Aggregate 3/4" (kg/m³)	Recommended Quantity Admixture (Liters/m³)	Fiber (%)
	140	204	130	165	780	1,070	1.20	–
Properties	Average		Standard deviation	Compressive strength (ksc)			Average	Standard deviation
Air content (%)[1]	3.2		0.2	7 days			94.12	14.52
Slump (mm)[2]	77.0		6.9	28 days			129.52	10.02
Initial setting time (min)[3]	166		1.4	90 days			181.02	3.99
Final setting time (min)[3]	298		24.9	365 days			219.84	2.41

Mix Symbol (MC280C240F105S75)	Cylindrical Strength (ksc)	Cement (kg/m³)	Fly Ash (kg/m³)	Water (kg/m³)	Sand (kg/m³)	Recycled Aggregate 3/4" (kg/m³)	Recommended Quantity Admixture (Liters/m³)	Fiber (%)
	140	184	116	165	782	1,067	1.30	–
Properties	Average		Standard deviation	Compressive strength (ksc)			Average	Standard deviation
Air content (%)[1]	3.0		0.3	7 days			115.98	22.01
Slump (mm)[2]	78.0		16.7	28 days			246.32	15.89

Properties	Average	Standard deviation	Compressive strength (ksc)	Average	Standard deviation
Initial setting time (min)[3]	186	8.7	90 days	294.56	13.62
Final setting time (min)[3]	328	44.8	365 days	319.66	4.46

Mix Symbol (MC280C240F105S75)	Cylindrical Strength (ksc)	Cement (kg/m³)	Fly Ash (kg/m³)	Water (kg/m³)	Sand (kg/m³)	Recycled Aggregate 3/4" (kg/m³)	Recommended Quantity Admixture (Liters/m³)	Fiber (%)
	140	230	54	165	782	1,067	1.30	–

Properties	Average	Standard deviation	Compressive strength (ksc)	Average	Standard deviation
Air content (%)[1]	3.1	0.1	7 days	85.77	23.54
Slump (mm)[2]	76.5	15.3	28 days	181.65	17.41
Initial setting time (min)[3]	136	14.9	90 days	264.14	5.88
Final setting time (min)[3]	318	14.7	365 days	291.85	2.65

Mix Symbol (MC280C257F112S100)	Cylindrical Strength (ksc)	Cement (kg/m³)	Fly Ash (kg/m³)	Water (kg/m³)	Sand (kg/m³)	Recycled Aggregate 3/4" (kg/m³)	Recommended Quantity Admixture (Liters/m³)	Fiber (%)
	180	334	0	190	769	1,039	1.40	–

Properties	Average	Standard deviation	Compressive strength (ksc)	Average	Standard deviation
Air content (%)[1]	3.3	0.3	7 days	80.98	16.51

(Continued)

TABLE 5.6 (Continued)
Typical Mix Proportions and Related Properties of RAC

Mix Symbol (MC280C257F112S100)	Cylindrical Strength (ksc)	Cement (kg/m³)	Fly Ash (kg/m³)	Water (kg/m³)	Sand (kg/m³)	Recycled Aggregate 3/4" (kg/m³)	Recommended Quantity Admixture (Liters/m³)	Fiber (%)
Slump (mm)[2]	80.0	14.6				28 days	171.65	10.36
Initial setting time (min)[3]	176	1.2				90 days	247.84	8.93
Final setting time (min)[3]	308	10.4				365 days	285.55	4.77

Mix Symbol (MC280C257F112S100)	Cylindrical Strength (ksc)	Cement (kg/m³)	Fly Ash (kg/m³)	Water (kg/m³)	Sand (kg/m³)	Recycled Aggregate 3/4" (kg/m³)	Recommended Quantity Admixture (Liters/m³)	Fiber (%)
	240	334	0	165	860	1,120	1.25	-
Properties	Average	Standard deviation				Compressive strength (ksc)	Average	Standard deviation
Air content (%)[1]	3.4	0.2				7 days	113.69	2.63
Slump (mm)[2]	80.5	8.3				28 days	142.11	1.41
Initial setting time (min)[3]	236	2.6				90 days	181.63	5.69
Final setting time (min)[3]	358	36.6				365 days	206.97	2.14

Mix Symbol (MC280C257F112S125)	Cylindrical Strength (ksc)	Cement (kg/m³)	Fly Ash (kg/m³)	Water (kg/m³)	Sand (kg/m³)	Recycled Aggregate 3/4" (kg/m³)	Recommended Quantity Admixture (Liters/m³)	Fiber (%)
	240	210	134	155	860	1,120	1.30	–
Properties	Average		Standard deviation	Compressive strength (ksc)			Average	Standard deviation
Air content (%)[1]	3.6		0.1	7 days			82.68	0.84
Slump (mm)[2]	79.0		8.7	28 days			191.41	1.03
Initial setting time (min)[3]	146		9.2	90 days			249.61	2.88
Final setting time (min)[3]	398		24.6	365 days			290.65	3.62

Mix Symbol (MC280C257F112S125)	Cylindrical Strength (ksc)	Cement (kg/m³)	Fly Ash (kg/m³)	Water (kg/m³)	Sand (kg/m³)	Recycled Aggregate 3/4" (kg/m³)	Recommended Quantity Admixture (Liters/m³)	Fiber (%)
	240	220	140	155	860	1,120	1.20	–
Properties	Average		Standard deviation	Compressive strength (ksc)			Average	Standard deviation
Air content (%)[1]	3.5		0.3	7 days			80.37	20.08
Slump (mm)[2]	81.0		12.7	28 days			182.19	17.63
Initial setting time (min)[3]	226		5.4	90 days			248.52	8.84
Final setting time (min)[3]	348		5.9	365 days			290.69	1.36

(Continued)

TABLE 5.6 (Continued)
Typical Mix Proportions and Related Properties of RAC

Mix Symbol (MC320C275F120S75)	Cylindrical Strength (ksc)	Cement (kg/m³)	Fly Ash (kg/m³)	Water (kg/m³)	Sand (kg/m³)	Recycled Aggregate 3/4" (kg/m³)	Recommended Quantity Admixture (Liters/m³)	Fiber (%)
Properties	240	198	126	155	860	1,120	1.40	–

Properties	Average	Standard deviation	Compressive strength (ksc)	Average	Standard deviation
Air content (%)[1]	1.7	0.2	7 days	81.52	23.36
Slump (mm)[2]	60.5	18.9	28 days	183.67	10.44
Initial setting time (min)[3]	156	12.7	90 days	267.44	11.98
Final setting time (min)[3]	388	32.2	365 days	301.73	5.28

Mix Symbol (MC280C345F0S75)	Cylindrical Strength (ksc)	Cement (kg/m³)	Fly Ash (kg/m³)	Water (kg/m³)	Sand (kg/m³)	Recycled Aggregate 3/4" (kg/m³)	Recommended Quantity Admixture (Liters/m³)	Fiber (%)
Properties	180	210	134	165	780	1,070	0.81	–

Properties	Average	Standard deviation	Compressive strength (ksc)	Average	Standard deviation
Air content (%)[1]	1.0	0.1	7 days	123.58	17.98
Slump (mm)[2]	87.5	4.8	28 days	209.36	12.61

(continuation of previous mix – properties)

Properties	Cylindrical Strength (ksc)	Average	Standard deviation	Compressive strength (ksc)	Average	Standard deviation
Initial setting time (min)[3]		148	12.8	90 days	338.51	13.55
Final setting time (min)[3]		398	11.0	365 days	374.52	2.58

Mix Symbol (MC280C275F70S75)	Cement (kg/m³)	Fly Ash (kg/m³)	Water (kg/m³)	Sand (kg/m³)	Recycled Aggregate 3/4" (kg/m³)	Recommended Quantity Admixture (Liters/m³)	Fiber (%)
	220	140	165	782	1,067	1.35	–

Properties	Cylindrical Strength (ksc)	Average	Standard deviation	Compressive strength (ksc)	Average	Standard deviation
Air content (%)[1]	180	1.5	0.3	7 days	115.63	24.74
Slump (mm)[2]		94.0	20.8	28 days	290.85	20.39
Initial setting time (min)[3]		158	7.6	90 days	308.62	10.94
Final setting time (min)[3]		319	2.5	365 days	380.32	5.99

Mix Symbol (MC280C211F144S75)	Cement (kg/m³)	Fly Ash (kg/m³)	Water (kg/m³)	Sand (kg/m³)	Recycled Aggregate 3/4" (kg/m³)	Recommended Quantity Admixture (Liters/m³)	Fiber (%)
	198	126	165	782	1,067	1.40	–

Properties	Cylindrical Strength (ksc)	Average	Standard deviation	Compressive strength (ksc)	Average	Standard deviation
Air content (%)[1]	180	1.1	0.2	7 days	83.23	13.99

(Continued)

TABLE 5.6 (Continued)
Typical Mix Proportions and Related Properties of RAC

Mix Symbol (MC280C211F144S75)	Cylindrical Strength (ksc)	Cement (kg/m³)	Fly Ash (kg/m³)	Water (kg/m³)	Sand (kg/m³)	Recycled Aggregate 3/4" (kg/m³)	Recommended Quantity Admixture (Liters/m³)	Fiber (%)
Slump (mm)[2]	88.0		6.0	28 days			118.84	11.14
Initial setting time (min)[3]	208		16.6	90 days			139.57	6.87
Final setting time (min)[3]	321		3.4	365 days			168.09	4.13

Mix Symbol (MC210C295F0S75)	Cylindrical Strength (ksc)	Cement (kg/m³)	Fly Ash (kg/m³)	Water (kg/m³)	Sand (kg/m³)	Recycled Aggregate 3/4" (kg/m³)	Recommended Quantity Admixture (Liters/m³)	Fiber (%)
Properties	180	250	60	165	769	1,039	1.25	–

Properties	Average		Standard deviation	Compressive strength (ksc)			Average	Standard deviation
Air content (%)[1]	1.6		0.2	7 days			104.82	7.14
Slump (mm)[2]	91.0		14.2	28 days			216.21	0.98
Initial setting time (min)[3]	128		3.6	90 days			309.64	1.76
Final setting time (min)[3]	272		32.9	365 days			359.82	5.95

Mix Symbol (MC210C280F0S75)	Cylindrical Strength (ksc)	Cement (kg/m³)	Fly Ash (kg/m³)	Water (kg/m³)	Sand (kg/m³)	Recycled Aggregate 3/4" (kg/m³)	Recommended Quantity Admixture (Liters/m³)	Fiber (%)
	200	196	124	155	870	1,135	0.50	–
Properties	Average		Standard deviation	Compressive strength (ksc)			Average	Standard deviation
Air content (%)[1]	1.2		0.3	7 days			65.36	4.87
Slump (mm)[2]	86.5		2.9	28 days			104.87	2.47
Initial setting time (min)[3]	138		9.8	90 days			143.25	1.63
Final setting time (min)[3]	298		26.8	365 days			157.89	0.58

Mix Symbol (MC210C320F0S100)	Cylindrical Strength (ksc)	Cement (kg/m³)	Fly Ash (kg/m³)	Water (kg/m³)	Sand (kg/m³)	Recycled Aggregate 3/4" (kg/m³)	Recommended Quantity Admixture (Liters/m³)	Fiber (%)
	200	204	130	155	870	1,135	0.55	–
Properties	Average		Standard deviation	Compressive strength (ksc)			Average	Standard deviation
Air content (%)[1]	1.7		0.3	7 days			101.68	7.11
Slump (mm)[2]	92.5		19.4	28 days			160.14	12.03
Initial setting time (min)[3]	178		1.7	90 days			191.47	6.49
Final setting time (min)[3]	252		9.2	365 days			220.53	1.44

(Continued)

TABLE 5.6 (Continued)
Typical Mix Proportions and Related Properties of RAC

Mix Symbol (MC240C345F0S100)	Cylindrical Strength (ksc)	Cement (kg/m³)	Fly Ash (kg/m³)	Water (kg/m³)	Sand (kg/m³)	Recycled Aggregate 3/4" (kg/m³)	Recommended Quantity Admixture (Liters/m³)	Fiber (%)
	210	184	116	155	870	1,135	0.60	–
Properties	Average		Standard deviation		Compressive strength (ksc)		Average	Standard deviation
Air content (%)[1]	1.3		0.1		7 days		65.73	10.49
Slump (mm)[2]	87.0		5.1		28 days		93.65	11.02
Initial setting time (min)[3]	198		17.4		90 days		131.59	6.47
Final setting time (min)[3]	355		10.7		365 days		153.04	3.29

Mix Symbol (MC280C370F0S100)	Cylindrical Strength (ksc)	Cement (kg/m³)	Fly Ash (kg/m³)	Water (kg/m³)	Sand (kg/m³)	Recycled Aggregate 3/4" (kg/m³)	Recommended Quantity Admixture (Liters/m³)	Fiber (%)
	210	230	54	155	870	1,135	0.70	–
Properties	Average		Standard deviation		Compressive strength (ksc)		Average	Standard deviation
Air content (%)[1]	1.8		0.1		7 days		59.14	3.65
Slump (mm)[2]	89.5		6.7		28 days		121.63	12.44

Initial setting time (min)[3]	118		14.9	90 days		183.06		6.74
Final setting time (min)[3]	364		41.7	365 days		202.41		2.56
Mix Symbol (MC280C410F0S175)	**Cylindrical Strength (ksc)**	**Cement (kg/m³)**	**Fly Ash (kg/m³)**	**Water (kg/m³)**	**Sand (kg/m³)**	**Recycled Aggregate 3/4″ (kg/m³)**	**Recommended Quantity Admixture (Liters/m³)**	**Fiber (%)**
Properties	240	250	60	155	860	1,120	1.40	–
	Average		Standard deviation	Compressive strength (ksc)		Average		Standard deviation
Air content (%)[1]	1.4		0.2	7 days		74.25		20.98
Slump (mm)[2]	93.5		10.9	28 days		147.56		16.02
Initial setting time (min)[3]	108		10.2	90 days		216.25		10.47
Final setting time (min)[3]	379		25.7	365 days		251.79		6.21
Mix Symbol (MC280C410F0S175)	**Cylindrical Strength (ksc)**	**Cement (kg/m³)**	**Fly Ash (kg/m³)**	**Water (kg/m³)**	**Sand (kg/m³)**	**Recycled Aggregate 3/4″ (kg/m³)**	**Recommended Quantity Admixture (Liters/m³)**	**Fiber (%)**
Properties	250	344	0	165	850	1,145	1.40	–
	Average		Standard deviation	Compressive strength (ksc)		Average		Standard deviation
Air content (%)[1]	1.9		0.2	7 days		129.51		17.47

(Continued)

TABLE 5.6 (Continued)
Typical Mix Proportions and Related Properties of RAC

Mix Symbol (MC280C410F0S175)	Cylindrical Strength (ksc)	Cement (kg/m³)	Fly Ash (kg/m³)	Water (kg/m³)	Sand (kg/m³)	Recycled Aggregate 3/4" (kg/m³)	Recommended Quantity Admixture (Liters/m³)	Fiber (%)
Slump (mm)[2]	90.5		13.5		28 days		147.44	15.62
Initial setting time (min)[3]	168		2.6		90 days		209.36	7.88
Final setting time (min)[3]	387		46.5		365 days		244.41	1.63

Mix Symbol (MC210C320F0S100)	Cylindrical Strength (ksc)	Cement (kg/m³)	Fly Ash (kg/m³)	Water (kg/m³)	Sand (kg/m³)	Recycled Aggregate 3/4" (kg/m³)	Recommended Quantity Admixture (Liters/m³)	Fiber (%)
	180	210	50	155	880	1,145	0.55	–

Properties	Average	Standard deviation	Compressive strength (ksc)	Average	Standard deviation
Air content (%)[1]	1.6	0.2	7 days	64.33	4.69
Slump (mm)[2]	91.0	14.2	28 days	78.41	3.16
Initial setting time (min)[3]	128	3.6	90 days	113.69	2.14
Final setting time (min)[3]	272	32.9	365 days	135.24	1.33

Mix Symbol (MC240C345F0S100)	Cylindrical Strength (ksc)	Cement (kg/m³)	Fly Ash (kg/m³)	Water (kg/m³)	Sand (kg/m³)	Recycled Aggregate 3/4" (kg/m³)	Recommended Quantity Admixture (Liters/m³)	Fiber (%)
Properties	210	310	0	165	870	1,135	0.60	–
	Average		Standard deviation	Compressive strength (ksc)			Average	Standard deviation
Air content (%)[1]	2.2		0.3	7 days			81.65	12.84
Slump (mm)[2]	83.5		4.8	28 days			118.48	9.67
Initial setting time (min)[3]	167		14.5	90 days			132.41	3.32
Final setting time (min)[3]	389		49.5	365 days			159.58	0.84

Mix Symbol (MC280C370F0S100)	Cylindrical Strength (ksc)	Cement (kg/m³)	Fly Ash (kg/m³)	Water (kg/m³)	Sand (kg/m³)	Recycled Aggregate 3/4" (kg/m³)	Recommended Quantity Admixture (Liters/m³)	Fiber (%)
Properties	250	226	144	155	850	1,145	0.70	–
	Average		Standard deviation	Compressive strength (ksc)			Average	Standard deviation
Air content (%)[1]	1.7		0.3	7 days			114.83	17.58
Slump (mm)[2]	92.5		19.4	28 days			185.03	11.36
Initial setting time (min)[3]	178		1.7	90 days			234.88	7.08
Final setting time (min)[3]	252		9.2	365 days			271.67	4.47

(Continued)

TABLE 5.6 (Continued)
Typical Mix Proportions and Related Properties of RAC

Mix Symbol (MC210C320F0S100)	Cylindrical Strength (ksc)	Cement (kg/m³)	Fly Ash (kg/m³)	Water (kg/m³)	Sand (kg/m³)	Recycled Aggregate 3/4" (kg/m³)	Recommended Quantity Admixture (Liters/m³)	Fiber (%)
	140	204	130	165	890	1,165	0.80	–

Properties	Average	Standard deviation	Compressive strength (ksc)	Average	Standard deviation
Air content (%)[1]	2.7	0.2	7 days	69.14	7.24
Slump (mm)[2]	83.0	10.5	28 days	104.89	6.33
Initial setting time (min)[3]	177	10.1	90 days	118.54	5.14
Final setting time (min)[3]	309	45.1	365 days	138.46	2.16

Mix Symbol (MC240C345F0S100)	Cylindrical Strength (ksc)	Cement (kg/m³)	Fly Ash (kg/m³)	Water (kg/m³)	Sand (kg/m³)	Recycled Aggregate 3/4" (kg/m³)	Recommended Quantity Admixture (Liters/m³)	Fiber (%)
	140	230	54	165	190	1,165	0.80	–

Properties	Average	Standard deviation	Compressive strength (ksc)	Average	Standard deviation
Air content (%)[1]	1.9	0.2	7 days	74.54	11.82
Slump (mm)[2]	90.5	13.5	28 days	98.93	3.67

(Continued from previous mix)

Properties	Average	Standard deviation	Compressive strength (ksc)	Average	Standard deviation
Initial setting time (min)[3]	168	2.6	90 days	134.89	5.77
Final setting time (min)[3]	387	46.5	365 days	150.22	2.10

Mix Symbol (MC280C370F0S100)	Cylindrical Strength (ksc)	Cement (kg/m³)	Fly Ash (kg/m³)	Water (kg/m³)	Sand (kg/m³)	Recycled Aggregate 3/4" (kg/m³)	Recommended Quantity Admixture (Liters/m³)	Fiber (%)
	250	216	138	155	850	1,145	0.90	–

Properties	Average	Standard deviation	Compressive strength (ksc)	Average	Standard deviation
Air content (%)[1]	2.8	0.1	7 days	94.14	20.11
Slump (mm)[2]	84.5	5.9	28 days	172.57	7.48
Initial setting time (min)[3]	247	18.9	90 days	236.28	5.97
Final setting time (min)[3]	319	30.2	365 days	281.09	1.33

Mix Symbol (MC210C255F65S100)	Cylindrical Strength (ksc)	Cement (kg/m³)	Fly Ash (kg/m³)	Water (kg/m³)	Sand (kg/m³)	Recycled Aggregate 3/4" (kg/m³)	Recommended Quantity Admixture (Liters/m³)	Fiber (%)
	140	190	44	155	800	1,005	1.20	–

Properties	Average	Standard deviation	Compressive strength (ksc)	Average	Standard deviation
Air content (%)[1]	2.0	0.3	7 days	61.36	6.48

(Continued)

TABLE 5.6 (Continued)
Typical Mix Proportions and Related Properties of RAC

Mix Symbol (MC210C255F65S100)	Cylindrical Strength (ksc)	Cement (kg/m³)	Fly Ash (kg/m³)	Water (kg/m³)	Sand (kg/m³)	Recycled Aggregate 3/4" (kg/m³)	Recommended Quantity Admixture (Liters/m³)	Fiber (%)
Slump (mm)[2]	81.5		2.6	28 days			94.56	3.11
Initial setting time (min)[3]	207		16.7	90 days			133.92	4.63
Final setting time (min)[3]	399		26.8	365 days			154.37	2.45
Mix Symbol (MC240C275F70S100)	**Cylindrical Strength (ksc)**	**Cement (kg/m³)**	**Fly Ash (kg/m³)**	**Water (kg/m³)**	**Sand (kg/m³)**	**Recycled Aggregate 3/4" (kg/m³)**	**Recommended Quantity Admixture (Liters/m³)**	**Fiber (%)**
Properties	180	284	0	165	810	1,005	130	–
	Average		Standard deviation	Compressive strength (ksc)			Average	Standard deviation
Air content (%)[1]	3.9		0.1	7 days			92.67	23.66
Slump (mm)[2]	79.5		18.4	28 days			173.26	0.78
Initial setting time (min)[3]	196		17.7	90 days			244.83	3.48
Final setting time (min)[3]	368		28.4	365 days			286.35	6.44

Mix Symbol (MC280C295F75S100)	Cylindrical Strength (ksc)	Cement (kg/m³)	Fly Ash (kg/m³)	Water (kg/m³)	Sand (kg/m³)	Recycled Aggregate 3/4" (kg/m³)	Recommended Quantity Admixture (Liters/m³)	Fiber (%)
	180	180	114	155	805	1,005	140	–

Properties	Average		Standard deviation	Compressive strength (ksc)			Average	Standard deviation
Air content (%)[1]	2.3		0.1	7 days			164.44	16.22
Slump (mm)[2]	84.0		9.4	28 days			179.87	7.36
Initial setting time (min)[3]	157		9.0	90 days			268.48	10.07
Final setting time (min)[3]	329		46.2	365 days			342.55	6.04

Mix Symbol (MC210C255F65S100)	Cylindrical Strength (ksc)	Cement (kg/m³)	Fly Ash (kg/m³)	Water (kg/m³)	Sand (kg/m³)	Recycled Aggregate 3/4" (kg/m³)	Recommended Quantity Admixture (Liters/m³)	Fiber (%)
	180	190	120	155	820	1,005	140	–

Properties	Average		Standard deviation	Compressive strength (ksc)			Average	Standard deviation
Air content (%)[1]	3.8		0.3	7 days			61.33	9.88
Slump (mm)[2]	77.5		19.3	28 days			119.58	4.61
Initial setting time (min)[3]	206		7.6	90 days			173.31	5.46
Final setting time (min)[3]	378		16.8	365 days			201.06	4.63

(Continued)

TABLE 5.6 (Continued)
Typical Mix Proportions and Related Properties of RAC

Mix Symbol (MC240C275F70S100)	Cylindrical Strength (ksc)	Cement (kg/m³)	Fly Ash (kg/m³)	Water (kg/m³)	Sand (kg/m³)	Recycled Aggregate 3/4" (kg/m³)	Recommended Quantity Admixture (Liters/m³)	Fiber (%)
	180	168	106	155	800	1,005	1.20	–
Properties	Average		Standard deviation	Compressive strength (ksc)		Average		Standard deviation
Air content (%)[1]	3.0		0.1	7 days		152.03		13.58
Slump (mm)[2]	87.0		5.1	28 days		269.77		24.71
Initial setting time (min)[3]	158		7.6	90 days		384.41		16.59
Final setting time (min)[3]	319		2.5	365 days		432.63		4.77

Mix Symbol (MC280C295F75S100)	Cylindrical Strength (ksc)	Cement (kg/m³)	Fly Ash (kg/m³)	Water (kg/m³)	Sand (kg/m³)	Recycled Aggregate 3/4" (kg/m³)	Recommended Quantity Admixture (Liters/m³)	Fiber (%)
	180	210	50	155	810	1,005	130	–
Properties	Average		Standard deviation	Compressive strength (ksc)		Average		Standard deviation
Air content (%)[1]	1.0		0.1	7 days		97.55		0.84
Slump (mm)[2]	91.0		14.2	28 days		140.59		3.26

Properties	Average	Standard deviation	Compressive strength (ksc)	Average	Standard deviation
Initial setting time (min)[3]	138	9.8	90 days	196.22	8.42
Final setting time (min)[3]	387	46.5	365 days	226.24	3.92

Mix Symbol (MC210C255F65S100)	Cylindrical Strength (ksc)	Cement (kg/m³)	Fly Ash (kg/m³)	Water (kg/m³)	Sand (kg/m³)	Recycled Aggregate 3/4" (kg/m³)	Recommended Quantity Admixture (Liters/m³)	Fiber (%)
	210	310	0	165	805	1,005	140	–

Properties	Average	Standard deviation	Compressive strength (ksc)	Average	Standard deviation
Air content (%)[1]	3.4	0.2	7 days	56.28	12.41
Slump (mm)[2]	88.0	6.0	28 days	112.63	10.36
Initial setting time (min)[3]	168	2.6	90 days	158.88	6.20
Final setting time (min)[3]	364	41.7	365 days	181.36	3.39

Mix Symbol (MC240C275F70S100)	Cylindrical Strength (ksc)	Cement (kg/m³)	Fly Ash (kg/m³)	Water (kg/m³)	Sand (kg/m³)	Recycled Aggregate 3/4" (kg/m³)	Recommended Quantity Admixture (Liters/m³)	Fiber (%)
	210	196	124	155	820	1,005	140	–

Properties	Average	Standard deviation	Compressive strength (ksc)	Average	Standard deviation
Air content (%)[1]	1.2	0.1	7 days	89.55	0.88

(Continued)

TABLE 5.6 (Continued)
Typical Mix Proportions and Related Properties of RAC

Mix Symbol (MC240C275F70S100)	Cylindrical Strength (ksc)	Cement (kg/m³)	Fly Ash (kg/m³)	Water (kg/m³)	Sand (kg/m³)	Recycled Aggregate 3/4" (kg/m³)	Recommended Quantity Admixture (Liters/m³)	Fiber (%)
Slump (mm)[2]	90.5		13.5	28 days			108.47	1.24
Initial setting time (min)[3]	178		1.7	90 days			149.53	0.65
Final setting time (min)[3]	252		9.2	365 days			181.23	2.95

Mix Symbol (MC280C295F75S100)	Cylindrical Strength (ksc)	Cement (kg/m³)	Fly Ash (kg/m³)	Water (kg/m³)	Sand (kg/m³)	Recycled Aggregate 3/4" (kg/m³)	Recommended Quantity Admixture (Liters/m³)	Fiber (%)
Properties	210	204	130	155	800	1,005	1.20	–

Properties	Average	Standard deviation		Compressive strength (ksc)	Average	Standard deviation
Air content (%)[1]	3.5	0.3		7 days	121.68	11.93
Slump (mm)[2]	89.5	6.7		28 days	180.14	7.01
Initial setting time (min)[3]	128	3.6		90 days	211.47	6.33
Final setting time (min)[3]	321	3.4		365 days	240.58	2.48

Mix Symbol (MC280C410F0S175)	Cylindrical Strength (ksc)	Cement (kg/m³)	Fly Ash (kg/m³)	Water (kg/m³)	Sand (kg/m³)	Recycled Aggregate 3/4" (kg/m³)	Recommended Quantity Admixture (Liters/m³)	Fiber (%)
	210	184	116	155	810	1,005	130	–
Properties	Average		Standard deviation	Compressive strength (ksc)		Average		Standard deviation
Air content (%)[1]	3.9		0.2	7 days		163.66		4.89
Slump (mm)[2]	94.0		20.8	28 days		298.63		0.49
Initial setting time (min)[3]	208		16.6	90 days		413.54		1.83
Final setting time (min)[3]	398		11.0	365 days		460.47		7.62

Mix Symbol (MC280C327F83S175)	Cylindrical Strength (ksc)	Cement (kg/m³)	Fly Ash (kg/m³)	Water (kg/m³)	Sand (kg/m³)	Recycled Aggregate 3/4" (kg/m³)	Recommended Quantity Admixture (Liters/m³)	Fiber (%)
	210	230	54	155	805	1,005	140	–
Properties	Average		Standard deviation	Compressive strength (ksc)		Average		Standard deviation
Air content (%)[1]	3.6		0.2	7 days		183.27		8.47
Slump (mm)[2]	86.5		2.9	28 days		289.67		3.62
Initial setting time (min)[3]	108		10.2	90 days		421.68		10.87
Final setting time (min)[3]	379		25.7	365 days		467.09		5.33

(Continued)

TABLE 5.6 (Continued)
Typical Mix Proportions and Related Properties of RAC

Mix Symbol (MC280C327F83S175)	Cylindrical Strength (ksc)	Cement (kg/m³)	Fly Ash (kg/m³)	Water (kg/m³)	Sand (kg/m³)	Recycled Aggregate 3/4" (kg/m³)	Recommended Quantity Admixture (Liters/m³)	Fiber (%)
Properties	240	334	0	165	820	1,005	140	–

Properties	Average	Standard deviation	Compressive strength (ksc)	Average	Standard deviation
Air content (%)[1]	4.0	0.3	7 days	174.47	12.20
Slump (mm)[2]	87.5	4.8	28 days	281.07	11.01
Initial setting time (min)[3]	118	14.9	90 days	386.93	13.59
Final setting time (min)[3]	355	10.7	365 days	465.25	8.47

Mix Symbol (MC280C327F83S175)	Cylindrical Strength (ksc)	Cement (kg/m³)	Fly Ash (kg/m³)	Water (kg/m³)	Sand (kg/m³)	Recycled Aggregate 3/4" (kg/m³)	Recommended Quantity Admixture (Liters/m³)	Fiber (%)
Properties	240	210	134	155	770	950	2.20	–

Properties	Average	Standard deviation	Compressive strength (ksc)	Average	Standard deviation
Air content (%)[1]	3.7	0.3	7 days	174.58	26.48
Slump (mm)[2]	92.5	19.4	28 days	299.62	21.66

(Continued from previous page)

Properties	Average	Standard deviation	Compressive strength (ksc)	Average	Standard deviation
Initial setting time (min)[3]	198	17.4	90 days	415.26	13.23
Final setting time (min)[3]	272	32.9	365 days	479.62	7.04

Mix Symbol (MC300C410F0S100)

Cylindrical Strength (ksc)	Cement (kg/m³)	Fly Ash (kg/m³)	Water (kg/m³)	Sand (kg/m³)	Recycled Aggregate 3/4" (kg/m³)	Recommended Quantity Admixture (Liters/m³)	Fiber (%)
240	220	140	155	800	1,010	0.50	–

Properties	Average	Standard deviation	Compressive strength (ksc)	Average	Standard deviation
Air content (%)[1]	1.1	0.3	7 days	101.64	14.18
Slump (mm)[2]	93.5	10.9	28 days	156.24	7.43
Initial setting time (min)[3]	148	12.8	90 days	192.57	2.63
Final setting time (min)[3]	298	26.8	365 days	213.25	4.44

Mix Symbol (MC280C275F70S75)

Cylindrical Strength (ksc)	Cement (kg/m³)	Fly Ash (kg/m³)	Water (kg/m³)	Sand (kg/m³)	Recycled Aggregate 3/4" (kg/m³)	Recommended Quantity Admixture (Liters/m³)	Fiber (%)
240	198	126	155	805	1,050	0.60	–

Properties	Average	Standard deviation	Compressive strength (ksc)	Average	Standard deviation
Air content (%)[1]	2.9	0.2	7 days	131.41	13.62

(Continued)

TABLE 5.6 (Continued)
Typical Mix Proportions and Related Properties of RAC

Mix Symbol (MC280C275F70S75)	Cylindrical Strength (ksc)	Cement (kg/m³)	Fly Ash (kg/m³)	Water (kg/m³)	Sand (kg/m³)	Recycled Aggregate 3/4" (kg/m³)	Recommended Quantity Admixture (Liters/m³)	Fiber (%)
Slump (mm)[2]		63.5	10.5		28 days		231.01	16.87
Initial setting time (min)[3]		143	5.8		90 days		325.62	13.02
Final setting time (min)[3]		350	49.8		365 days		371.44	7.14

Mix Symbol (MC320C315F80S75)	Cylindrical Strength (ksc)	Cement (kg/m³)	Fly Ash (kg/m³)	Water (kg/m³)	Sand (kg/m³)	Recycled Aggregate 3/4" (kg/m³)	Recommended Quantity Admixture (Liters/m³)	Fiber (%)
	240	250	60	155	790	1,050	0.60	–

Properties	Average	Standard deviation		Compressive strength (ksc)		Average	Standard deviation
Air content (%)[1]	3.1	0.1		7 days		112.67	20.11
Slump (mm)[2]	61.5	3.9		28 days		193.21	7.48
Initial setting time (min)[3]	103	1.3		90 days		264.83	5.97
Final setting time (min)[3]	390	17.8		365 days		306.35	1.33

Mix Symbol (MC280C275F70S75)	Cylindrical Strength (ksc)	Cement (kg/m³)	Fly Ash (kg/m³)	Water (kg/m³)	Sand (kg/m³)	Recycled Aggregate 3/4" (kg/m³)	Recommended Quantity Admixture (Liters/m³)	Fiber (%)
Properties	250	344	0	165	811	1,097	0.60	–
	Average		Standard deviation	Compressive strength (ksc)			Average	Standard deviation
Air content (%)[1]	3.0		0.3	7 days			112.84	17.58
Slump (mm)[2]	64.5		11.7	28 days			193.25	11.36
Initial setting time (min)[3]	153		6.9	90 days			264.84	7.08
Final setting time (min)[3]	355		7.7	365 days			303.03	4.47

Mix Symbol (MC320C315F80S75)	Cylindrical Strength (ksc)	Cement (kg/m³)	Fly Ash (kg/m³)	Water (kg/m³)	Sand (kg/m³)	Recycled Aggregate 3/4" (kg/m³)	Recommended Quantity Admixture (Liters/m³)	Fiber (%)
Properties	250	216	138	155	793	1,072	0.70	–
	Average		Standard deviation	Compressive strength (ksc)			Average	Standard deviation
Air content (%)[1]	1.2		0.2	7 days			114.14	7.14
Slump (mm)[2]	60.5		4.2	28 days			192.53	13.93
Initial setting time (min)[3]	113		2.4	90 days			256.28	8.55
Final setting time (min)[3]	395		44.1	365 days			301.09	4.26

(Continued)

TABLE 5.6 (Continued)
Typical Mix Proportions and Related Properties of RAC

Mix Symbol (MC280C345F0S75)	Cylindrical Strength (ksc)	Cement (kg/m³)	Fly Ash (kg/m³)	Water (kg/m³)	Sand (kg/m³)	Recycled Aggregate 3/4" (kg/m³)	Recommended Quantity Admixture (Liters/m³)	Fiber (%)
Properties	250	226	144	155	811	1,097	0.60	–
	Average		Standard deviation	Compressive strength (ksc)			Average	Standard deviation
Air content (%)[1]	2.4		0.2	7 days			94.36	6.21
Slump (mm)[2]	54.0		9.4	28 days			164.52	7.44
Initial setting time (min)[3]	161		6.9	90 days			243.20	8.99
Final setting time (min)[3]	290		8.6	365 days			258.43	5.25

Mix Symbol (MC280C275F70S75)	Cylindrical Strength (ksc)	Cement (kg/m³)	Fly Ash (kg/m³)	Water (kg/m³)	Sand (kg/m³)	Recycled Aggregate 3/4" (kg/m³)	Recommended Quantity Admixture (Liters/m³)	Fiber (%)
Properties	250	204	130	155	811	1,097	1.20	–
	Average		Standard deviation	Compressive strength (ksc)			Average	Standard deviation
Air content (%)[1]	2.7		0.3	7 days			93.25	20.14
Slump (mm)[2]	70.0		7.9	28 days			160.23	15.23

(Continuation of preceding mix design)

Property	Average	Standard deviation
Initial setting time (min)[3]	154	18.1
Final setting time (min)[3]	316	22.6

Compressive strength (ksc)	Average	Standard deviation
90 days	224.88	7.48
365 days	267.48	5.32

Mix Symbol (MC320C455F0S100)

Cylindrical Strength (ksc)	Cement (kg/m³)	Fly Ash (kg/m³)	Water (kg/m³)	Sand (kg/m³)	Recycled Aggregate 3/4" (kg/m³)	Recommended Quantity Admixture (Liters/m³)	Fiber (%)
250	260	62	155	782	1,016	0.70	–

Properties	Average	Standard deviation
Air content (%)[1]	2.6	0.1
Slump (mm)[2]	71.0	14.7
Initial setting time (min)[3]	204	17.9
Final setting time (min)[3]	306	24.8

Compressive strength (ksc)	Average	Standard deviation
7 days	105.80	14.63
28 days	166.59	2.47
90 days	229.32	8.22
365 days	264.83	1.93

Mix Symbol (MC320C386F69S100)

Cylindrical Strength (ksc)	Cement (kg/m³)	Fly Ash (kg/m³)	Water (kg/m³)	Sand (kg/m³)	Recycled Aggregate 3/4" (kg/m³)	Recommended Quantity Admixture (Liters/m³)	Fiber (%)
280	360	0	165	782	1,016	0.80	–

Properties	Average	Standard deviation
Air content (%)[1]	2.5	0.3

Compressive strength (ksc)	Average	Standard deviation
7 days	99.52	7.44

(Continued)

TABLE 5.6 (Continued)
Typical Mix Proportions and Related Properties of RAC

Mix Symbol (MC320C386F69S100)	Cylindrical Strength (ksc)	Cement (kg/m³)	Fly Ash (kg/m³)	Water (kg/m³)	Sand (kg/m³)	Recycled Aggregate 3/4" (kg/m³)	Recommended Quantity Admixture (Liters/m³)	Fiber (%)
Slump (mm)[2]	70.5		11.7		28 days		160.36	12.93
Initial setting time (min)[3]	194		10.8		90 days		226.32	8.25
Final setting time (min)[3]	296		23.4		365 days		562.83	4.67
Mix Symbol (MC320C435F0S100)	**Cylindrical Strength (ksc)**	**Cement (kg/m³)**	**Fly Ash (kg/m³)**	**Water (kg/m³)**	**Sand (kg/m³)**	**Recycled Aggregate 3/4" (kg/m³)**	**Recommended Quantity Admixture (Liters/m³)**	**Fiber (%)**
Properties	280	226	144	155	782	1,016	0.70	–
	Average		Standard deviation		Compressive strength (ksc)		Average	Standard deviation
Air content (%)[1]	3.8		0.1		7 days		134.83	20.01
Slump (mm)[2]	95.0		5.7		28 days		205.03	14.99
Initial setting time (min)[3]	239		1.4		90 days		254.88	9.82
Final setting time (min)[3]	355		14.3		365 days		291.67	4.67

Mix Symbol (MC320C369F66S100)	Cylindrical Strength (ksc)	Cement (kg/m³)	Fly Ash (kg/m³)	Water (kg/m³)	Sand (kg/m³)	Recycled Aggregate 3/4" (kg/m³)	Recommended Quantity Admixture (Liters/m³)	Fiber (%)
	280	234	150	155	782	1,016	1.40	–

Properties	Average	Standard deviation	Compressive strength (ksc)	Average	Standard deviation
Air content (%)[1]	3.7	0.3	7 days	113.20	23.66
Slump (mm)[2]	94.5	14.2	28 days	218.56	5.78
Initial setting time (min)[3]	216	14.6	90 days	259.47	3.48
Final setting time (min)[3]	279	41.7	365 days	311.53	6.44

Mix Symbol (MC240C267F68S75)	Cylindrical Strength (ksc)	Cement (kg/m³)	Fly Ash (kg/m³)	Water (kg/m³)	Sand (kg/m³)	Recycled Aggregate 3/4" (kg/m³)	Recommended Quantity Admixture (Liters/m³)	Fiber (%)
	140	260	20	165	740	990	1.30	–

Properties	Average	Standard deviation	Compressive strength (ksc)	Average	Standard deviation
Air content (%)[1]	2.2	0.1	7 days	112.89	30.01
Slump (mm)[2]	51.5	0.8	28 days	233.36	14.55

(Continued)

TABLE 5.6 (Continued)
Typical Mix Proportions and Related Properties of RAC

Mix Symbol (MC240C267F68S75)	Cylindrical Strength (ksc)	Cement (kg/m³)	Fly Ash (kg/m³)	Water (kg/m³)	Sand (kg/m³)	Recycled Aggregate 3/4" (kg/m³)	Recommended Quantity Admixture (Liters/m³)	Fiber (%)
Initial setting time (min)[3]	133		10.2		90 days		349.54	8.74
Final setting time (min)[3]	287		21.5		365 days		374.82	2.63

Mix Symbol (MC280C410F0S175)	Cylindrical Strength (ksc)	Cement (kg/m³)	Fly Ash (kg/m³)	Water (kg/m³)	Sand (kg/m³)	Recycled Aggregate 3/4" (kg/m³)	Recommended Quantity Admixture (Liters/m³)	Fiber (%)
	140	166	54	155	811	1,098	2.00	–
Properties	Average		Standard deviation		Compressive strength (ksc)		Average		Standard deviation
Air content (%)[1]	2.3		0.1		7 days		108.36	31.52
Slump (mm)[2]	150.5		5.3		28 days		233.63	14.69
Initial setting time (min)[3]	117		4.7		90 days		328.47	12.33
Final setting time (min)[3]	264		34.8		365 days		369.22	4.96

Mix Symbol (MC280C285F125S175) Properties	Cylindrical Strength (ksc)	Cement (kg/m³)	Fly Ash (kg/m³)	Water (kg/m³)	Sand (kg/m³)	Recycled Aggregate 3/4" (kg/m³)	Recommended Quantity Admixture (Liters/m³)	Fiber (%)
Properties	140	174	50	155	811	1,098	2.00	–
	Average		Standard deviation	Compressive strength (ksc)			Average	Standard deviation
Air content (%)[1]	2.4		0.2	7 days			154.47	12.20
Slump (mm)[2]	188.5		13.7	28 days			261.07	11.01
Initial setting time (min)[3]	125		12.9	90 days			364.24	16.35
Final setting time (min)[3]	275		14.7	365 days			412.59	8.77

Mix Symbol (MC320C460F0S175) Properties	Cylindrical Strength (ksc)	Cement (kg/m³)	Fly Ash (kg/m³)	Water (kg/m³)	Sand (kg/m³)	Recycled Aggregate 3/4" (kg/m³)	Recommended Quantity Admixture (Liters/m³)	Fiber (%)
Properties	280	214	136	155	870	1,135	2.30	–
	Average		Standard deviation	Compressive strength (ksc)			Average	Standard deviation
Air content (%)[1]	2.6		0.2	7 days			103.04	6.81
Slump (mm)[2]	190.5		1.2	28 days			121.69	12.58

(Continued)

TABLE 5.6 (Continued)
Typical Mix Proportions and Related Properties of RAC

Mix Symbol (MC320C460F0S175)	Cylindrical Strength (ksc)	Cement (kg/m³)	Fly Ash (kg/m³)	Water (kg/m³)	Sand (kg/m³)	Recycled Aggregate 3/4" (kg/m³)	Recommended Quantity Admixture (Liters/m³)	Fiber (%)
Initial setting time (min)[3]	203			90 days			194.11	7.66
Final setting time (min)[3]	325			365 days			212.77	1.07
Mix Symbol (MC320C320F140S175)	**Cylindrical Strength (ksc)**	**Cement (kg/m³)**	**Fly Ash (kg/m³)**	**Water (kg/m³)**	**Sand (kg/m³)**	**Recycled Aggregate 3/4" (kg/m³)**	**Recommended Quantity Admixture (Liters/m³)**	**Fiber (%)**
Properties	300	240	154	155	740	990	2.33	–
	Average		Standard deviation	Compressive strength (ksc)			Average	Standard deviation
Air content (%)[1]	2.5		0.1	7 days			147.46	24.77
Slump (mm)[2]	187.0		24.5	28 days			275.47	16.32
Initial setting time (min)[3]	196		6.6	90 days			366.93	7.47
Final setting time (min)[3]	303		33.3	365 days			445.25	3.62

Mix Symbol (MC240C267F68S75)	Cylindrical Strength (ksc)	Cement (kg/m³)	Fly Ash (kg/m³)	Water (kg/m³)	Sand (kg/m³)	Recycled Aggregate 3/4" (kg/m³)	Recommended Quantity Admixture (Liters/m³)	Fiber (%)
	240	282	68	170	811	1,098	1.36	–
Properties	Average		Standard deviation	Compressive strength (ksc)			Average	Standard deviation
Air content (%)[1]	2.1		0.2	7 days			63.62	10.69
Slump (mm)[2]	60.5		6.8	28 days			106.27	7.60
Initial setting time (min)[3]	124		16.5	90 days			146.58	1.47
Final setting time (min)[3]	256		17.0	365 days			151.36	3.66

Mix Symbol (MC240C320F0S75)	Cylindrical Strength (ksc)	Cement (kg/m³)	Fly Ash (kg/m³)	Water (kg/m³)	Sand (kg/m³)	Recycled Aggregate 3/4" (kg/m³)	Recommended Quantity Admixture (Liters/m³)	Fiber (%)
	240	335	0	170	811	1,098	1.30	–
Properties	Average		Standard deviation	Compressive strength (ksc)			Average	Standard deviation
Air content (%)[1]	1.9		0.2	7 days			53.69	9.54
Slump (mm)[2]	67.5		10.5	28 days			80.89	7.32

(Continued)

TABLE 5.6 (Continued)
Typical Mix Proportions and Related Properties of RAC

Mix Symbol (MC240C320F0S75)	Cylindrical Strength (ksc)	Cement (kg/m³)	Fly Ash (kg/m³)	Water (kg/m³)	Sand (kg/m³)	Recycled Aggregate 3/4" (kg/m³)	Recommended Quantity Admixture (Liters/m³)	Fiber (%)
Initial setting time (min)[3]	174		17.2	90 days			117.64	5.66
Final setting time (min)[3]	276		21.7	365 days			130.36	2.94

Mix Symbol (MC240C247F63S75)	Cylindrical Strength (ksc)	Cement (kg/m³)	Fly Ash (kg/m³)	Water (kg/m³)	Sand (kg/m³)	Recycled Aggregate 3/4" (kg/m³)	Recommended Quantity Admixture (Liters/m³)	Fiber (%)
	240	362	63	155	870	1,135	1.26	–

Properties	Average		Standard deviation	Compressive strength (ksc)			Average	Standard deviation
Air content (%)[1]	2.2		0.3	7 days			55.32	0.84
Slump (mm)[2]	67.0		5.4	28 days			103.62	2.36
Initial setting time (min)[3]	134		15.1	90 days			131.25	4.87
Final setting time (min)[3]	266		17.9	365 days			159.53	1.69

Mix Symbol (MC240C267F68S75)	Cylindrical Strength (ksc)	Cement (kg/m³)	Fly Ash (kg/m³)	Water (kg/m³)	Sand (kg/m³)	Recycled Aggregate 3/4" (kg/m³)	Recommended Quantity Admixture (Liters/m³)	Fiber (%)
Properties	140	154	60	155	850	1,105	1.36	–
	Average		Standard deviation	Compressive strength (ksc)			Average	Standard deviation
Air content (%)[1]	2.0		0.3	7 days			120.36	13.56
Slump (mm)[2]	68.0		15.3	28 days			225.19	10.05
Initial setting time (min)[3]	184		9.9	90 days			291.18	9.45
Final setting time (min)[3]	286		20.9	365 days			337.08	5.63

Mix Symbol (MC240C284F51S75)	Cylindrical Strength (ksc)	Cement (kg/m³)	Fly Ash (kg/m³)	Water (kg/m³)	Sand (kg/m³)	Recycled Aggregate 3/4" (kg/m³)	Recommended Quantity Admixture (Liters/m³)	Fiber (%)
Properties	140	335	50	155	860	1,120	1.36	–
	Average		Standard deviation	Compressive strength (ksc)			Average	Standard deviation
Air content (%)[1]	2.4		0.2	7 days			73.25	6.41
Slump (mm)[2]	69.0		2.9	28 days			117.82	5.33
Initial setting time (min)[3]	114		12.7	90 days			186.13	4.01
Final setting time (min)[3]	346		33.4	365 days			199.88	2.89

TABLE 5.7

Chemical Compositions and Physical Properties of Type 1 Portland Cements and FA

Chemical Compositions (% by Mass)	SiO_2	Al_2O_3	Fe_2O_3	CaO	MgO	Na_2O	SO_3	Free CaO
Type 1 Portland Cement	20.84	5.22	3.20	66.28	1.24	0.10	2.41	0.99
FA	42.10	21.80	11.22	13.56	2.41	2.90	1.88	1.44

Physical Properties	Type 1 Portland Cement	FA
Loss on Ignition (%)	0.96	2.33
Moisture Content (%)	0.19	1.50
Blaine Surface Area (cm²/g)		
Fineness (Particle Size, % Retained)	3,200	2,850
− ≥ 75 μm	0.50	0.56
− 75 μm	5.25	8.25
− 45 μm	3.60	4.76
− ≤ 36 μm	90.62	86.43
Fineness (Retained) on 45 Micron		
(No. 325)	5.75	4.90
Water Requirement (%)	100	97
Bulk Density (kg/l)	1.03	0.51
Specific Gravity	3.15	2.13

- River sand with gradation conforming to the ASTM C33 (ASTM C33/ C33M–1601 standard specification for concrete aggregates).
- Crushed lime recycled aggregate rock with gradation conforming to the ASTM C33 (ASTM C33/C33M–1601 standard specification for RAC aggregates).
- The chemical admixture used superplasticizer that conforms to the ASTM C494 (ASTM C494/C494M–16 Standard specification for chemical admixtures for concrete); that is, the superplasticizer had a recommended dosage rate of cementitious materials (per 100 of a kilogram of cementitious materials).

5.2.4 SELF-COMPACTING RAC

Self-compacting RAC, often known as SCC, is a fundamentally unique kind of contemporary RAC widely regarded as the most promising development in concrete engineering. This is because SCC offers a number of benefits over conventional RAC. The SCC may be substantially compressed on its gravitational weight and does not need extra compaction efforts or energy. These three primary behaviours characterize the SCC including the capability of flowing under its weight without experiencing vibration, the capability of flowing through extremely crowded reinforcement under its weight, and the capability of being homogenous without the aggregates becoming segregated.

The capacity of self-consolidating RAC to flow more efficiently than traditional RAC is the primary distinction between the two types of RAC. The bleeding in conventional RAC is rather considerable, as shown by the slump test results conducted in accordance with ASTM C143, which show that the bleeding is more than 200 mm. Because SCC has a slump flow of more than 600 mm per ASTM C 1611 (2020), it produces a high degree of cohesiveness, and allows the RAC to flow into the casing without needing vibration to be applied. In addition, the viscosity of SCC is high enough to overcome the resistance caused by the coarse aggregates' friction with one another. When the RAC is poured into the casing, this feature ensures that the aggregates do not get separated, which would otherwise impede the flow of the RAC. The coarse aggregates and the mortar must not get separated in the SCC.

*   Filling ability: SCC can flow through heavily congested reinforcement under its weight, as well as the ability to become homogeneous without the aggregates becoming segregated, as stated by the European Federation of National Associations Representing Producers and Applicators of Specialist Building Products for Concrete (EFNARC) and the American Concrete Institute (ACI). Consequently, the new SCC can flow under its weight and through highly packed reinforcement. Other important parameters include the effect on deformability, which refers to the reduction of internal friction between particles. This can be accomplished by either lowering the surface tension through a superplasticizer or increasing the volume of coarse and fine aggregates, as well as by increasing the volume of the paste to improve its filling ability. Well-graded cement and powder can maintain a high water-to-cement ratio. This decreases inter-particle friction and makes the paste less sticky, both of which contribute to a reduction in the amount of segregation that occurs and helps to prevent excessive bleeding. Some of the bleed water that is produced makes its way to the top surface of the RAC, while other bleed water stays trapped in bleed channels and beneath various impediments, such as aggregate and reinforcement. The compressive strength and the durability of RAC are impacted by bleeding, and the water-to-cement ratio in this instance is relatively high. Both of these factors contribute to the overall quality of the RAC. Through either their physical or chemical impacts, the use of fine filler has the potential to improve several elements of cement-based systems. Because of the particles' tiny size, several physical phenomena are linked with them. These effects may enhance the packing density of powder and minimize interstitial spaces, which leads to a decrease in the quantity of water entrapped in the system. It has been claimed that using a constantly graded skeleton of powder may cut down on the amount of powder necessary to guarantee that the RAC has sufficient deformability. However, an excessive amount of tiny particles might lead to a significant increase in the specific surface area of the powder. Consequently, a greater quantity of water to obtain the desired consistency. In both binary and ternary systems, the addition of mineral additives is an essential step. However, the benefits of a specific material might sometimes make up for the drawbacks of that material. For instance, while the presence

of a substance with a high water-absorption effect has a detrimental impact on RAC, such a material would also contribute to the RAC's development of compact strength and endurance.

- Passing ability: The ability to flow through complicated reinforcing structures that are close together is related to the flowability of RAC in confined places like molds that include reinforced RAC. When choosing the coarse aggregates' size and form and the RAC's mortar volume, it is necessary to consider the critical characteristics of the space and layout of the reinforced structure. These criteria must be taken into consideration. If the arrangement of the reinforcing structures is particularly dense, the quantity of paste included in the RAC has to be increased correspondingly to the coarse particles. According to the definition provided by the Technical Committee of the Reunion Internationale des Laboratoires Experts des Matériaux, Systèmes de Construction et Ouvrages (RILEM) (Wardeh et al., 2014), flowability is the capacity of a material to pass through a variety of obstacles and narrow sections in the formwork and closely spaced reinforcing bars without segregation or blocking. Flowability also refers to the capacity of a material to flow through tight openings, such as spaces between steel reinforcing bars.
- Resistance to segregation: The capacity of SCC to become homogenous without its aggregates being separated is the most important characteristic of this material. High fluidity and enough stability are the minimum requirements for SCC to be considered acceptable. Stability can be broken down into two categories: dynamic and static. In terms of stability, dynamic stability refers to the resistance of RAC to the separation of constituents during transport, placement, and the casting process, and (ii) static stability refers to the RAC's resistance to bleeding, segregation, and settlement after casting while the RAC is still in a plastic state. Static segregation occurs in RAC when the yield stress of the suspending matrix is not adequate to maintain the weight of the aggregate after its buoyancy is subtracted from the total weight of the aggregate. The main aspects of segregation are (i) the bleeding of water, (ii) the separation of paste and aggregate, (iii) the separation of coarse aggregate that leads to obstructions, and (iv) the lack of consistency in the distribution of air pores. Either the amount of powder on the surface or the amount of water in the mixture may be decreased to alleviate the issue of water and aggregates being separated from one another.

Self-Compacting Recycled Aggregate Concrete Materials

- Type 1 hydraulic Portland cements conforming to ASTM C150 (ASTM C150/C150M–17 Standard specifications for Portland cement) were used throughout concrete mixtures. Their chemical compositions and physical properties are shown in Tables 5.7 and 5.8.
- The ASTM C618 (ASTM C618–15 Standard specification for coal fly ash and raw or calcined natural pozzolan for use in concrete) classifies the PFA as low calcium (Type F).

TABLE 5.8

Typical Mix Proportions and Related Properties of RAC

Mix Symbol (SC100C245F0S75)	Cylindrical Strength (ksc)	Cement (kg/m³)	Fly Ash (kg/m³)	Water (kg/m³)	Sand (kg/m³)	Recycled Aggregate 3/4" (kg/m³)	Recommended Quantity admixture (liters/m³)	Fiber (%)
Properties	95	166	104	155	800	1,110	0.50	–
	Average		Standard deviation	Compressive strength (ksc)			Average	Standard deviation
Air content (%)[1]	2.9		0.2	7 days			30.70	7.00
Slump (mm)[2]	56.5		13.7	28 days			55.30	3.05
Initial setting time (min)[3]	137		14.6	90 days			80.20	2.10
Final setting time (min)[3]	269		28.8	365 days			100.10	1.85

Mix Symbol (SC100C170F75S75)	Cylindrical Strength (ksc)	Cement (kg/m³)	Fly Ash (kg/m³)	Water (kg/m³)	Sand (kg/m³)	Recycled Aggregate 3/4" (kg/m³)	Recommended Quantity Admixture (Liters/m³)	Fiber (%)
Properties	90	174	110	155	820	1,100	1.00	–
	Average		Standard deviation	Compressive strength (ksc)			Average	Standard deviation
Air content (%)[1]	3.0		0.3	7 days			43.40	9.05

(Continued)

TABLE 5.8 (Continued)
Typical Mix Proportions and Related Properties of RAC

Mix Symbol (SC100C170F75S75)	Cylindrical Strength (ksc)	Cement (kg/m³)	Fly Ash (kg/m³)	Water (kg/m³)	Sand (kg/m³)	Recycled Aggregate 3/4" (kg/m³)	Recommended Quantity Admixture (Liters/m³)	Fiber (%)
Slump (mm)[2]	79.0		3.2	28 days			78.05	7.30
Initial setting time (min)[3]	217		16.4	90 days			90.70	3.98
Final setting time (min)[3]	289		45.8	365 days			100.05	2.00

Mix Symbol (SC150C270F0S75)	Cylindrical Strength (ksc)	Cement (kg/m³)	Fly Ash (kg/m³)	Water (kg/m³)	Sand (kg/m³)	Recycled Aggregate 3/4" (kg/m³)	Recommended Quantity Admixture (Liters/m³)	Fiber (%)
	140	260	0	165	829	1,123	0.55	–
Properties	Average		Standard deviation	Compressive strength (ksc)		Average		Standard deviation
Air content (%)[1]	3.2		0.3	7 days			46.03	11.82
Slump (mm)[2]	78.5		1.9	28 days			88.56	3.67
Initial setting time (min)[3]	247		9.5	90 days			129.88	5.77
Final setting time (min)[3]	365		22.7	365 days			151.36	2.10

Mix Symbol (SC150C188F82S75)	Cylindrical Strength (ksc)	Cement (kg/m³)	Fly Ash (kg/m³)	Water (kg/m³)	Sand (kg/m³)	Recycled Aggregate 3/4" (kg/m³)	Recommended Quantity Admixture (Liters/m³)	Fiber (%)
	140	154	96	155	818	1,107	1.10	—
Properties	Average		Standard deviation	Compressive strength (ksc)			Average	Standard deviation
Air content (%)[1]	3.3		0.1	7 days			61.36	0.84
Slump (mm)[2]	80.0		8.4	28 days			94.56	2.36
Initial setting time (min)[3]	126		3.4	90 days			133.92	4.87
Final setting time (min)[3]	333		6.1	365 days			154.23	0.69

Mix Symbol (SC180C295F0S75)	Cylindrical Strength (ksc)	Cement (kg/m³)	Fly Ash (kg/m³)	Water (kg/m³)	Sand (kg/m³)	Recycled Aggregate 3/4" (kg/m³)	Recommended Quantity Admixture (Liters/m³)	Fiber (%)
	140	190	44	155	820	1,110	0.60	—
Properties	Average		Standard deviation	Compressive strength (ksc)			Average	Standard deviation
Air content (%)[1]	2.5		0.3	7 days			63.62	4.82
Slump (mm)[2]	53.5		8.4	28 days			106.27	7.24
Initial setting time (min)[3]	171		8.1	90 days			146.58	3.06
Final setting time (min)[3]	295		2.2	365 days			151.77	1.84

(Continued)

TABLE 5.8 (Continued)
Typical Mix Proportions and Related Properties of RAC

Mix Symbol (SC180C205F90S575)	Cylindrical Strength (ksc)	Cement (kg/m³)	Fly Ash (kg/m³)	Water (kg/m³)	Sand (kg/m³)	Recycled Aggregate 3/4" (kg/m³)	Recommended Quantity Admixture (Liters/m³)	Fiber (%)
	140	200	20	165	808	1,094	1.20	–
Properties	Average		Standard deviation		Compressive strength (ksc)		Average	Standard deviation
Air content (%)[1]	2.7		0.3		7 days		55.32	10.69
Slump (mm)[2]	55.5		20.3		28 days		103.62	7.60
Initial setting time (min)[3]	181		9.2		90 days		131.99	1.47
Final setting time (min)[3]	255		9.4		365 days		159.81	3.66

Mix Symbol (SC210C320F0S575)	Cylindrical Strength (ksc)	Cement (kg/m³)	Fly Ash (kg/m³)	Water (kg/m³)	Sand (kg/m³)	Recycled Aggregate 3/4" (kg/m³)	Recommended Quantity Admixture (Liters/m³)	Fiber (%)
	180	168	106	155	811	1,098	0.65	–
Properties	Average		Standard deviation		Compressive strength (ksc)		Average	Standard deviation
Air content (%)[1]	1.5		0.1		7 days		103.04	12.84
Slump (mm)[2]	60.0		8.6		28 days		121.69	9.67

Properties	Average	Standard deviation	Compressive strength (ksc)	Average	Standard deviation
Initial setting time (min)[3]	192	18.7	90 days	194.11	3.32
Final setting time (min)[3]	315	41.1	365 days	212.77	0.84

Mix Symbol (SC210C223F97S75)	Cylindrical Strength (ksc)	Cement (kg/m³)	Fly Ash (kg/m³)	Water (kg/m³)	Sand (kg/m³)	Recycled Aggregate 3/4" (kg/m³)	Recommended Quantity Admixture (Liters/m³)	Fiber (%)
Properties	180	210	50	155	798	1,080	1.30	–

Properties	Average	Standard deviation	Compressive strength (ksc)	Average	Standard deviation
Air content (%)[1]	1.0	0.1	7 days	112.85	22.48
Slump (mm)[2]	74.5	17.1	28 days	219.43	21.06
Initial setting time (min)[3]	175	1.5	90 days	277.78	7.56
Final setting time (min)[3]	317	47.6	365 days	322.59	4.22

Mix Symbol (SC240C345F0S75)	Cylindrical Strength (ksc)	Cement (kg/m³)	Fly Ash (kg/m³)	Water (kg/m³)	Sand (kg/m³)	Recycled Aggregate 3/4" (kg/m³)	Recommended Quantity Admixture (Liters/m³)	Fiber (%)
Properties	210	310	0	165	803	1,086	0.70	–

Properties	Average	Standard deviation	Compressive strength (ksc)	Average	Standard deviation
Air content (%)[1]	1.2	0.3	7 days	61.33	4.25

(Continued)

TABLE 5.8 (Continued)
Typical Mix Proportions and Related Properties of RAC

Mix Symbol (SC240C345F0S575)	Cylindrical Strength (ksc)	Cement (kg/m³)	Fly Ash (kg/m³)	Water (kg/m³)	Sand (kg/m³)	Recycled Aggregate 3/4" (kg/m³)	Recommended Quantity Admixture (Liters/m³)	Fiber (%)
Slump (mm)[2]	71.5		15.1	28 days			119.04	8.59
Initial setting time (min)[3]	155		2.4	90 days			173.59	7.32
Final setting time (min)[3]	357		21.7	365 days			201.77	2.58

Mix Symbol (SC240C240F105S575)	Cylindrical Strength (ksc)	Cement (kg/m³)	Fly Ash (kg/m³)	Water (kg/m³)	Sand (kg/m³)	Recycled Aggregate 3/4" (kg/m³)	Recommended Quantity Admixture (Liters/m³)	Fiber (%)
	210	196	124	155	788	1,066	1.40	–
Properties	Average	Standard deviation	Compressive strength (ksc)				Average	Standard deviation
Air content (%)[1]	1.0	0.3	7 days				120.36	13.56
Slump (mm)[2]	76.0	16.1	28 days				225.19	10.05
Initial setting time (min)[3]	165	15.7	90 days				281.26	7.60
Final setting time (min)[3]	307	28.1	365 days				335.53	4.08

Mix Symbol (SC250C355F0S75)	Cylindrical Strength (ksc)	Cement (kg/m³)	Fly Ash (kg/m³)	Water (kg/m³)	Sand (kg/m³)	Recycled Aggregate 3/4" (kg/m³)	Recommended Quantity Admixture (Liters/m³)	Fiber (%)
	210	204	130	155	799	1,081	0.72	–

Properties	Average	Standard deviation	Compressive strength (ksc)	Average	Standard deviation (%)
Air content (%)[1]	2.2	0.1	7 days	135.98	5.48
Slump (mm)[2]	51.5	0.8	28 days	257.14	4.21
Initial setting time (min)[3]	133	10.2	90 days	316.87	3.99
Final setting time (min)[3]	287	21.5	365 days	361.96	2.63

Mix Symbol (SC250C247F108S75)	Cylindrical Strength (ksc)	Cement (kg/m³)	Fly Ash (kg/m³)	Water (kg/m³)	Sand (kg/m³)	Recycled Aggregate 3/4" (kg/m³)	Recommended Quantity Admixture (Liters/m³)	Fiber (%)
	210	184	116	155	784	1,061	1.44	–

Properties	Average	Standard deviation	Compressive strength (ksc)	Average	Standard deviation (%)
Air content (%)[1]	2.3	0.2	7 days	120.73	7.56
Slump (mm)[2]	53.5	1.2	28 days	241.38	2.65
Initial setting time (min)[3]	140	1.4	90 days	325.66	1.99
Final setting time (min)[3]	291	22.5	365 days	358.14	0.87

(Continued)

TABLE 5.8 (Continued)
Typical Mix Proportions and Related Properties of RAC

Mix Symbol (SC280C370F0S75)	Cylindrical Strength (ksc)	Cement (kg/m³)	Fly Ash (kg/m³)	Water (kg/m³)	Sand (kg/m³)	Recycled Aggregate 3/4" (kg/m³)	Recommended Quantity Admixture (Liters/m³)	Fiber (%)
	210	230	54	155	794	1,074	0.75	–
Properties	Average		Standard deviation	Compressive strength (ksc)			Average	Standard deviation
Air content (%)[1]	3.2		0.1	7 days			93.26	20.01
Slump (mm)[2]	50.5		14.5	28 days			198.44	14.99
Initial setting time (min)[3]	188		11.4	90 days			239.09	9.82
Final setting time (min)[3]	387		23.2	365 days			291.54	4.67

Mix Symbol (SC280C258F112S75)	Cylindrical Strength (ksc)	Cement (kg/m³)	Fly Ash (kg/m³)	Water (kg/m³)	Sand (kg/m³)	Recycled Aggregate 3/4" (kg/m³)	Recommended Quantity Admixture (Liters/m³)	Fiber (%)
	240	334	0	165	778	1,053	1.50	–
Properties	Average		Standard deviation	Compressive strength (ksc)			Average	Standard deviation
Air content (%)[1]	1.3		0.2	7 days			119.25	26.74
Slump (mm)[2]	50.5		23.5	28 days			215.89	15.23

	Average	Standard deviation	Compressive strength (ksc)	Average	Standard deviation
Initial setting time (min)[3]	121	2.7	90 days	305.98	11.28
Final setting time (min)[3]	250	10.9	365 days	369.52	7.54

Mix Symbol (SC300C395F0S75)	Cylindrical Strength (ksc)	Cement (kg/m³)	Fly Ash (kg/m³)	Water (kg/m³)	Sand (kg/m³)	Recycled Aggregate 3/4" (kg/m³)	Recommended Quantity Admixture (Liters/m³)	Fiber (%)
Properties	240	210	134	155	785	1,062	0.80	–

	Average	Standard deviation	Compressive strength (ksc)	Average	Standard deviation
Air content (%)[1]	3.3	0.2	7 days	149.84	19.63
Slump (mm)[2]	52.5	11.8	28 days	267.14	17.54
Initial setting time (min)[3]	141	3.5	90 days	327.95	16.22
Final setting time (min)[3]	270	25.7	365 days	346.18	15.01

Mix Symbol (SC300C275F120S75)	Cylindrical Strength (ksc)	Cement (kg/m³)	Fly Ash (kg/m³)	Water (kg/m³)	Sand (kg/m³)	Recycled Aggregate 3/4" (kg/m³)	Recommended Quantity Admixture (Liters/m³)	Fiber (%)
Properties	240	220	140	155	769	1,039	1.60	–

	Average	Standard deviation	Compressive strength (ksc)	Average	Standard deviation
Air content (%)[1]	2.2	0.1	7 days	137.96	3.15

(Continued)

TABLE 5.8 (Continued)
Typical Mix Proportions and Related Properties of RAC

Mix Symbol (SC300C275F120S75)	Cylindrical Strength (ksc)	Cement (kg/m³)	Fly Ash (kg/m³)	Water (kg/m³)	Sand (kg/m³)	Recycled Aggregate 3/4" (kg/m³)	Recommended Quantity Admixture (Liters/m³)	Fiber (%)
Slump (mm)[2]	51.5		0.8	28 days			278.63	2.77
Initial setting time (min)[3]	133		10.2	90 days			339.11	1.62
Final setting time (min)[3]	287		21.5	365 days			347.98	1.04

Mix Symbol (SC320C420F0S75)	Cylindrical Strength (ksc)	Cement (kg/m³)	Fly Ash (kg/m³)	Water (kg/m³)	Sand (kg/m³)	Recycled Aggregate 3/4" (kg/m³)	Recommended Quantity Admixture (Liters/m³)	Fiber (%)
	240	198	126	155	758	1,068	0.85	–

Properties		Average	Standard deviation		Compressive strength (ksc)		Average	Standard deviation
Air content (%)[1]		2.0	0.2		7 days		97.98	15.47
Slump (mm)[2]		49.5	20.7		28 days		232.67	19.32
Initial setting time (min)[3]		101	7.9		90 days		269.48	10.03
Final setting time (min)[3]		374	10.9		365 days		305.55	5.48

Mix Symbol (SC320C293F127S75)	Cylindrical Strength (ksc)	Cement (kg/m³)	Fly Ash (kg/m³)	Water (kg/m³)	Sand (kg/m³)	Recycled Aggregate 3/4" (kg/m³)	Recommended Quantity Admixture (Liters/m³)	Fiber (%)
	240	250	60	155	741	1,044	1.70	–

Properties	Average	Standard deviation	Compressive strength (ksc)		Average	Standard deviation
Air content (%)[1]	2.1	0.3	7 days		201.54	29.51
Slump (mm)[2]	50.0	4.2	28 days		287.55	22.66
Initial setting time (min)[3]	110	19.2	90 days		316.74	14.41
Final setting time (min)[3]	256	38.4	365 days		281.59	3.57

Mix Symbol (SC350C445F0S75)	Cylindrical Strength (ksc)	Cement (kg/m³)	Fly Ash (kg/m³)	Water (kg/m³)	Sand (kg/m³)	Recycled Aggregate 3/4" (kg/m³)	Recommended Quantity Admixture (Liters/m³)	Fiber (%)
	250	344	100	165	749	1,055	0.90	–

Properties	Average	Standard deviation	Compressive strength (ksc)		Average	Standard deviation
Air content (%)[1]	3.2	0.1	7 days		193.67	14.82
Slump (mm)[2]	50.5	14.5	28 days		233.33	23.02
Initial setting time (min)[3]	188	11.4	90 days		325.29	7.47
Final setting time (min)[3]	387	23.2	365 days		369.22	4.58

(*Continued*)

TABLE 5.8 (Continued)
Typical Mix Proportions and Related Properties of RAC

Mix Symbol (SC350C310F135S75)	Cylindrical Strength (ksc)	Cement (kg/m³)	Fly Ash (kg/m³)	Water (kg/m³)	Sand (kg/m³)	Recycled Aggregate 3/4″ (kg/m³)	Recommended Quantity Admixture (Liters/m³)	Fiber (%)
	250	216	138	155	732	1,030	1.80	–

Properties	Average		Standard deviation	Compressive strength (ksc)			Average	Standard deviation
Air content (%)[1]	3.4		0.3	7 days			205.43	33.25
Slump (mm)[2]	53.0		7.6	28 days			266.93	24.17
Initial setting time (min)[3]	191		17.2	90 days			389.84	13.59
Final setting time (min)[3]	275		15.9	365 days			413.65	8.47

Mix Symbol (SC100C245F0S75)	Cylindrical Strength (ksc)	Cement (kg/m³)	Fly Ash (kg/m³)	Water (kg/m³)	Sand (kg/m³)	Recycled Aggregate 3/4″ (kg/m³)	Recommended Quantity Admixture (Liters/m³)	Fiber (%)
	90	190	120	155	857	1,115	0.50	–

Properties	Average		Standard deviation	Compressive strength (ksc)			Average	Standard deviation
Air content (%)[1]	2.7		0.3	7 days			20.80	5.90
Slump (mm)[2]	56.0		4.2	28 days			55.11	4.00

Continued — test results (previous mix):

Properties	Average	Standard deviation	Compressive strength (ksc)	Average	Standard deviation
Initial setting time (min)[3]	110	9.6	90 days	90.00	3.25
Final setting time (min)[3]	287	19.2	365 days	101.44	1.00

Mix design:

Mix Symbol (SC150C270F0S75)	Cylindrical Strength (ksc)	Cement (kg/m³)	Fly Ash (kg/m³)	Water (kg/m³)	Sand (kg/m³)	Recycled Aggregate 3/4" (kg/m³)	Recommended Quantity Admixture (Liters/m³)	Fiber (%)
Properties	140	166	104	155	829	1,123	0.55	–

Test results:

Properties	Average	Standard deviation	Compressive strength (ksc)	Average	Standard deviation
Air content (%)[1]	3.0	0.1	7 days	58.67	7.56
Slump (mm)[2]	58.0	22.6	28 days	114.63	4.11
Initial setting time (min)[3]	102	6.8	90 days	155.20	3.26
Final setting time (min)[3]	335	8.7	365 days	180.32	2.10

Mix design:

Mix Symbol (SC180C295F0S75)	Cylindrical Strength (ksc)	Cement (kg/m³)	Fly Ash (kg/m³)	Water (kg/m³)	Sand (kg/m³)	Recycled Aggregate 3/4" (kg/m³)	Recommended Quantity Admixture (Liters/m³)	Fiber (%)
Properties	140	174	110	155	820	1,110	0.60	–

Test results:

Properties	Average	Standard deviation	Compressive strength (ksc)	Average	Standard deviation
Air content (%)[1]	1.3	0.2	7 days	38.67	4.88

(Continued)

TABLE 5.8 (Continued)
Typical Mix Proportions and Related Properties of RAC

Mix Symbol (SC180C295F0575)	Cylindrical Strength (ksc)	Cement (kg/m³)	Fly Ash (kg/m³)	Water (kg/m³)	Sand (kg/m³)	Recycled Aggregate 3/4" (kg/m³)	Recommended Quantity Admixture (Liters/m³)	Fiber (%)
							Average	
Slump (mm)[2]	54.5		5.9	28 days			94.63	13.62
Initial setting time (min)[3]	157		9.0	90 days			146.47	4.78
Final setting time (min)[3]	359		28.0	365 days			160.23	1.06

Mix Symbol (SC210C320F0575)	Cylindrical Strength (ksc)	Cement (kg/m³)	Fly Ash (kg/m³)	Water (kg/m³)	Sand (kg/m³)	Recycled Aggregate 3/4" (kg/m³)	Recommended Quantity Admixture (Liters/m³)	Fiber (%)
Properties	140	154	96	155	811	1,098	0.65	–
	Average		Standard deviation	Compressive strength (ksc)			Average	Standard deviation
Air content (%)[1]	2.5		0.1	7 days			69.55	12.41
Slump (mm)[2]	57.5		5.8	28 days			88.47	10.36
Initial setting time (min)[3]	109		15.8	90 days			138.88	1.65
Final setting time (min)[3]	273		13.3	365 days			161.36	2.95

Mix Symbol (SC240C345F0S75)	Cylindrical Strength (ksc)	Cement (kg/m³)	Fly Ash (kg/m³)	Water (kg/m³)	Sand (kg/m³)	Recycled Aggregate 3/4" (kg/m³)	Recommended Quantity Admixture (Liters/m³)	Fiber (%)
	140	190	44	155	803	1,086	0.60	–
Properties	Average		Standard deviation	Compressive strength (ksc)			Average	Standard deviation
Air content (%)[1]	1.7		0.3	7 days			113.69	12.28
Slump (mm)[2]	57.0		2.4	28 days			142.16	7.96
Initial setting time (min)[3]	132		7.6	90 days			183.66	3.99
Final setting time (min)[3]	325		42.8	365 days			202.47	2.41

Mix Symbol (SC250C355F0S75)	Cylindrical Strength (ksc)	Cement (kg/m³)	Fly Ash (kg/m³)	Water (kg/m³)	Sand (kg/m³)	Recycled Aggregate 3/4" (kg/m³)	Recommended Quantity Admixture (Liters/m³)	Fiber (%)
	180	284	0	165	799	1,081	0.70	–
Properties	Average		Standard deviation	Compressive strength (ksc)			Average	Standard deviation
Air content (%)[1]	2.6		0.2	7 days			94.12	12.63
Slump (mm)[2]	51.0		8.4	28 days			129.52	11.41
Initial setting time (min)[3]	106		4.8	90 days			186.82	4.44
Final setting time (min)[3]	326		9.8	365 days			211.47	0.94

(Continued)

TABLE 5.8 (Continued)
Typical Mix Proportions and Related Properties of RAC

Mix Symbol (SC280C370F0S75)	Cylindrical Strength (ksc)	Cement (kg/m³)	Fly Ash (kg/m³)	Water (kg/m³)	Sand (kg/m³)	Recycled Aggregate 3/4" (kg/m³)	Recommended Quantity Admixture (Liters/m³)	Fiber (%)
	180	180	114	155	794	1,074	0.60	–
Properties	Average		Standard deviation	Compressive strength (ksc)			Average	Standard deviation
Air content (%)[1]	1.6		0.2	7 days			80.98	16.51
Slump (mm)[2]	57.5		5.6	28 days			171.65	10.36
Initial setting time (min)[3]	122		20.0	90 days			254.33	11.62
Final setting time (min)[3]	320		10.5	365 days			305.81	5.58

Mix Symbol (SC300C395F0S75)	Cylindrical Strength (ksc)	Cement (kg/m³)	Fly Ash (kg/m³)	Water (kg/m³)	Sand (kg/m³)	Recycled Aggregate 3/4" (kg/m³)	Recommended Quantity Admixture (Liters/m³)	Fiber (%)
	180	190	120	155	785	1,062	0.70	–
Properties	Average		Standard deviation	Compressive strength (ksc)			Average	Standard deviation
Air content (%)[1]	2.6		0.2	7 days			137.32	18.73
Slump (mm)[2]	55.0		15.2	28 days			277.47	20.36

Initial setting time (min)[3]	101		4.2	90 days		432.66		13.22
Final setting time (min)[3]	250		30.2	365 days		483.14		4.79
Mix Symbol (SC320C420F0S75)	**Cylindrical Strength (ksc)**	**Cement (kg/m³)**	**Fly Ash (kg/m³)**	**Water (kg/m³)**	**Sand (kg/m³)**	**Recycled Aggregate 3/4″ (kg/m³)**	**Recommended Quantity Admixture (Liters/m³)**	**Fiber (%)**
Properties	250	260	62	155	758	1,068	0.70	–
	Average		Standard deviation	Compressive strength (ksc)		Average		Standard deviation
Air content (%)[1]	1.4		0.3	7 days		191.25		22.40
Slump (mm)[2]	59.5		4.5	28 days		309.87		23.68
Initial setting time (min)[3]	186		14.5	90 days		361.82		15.36
Final setting time (min)[3]	278		15.7	365 days		405.64		8.64
Mix Symbol (SC350C445F0S75)	**Cylindrical Strength (ksc)**	**Cement (kg/m³)**	**Fly Ash (kg/m³)**	**Water (kg/m³)**	**Sand (kg/m³)**	**Recycled Aggregate 3/4″ (kg/m³)**	**Recommended Quantity Admixture (Liters/m³)**	**Fiber (%)**
Properties	280	360	0	165	749	1,055	0.72	–
	Average		Standard deviation	Compressive strength (ksc)		Average		Standard deviation
Air content (%)[1]	2.1		0.3	7 days		145.48		0.45

(*Continued*)

TABLE 5.8 (Continued)
Typical Mix Proportions and Related Properties of RAC

Mix Symbol (SC350C445F0S75)	Cylindrical Strength (ksc)	Cement (kg/m³)	Fly Ash (kg/m³)	Water (kg/m³)	Sand (kg/m³)	Recycled Aggregate 3/4" (kg/m³)	Recommended Quantity Admixture (Liters/m³)	Fiber (%)
Slump (mm)[2]	55.5		4.7		28 days		235.47	3.92
Initial setting time (min)[3]	165		19.2		90 days		328.54	8.16
Final setting time (min)[3]	255		42.9		365 days		361.43	9.54

Mix Symbol (SC380C470F0S75)	Cylindrical Strength (ksc)	Cement (kg/m³)	Fly Ash (kg/m³)	Water (kg/m³)	Sand (kg/m³)	Recycled Aggregate 3/4" (kg/m³)	Recommended Quantity Admixture (Liters/m³)	Fiber (%)
	280	226	144	155	741	1,043	0.70	–
Properties	Average		Standard deviation		Compressive strength (ksc)		Average	Standard deviation
Air content (%)[1]	2.7		0.1		7 days		155.75	27.30
Slump (mm)[2]	54.5		1.9		28 days		286.25	22.45
Initial setting time (min)[3]	110		10.8		90 days		400.57	16.45
Final setting time (min)[3]	344		3.7		365 days		424.52	6.65

Mix Symbol (SC100C170F75S75)	Cylindrical Strength (ksc)	Cement (kg/m³)	Fly Ash (kg/m³)	Water (kg/m³)	Sand (kg/m³)	Recycled Aggregate 3/4" (kg/m³)	Recommended Quantity Admixture (Liters/m³)	Fiber (%)
	90	175	70	160	847	1,101	0.80	–
Properties	Average		Standard deviation	Compressive strength (ksc)			Average	Standard deviation
Air content (%)[1]	2.4		0.3	7 days			36.28	12.41
Slump (mm)[2]	55.5		9.3	28 days			92.63	10.36
Initial setting time (min)[3]	247		12.2	90 days			138.88	6.20
Final setting time (min)[3]	312		10.9	365 days			161.36	3.39

Mix Symbol (SC150C188F82S75)	Cylindrical Strength (ksc)	Cement (kg/m³)	Fly Ash (kg/m³)	Water (kg/m³)	Sand (kg/m³)	Recycled Aggregate 3/4" (kg/m³)	Recommended Quantity Admixture (Liters/m³)	Fiber (%)
	140	269	80	165	818	1,110	0.70	–
Properties	Average		Standard deviation	Compressive strength (ksc)			Average	Standard deviation
Air content (%)[1]	2.3		0.2	7 days			62.56	12.14
Slump (mm)[2]	53.5		1.2	28 days			119.63	13.99
Initial setting time (min)[3]	140		1.4	90 days			177.52	0.47
Final setting time (min)[3]	291		22.5	365 days			204.21	3.59

(Continued)

TABLE 5.8 (Continued)
Typical Mix Proportions and Related Properties of RAC

Mix Symbol (SC180C205F90S75)	Cylindrical Strength (ksc)	Cement (kg/m³)	Fly Ash (kg/m³)	Water (kg/m³)	Sand (kg/m³)	Recycled Aggregate 3/4" (kg/m³)	Recommended Quantity Admixture (Liters/m³)	Fiber (%)
	140	165	114	155	808	1,094	1.15	–

Properties	Average	Standard deviation	Compressive strength (ksc)	Average	Standard deviation
Air content (%)[1]	2.5	0.1	7 days	59.14	12.01
Slump (mm)[2]	58.0	4.7	28 days	121.63	9.57
Initial setting time (min)[3]	145	7.1	90 days	183.09	4.11
Final setting time (min)[3]	300	24.7	365 days	202.41	2.58

Mix Symbol (SC210C223F97S75)	Cylindrical Strength (ksc)	Cement (kg/m³)	Fly Ash (kg/m³)	Water (kg/m³)	Sand (kg/m³)	Recycled Aggregate 3/4" (kg/m³)	Recommended Quantity Admixture (Liters/m³)	Fiber (%)
	180	181	124	155	803	1,080	1.20	–

Properties	Average	Standard deviation	Compressive strength (ksc)	Average	Standard deviation
Air content (%)[1]	2.6	0.3	7 days	69.55	0.88
Slump (mm)[2]	60.0	9.4	28 days	88.47	1.24

[Continued from previous mix]

Properties	Average	Standard deviation	Compressive strength (ksc)	Average	Standard deviation
Initial setting time (min)[3]	229	1.2	90 days	129.05	0.65
Final setting time (min)[3]	273	24.1	365 days	161.23	2.95

Mix Symbol (SC240C240F105S75)

Cement (kg/m³)	Fly Ash (kg/m³)	Water (kg/m³)	Sand (kg/m³)	Recycled Aggregate 3/4" (kg/m³)	Recommended Quantity Admixture (Liters/m³)	Fiber (%)
195	134	155	788	1,066	1.05	–

Properties	Cylindrical Strength (ksc)	Average	Compressive strength (ksc)	Average	Standard deviation
	210		7 days	81.64	14.18
Air content (%)[1]		3.5	28 days	136.24	7.43
Slump (mm)[2]		65.0	90 days	172.57	2.63
Initial setting time (min)[3]		108	365 days	193.25	4.44
Final setting time (min)[3]		265			

Mix Symbol (SC250C247F108S75)

Cement (kg/m³)	Fly Ash (kg/m³)	Water (kg/m³)	Sand (kg/m³)	Recycled Aggregate 3/4" (kg/m³)	Recommended Quantity Admixture (Liters/m³)	Fiber (%)
205	140	155	784	1,061	1.40	–

Properties	Cylindrical Strength (ksc)	Average	Compressive strength (ksc)	Average	Standard deviation
	210		7 days	132.03	13.58
Air content (%)[1]					

(Continued)

TABLE 5.8 (Continued)
Typical Mix Proportions and Related Properties of RAC

Mix Symbol (SC250C247F108S75)	Cylindrical Strength (ksc)	Cement (kg/m³)	Fly Ash (kg/m³)	Water (kg/m³)	Sand (kg/m³)	Recycled Aggregate 3/4" (kg/m³)	Recommended Quantity Admixture (Liters/m³)	Fiber (%)
	Average		Standard deviation				Average	Standard deviation
Slump (mm)[2]	2.7		0.1	28 days			249.77	24.71
Initial setting time (min)[3]	62.0		5.5	90 days			364.41	16.59
Final setting time (min)[3]	163		0.9	365 days			412.63	4.77

Mix Symbol (SC280C258F112S75)	Cylindrical Strength (ksc)	Cement (kg/m³)	Fly Ash (kg/m³)	Water (kg/m³)	Sand (kg/m³)	Recycled Aggregate 3/4" (kg/m³)	Recommended Quantity Admixture (Liters/m³)	Fiber (%)
Properties	210	183	126	155	778	1,052	1.25	–
	Average		Standard deviation	Compressive strength (ksc)			Average	Standard deviation
Air content (%)[1]				7 days			154.47	12.20
Slump (mm)[2]	3.8		0.2	28 days			261.07	11.01
Initial setting time (min)[3]	60.5		18.3	90 days			366.93	13.59
Final setting time (min)[3]	177		11.3	365 days			446.25	8.47

Mix Symbol (SC300C275F120S75)	Cylindrical Strength (ksc)	Cement (kg/m³)	Fly Ash (kg/m³)	Water (kg/m³)	Sand (kg/m³)	Recycled Aggregate 3/4" (kg/m³)	Recommended Quantity Admixture (Liters/m³)	Fiber (%)
Properties	210	235	60	155	769	1,039	1.50	–

Properties	Average	Standard deviation	Compressive strength (ksc)	Average	Standard deviation
Air content (%)[1]	1.5	0.1	7 days	185.99	10.93
Slump (mm)[2]	67.5	10.5	28 days	243.47	4.87
Initial setting time (min)[3]	174	17.2	90 days	341.52	2.69
Final setting time (min)[3]	256	17.0	365 days	374.77	1.55

Mix Symbol (SC320C293F127S75)	Cylindrical Strength (ksc)	Cement (kg/m³)	Fly Ash (kg/m³)	Water (kg/m³)	Sand (kg/m³)	Recycled Aggregate 3/4" (kg/m³)	Recommended Quantity Admixture (Liters/m³)	Fiber (%)
Properties	240	345	80	165	741	1,044	1.50	–

Properties	Average	Standard deviation	Compressive strength (ksc)	Average	Standard deviation
Air content (%)[1]	1.9	0.3	7 days	107.86	27.98
Slump (mm)[2]	66.5	1.7	28 days	209.33	21.86
Initial setting time (min)[3]	104	11.8	90 days	305.14	13.02
Final setting time (min)[3]	326	15.1	365 days	371.51	7.58

(Continued)

TABLE 5.8 (Continued)
Typical Mix Proportions and Related Properties of RAC

Mix Symbol (SC350C310F135S75)	Cylindrical Strength (ksc)	Cement (kg/m³)	Fly Ash (kg/m³)	Water (kg/m³)	Sand (kg/m³)	Recycled Aggregate 3/4" (kg/m³)	Recommended Quantity Admixture (Liters/m³)	Fiber (%)
Properties	250	265	66	155	732	1,030	1.70	–
	Average		Standard deviation	Compressive strength (ksc)			Average	Standard deviation
Air content (%)[1]	2.6		0.2	7 days			143.55	21.59
Slump (mm)[2]	64.0		19.8	28 days			172.66	19.25
Initial setting time (min)[3]	163		10.7	90 days			364.51	8.77
Final setting time (min)[3]	385		36.7	365 days			295.47	5.63

Mix Symbol (SC380C328F142S75)	Cylindrical Strength (ksc)	Cement (kg/m³)	Fly Ash (kg/m³)	Water (kg/m³)	Sand (kg/m³)	Recycled Aggregate 3/4" (kg/m³)	Recommended Quantity Admixture (Liters/m³)	Fiber (%)
Properties	280	369	0	165	722	1,016	1.50	–
	Average		Standard deviation	Compressive strength (ksc)			Average	Standard deviation
Air content (%)[1]	2.7		0.3	7 days			132.56	32.14
Slump (mm)[2]	63.0		6.1	28 days			249.63	25.77

(Continued from previous mix)

Properties	Average	Standard deviation	Compressive strength (ksc)	Average	Standard deviation
Initial setting time (min)[3]	133	4.7	90 days	367.14	16.52
Final setting time (min)[3]	375	18.8	365 days	413.27	9.73

Mix Symbol (SC100C170F75S75)	Cylindrical Strength (ksc)	Cement (kg/m³)	Fly Ash (kg/m³)	Water (kg/m³)	Sand (kg/m³)	Recycled Aggregate 3/4" (kg/m³)	Recommended Quantity Admixture (Liters/m³)	Fiber (%)
	100	185	75	170	847	1,100	1.00	–

Properties	Average	Standard deviation	Compressive strength (ksc)	Average	Standard deviation
Air content (%)[1]	3.1	0.2	7 days	38.67	7.56
Slump (mm)[2]	63.5	10.5	28 days	94.37	4.11
Initial setting time (min)[3]	143	5.8	90 days	135.23	3.26
Final setting time (min)[3]	390	17.8	365 days	160.32	2.10

Mix Symbol (SC150C188F82S75)	Cylindrical Strength (ksc)	Cement (kg/m³)	Fly Ash (kg/m³)	Water (kg/m³)	Sand (kg/m³)	Recycled Aggregate 3/4" (kg/m³)	Recommended Quantity Admixture (Liters/m³)	Fiber (%)
	150	203	82	170	818	1,107	1.00	–

Properties	Average	Standard deviation	Compressive strength (ksc)	Average	Standard deviation
Air content (%)[1]	3.0	0.1	7 days	37.69	8.32

(Continued)

TABLE 5.8 (Continued)
Typical Mix Proportions and Related Properties of RAC

Mix Symbol (SC150C188F82S75)	Cylindrical Strength (ksc)	Cement (kg/m³)	Fly Ash (kg/m³)	Water (kg/m³)	Sand (kg/m³)	Recycled Aggregate 3/4" (kg/m³)	Recommended Quantity Admixture (Liters/m³)	Fiber (%)
Slump (mm)[2]	66.0		1.1	28 days			90.43	6.41
Initial setting time (min)[3]	193		16.7	90 days			146.37	2.88
Final setting time (min)[3]	355		7.7	365 days			160.28	3.21

Mix Symbol (SC180C205F90S75)	Cylindrical Strength (ksc)	Cement (kg/m³)	Fly Ash (kg/m³)	Water (kg/m³)	Sand (kg/m³)	Recycled Aggregate 3/4" (kg/m³)	Recommended Quantity Admixture (Liters/m³)	Fiber (%)
	140	195	50	155	800	1,090	1.10	–
Properties	Average		Standard deviation	Compressive strength (ksc)			Average	Standard deviation
Air content (%)[1]	2.9		0.3	7 days			77.55	0.84
Slump (mm)[2]	61.5		3.9	28 days			120.87	3.26
Initial setting time (min)[3]	103		1.3	90 days			176.24	8.42
Final setting time (min)[3]	365		3.5	365 days			206.97	3.92

Mix Symbol (SC210C223F97S75)	Cylindrical Strength (ksc)	Cement (kg/m³)	Fly Ash (kg/m³)	Water (kg/m³)	Sand (kg/m³)	Recycled Aggregate 3/4" (kg/m³)	Recommended Quantity Admixture (Liters/m³)	Fiber (%)
	180	295	90	165	790	1,070	1.20	–

Properties	Average	Standard deviation	Compressive strength (ksc)	Average	Standard deviation
Air content (%)[1]	2.8	0.2	7 days	101.68	11.93
Slump (mm)[2]	62.0	5.8	28 days	160.14	7.01
Initial setting time (min)[3]	123	3.6	90 days	191.47	6.33
Final setting time (min)[3]	380	9.5	365 days	220.58	2.48

Mix Symbol (SC240C240F105S75)	Cylindrical Strength (ksc)	Cement (kg/m³)	Fly Ash (kg/m³)	Water (kg/m³)	Sand (kg/m³)	Recycled Aggregate 3/4" (kg/m³)	Recommended Quantity Admixture (Liters/m³)	Fiber (%)
	180	189	130	155	780	1,060	1.30	–

Properties	Average	Standard deviation	Compressive strength (ksc)	Average	Standard deviation
Air content (%)[1]	3.6	0.2	7 days	85.91	14.63
Slump (mm)[2]	60.5	12.7	28 days	146.59	2.47
Initial setting time (min)[3]	182	9.5	90 days	209.32	8.22
Final setting time (min)[3]	315	41.1	365 days	244.84	1.93

(Continued)

TABLE 5.8 (Continued)
Typical Mix Proportions and Related Properties of RAC

Mix Symbol (SC250C247F108S75)	Cylindrical Strength (ksc)	Cement (kg/m³)	Fly Ash (kg/m³)	Water (kg/m³)	Sand (kg/m³)	Recycled Aggregate 3/4" (kg/m³)	Recommended Quantity Admixture (Liters/m³)	Fiber (%)
	240	252	100	165	780	1,060	1.40	–
Properties	Average		Standard deviation	Compressive strength (ksc)			Average	Standard deviation
Air content (%)[1]	3.5		0.1	7 days			68.31	4.88
Slump (mm)[2]	58.0		6.5	28 days			120.26	13.62
Initial setting time (min)[3]	142		19.8	90 days			178.63	4.78
Final setting time (min)[3]	335		25.4	365 days			200.66	1.06

Mix Symbol (SC280C258F112S75)	Cylindrical Strength (ksc)	Cement (kg/m³)	Fly Ash (kg/m³)	Water (kg/m³)	Sand (kg/m³)	Recycled Aggregate 3/4" (kg/m³)	Recommended Quantity Admixture (Liters/m³)	Fiber (%)
	250	355	100	165	778	1,053	1.00	–
Properties	Average		Standard deviation	Compressive strength (ksc)			Average	Standard deviation
Air content (%)[1]	3.3		0.1	7 days			102.52	14.83
Slump (mm)[2]	65.5		7.4	28 days			214.18	20.33

(Continued from previous mix)

Property	Value	Standard deviation
Initial setting time (min)[3]	183	15.0
Final setting time (min)[3]	360	37.4

Compressive strength (ksc)	Average	Standard deviation
90 days	304.14	6.59
365 days	349.22	5.08

Mix Symbol (SC300C275F120S75)

Cement (kg/m³)	Fly Ash (kg/m³)	Water (kg/m³)	Sand (kg/m³)	Recycled Aggregate 3/4" (kg/m³)	Recommended Quantity Admixture (Liters/m³)	Fiber (%)
217	148	155	769	1,039	1.50	–

Properties

Cylindrical Strength (ksc)		
250		

Compressive strength (ksc)	Average	Standard deviation
7 days	112.55	27.98
28 days	249.28	21.86
90 days	324.51	7.41
365 days	353.25	5.33

Property	Average	Standard deviation
Air content (%)[1]	3.8	0.3
Slump (mm)[2]	65.0	18.9
Initial setting time (min)[3]	173	11.2
Final setting time (min)[3]	375	18.8

Mix Symbol (SC320C293F127S75)

Cement (kg/m³)	Fly Ash (kg/m³)	Water (kg/m³)	Sand (kg/m³)	Recycled Aggregate 3/4" (kg/m³)	Recommended Quantity Admixture (Liters/m³)	Fiber (%)
225	154	155	741	1,044	1.55	–

Properties

Cylindrical Strength (ksc)		
280		

Compressive strength (ksc)	Average	Standard deviation
7 days	107.82	18.84

Property	Average	Standard deviation
Air content (%)[1]	3.4	0.2

(Continued)

TABLE 5.8 (Continued)
Typical Mix Proportions and Related Properties of RAC

Mix Symbol (SC320C293F127S75)	Cylindrical Strength (ksc)	Cement (kg/m³)	Fly Ash (kg/m³)	Water (kg/m³)	Sand (kg/m³)	Recycled Aggregate 3/4" (kg/m³)	Recommended Quantity Admixture (Liters/m³)	Fiber (%)
Properties	Average		Standard deviation	Compressive strength (ksc)			Average	Standard deviation
Slump (mm)[2]	66.0		1.1	28 days			209.33	20.44
Initial setting time (min)[3]	193		16.7	90 days			306.74	10.32
Final setting time (min)[3]	365		3.5	365 days			380.15	5.58

Mix Symbol (SC350C310F135S75)	Cylindrical Strength (ksc)	Cement (kg/m³)	Fly Ash (kg/m³)	Water (kg/m³)	Sand (kg/m³)	Recycled Aggregate 3/4" (kg/m³)	Recommended Quantity Admixture (Liters/m³)	Fiber (%)
	280	235	160	155	732	1,030	1.40	–
Properties	Average		Standard deviation	Compressive strength (ksc)			Average	Standard deviation
Air content (%)[1]	1.5		0.1	7 days			124.14	22.02
Slump (mm)[2]	74.5		10.1	28 days			216.84	14.77
Initial setting time (min)[3]	104		15.7	90 days			299.66	14.77
Final setting time (min)[3]	378		33.5	365 days			353.28	4.84

Mix Symbol (SC380C328F142S75)	Cylindrical Strength (ksc)	Cement (kg/m³)	Fly Ash (kg/m³)	Water (kg/m³)	Sand (kg/m³)	Recycled Aggregate 3/4" (kg/m³)	Recommended Quantity Admixture (Liters/m³)	Fiber (%)
Properties	250	225	154	155	722	1,016	1.45	–
	Average		Standard deviation	Compressive strength (ksc)			Average	Standard deviation
Air content (%)[1]	3.5		0.3	7 days			185.99	0.93
Slump (mm)[2]	62.0		5.8	28 days			243.47	4.87
Initial setting time (min)[3]	123		3.6	90 days			342.48	14.24
Final setting time (min)[3]	380		9.5	365 days			369.88	8.46

Mix Symbol (SC320C335F85S75)	Cylindrical Strength (ksc)	Cement (kg/m³)	Fly Ash (kg/m³)	Water (kg/m³)	Sand (kg/m³)	Recycled Aggregate 3/4" (kg/m³)	Recommended Quantity Admixture (Liters/m³)	Fiber (%)
Properties	250	205	140	155	755	1,060	1.50	–
	Average		Standard deviation	Compressive strength (ksc)			Average	Standard deviation
Air content (%)[1]	1.6		0.2	7 days			111.83	30.41
Slump (mm)[2]	72.0		9.7	28 days			226.47	23.55
Initial setting time (min)[3]	114		7.6	90 days			329.84	15.88
Final setting time (min)[3]	374		28.7	365 days			368.27	4.96

(Continued)

TABLE 5.8 (Continued)
Typical Mix Proportions and Related Properties of RAC

Mix Symbol (SC240C240F105S75)	Cylindrical Strength (ksc)	Cement (kg/m³)	Fly Ash (kg/m³)	Water (kg/m³)	Sand (kg/m³)	Recycled Aggregate 3/4" (kg/m³)	Recommended Quantity Admixture (Liters/m³)	Fiber (%)
	240	255	105	155	815	1,100	1.30	–
Properties	Average		Standard deviation	Compressive strength (ksc)			Average	Standard deviation
Air content (%)[1]	3.6		0.1	7 days			145.14	12.48
Slump (mm)[2]	63.0		6.1	28 days			269.52	22.36
Initial setting time (min)[3]	133		4.7	90 days			360.06	13.62
Final setting time (min)[3]	385		36.7	365 days			411.59	6.99

Mix Symbol (SC280C258F112S75)	Cylindrical Strength (ksc)	Cement (kg/m³)	Fly Ash (kg/m³)	Water (kg/m³)	Sand (kg/m³)	Recycled Aggregate 3/4" (kg/m³)	Recommended Quantity Admixture (Liters/m³)	Fiber (%)
	280	273	112	155	800	1,095	1.40	–
Properties	Average		Standard deviation	Compressive strength (ksc)			Average	Standard deviation
Air content (%)[1]	1.1		0.2	7 days			141.52	29.84
Slump (mm)[2]	62.5		18.6	28 days			251.31	16.54

Properties	Average	Standard deviation	Compressive strength (ksc)	Average	Standard deviation
Initial setting time (min)[3]	110	5.4	90 days	368.02	13.30
Final setting time (min)[3]	366	23.9	365 days	406.84	9.47

Mix Symbol (SC240C240F105S75)	Cylindrical Strength (ksc)	Cement (kg/m³)	Fly Ash (kg/m³)	Water (kg/m³)	Sand (kg/m³)	Recycled Aggregate 3/4" (kg/m³)	Recommended Quantity Admixture (Liters/m³)	Fiber (%)
	180	169	116	155	815	1,100	1.20	–

Properties	Average	Standard deviation	Compressive strength (ksc)	Average	Standard deviation
Air content (%)[1]	3.7	0.2	7 days	131.36	32.48
Slump (mm)[2]	64.0	19.8	28 days	254.48	24.71
Initial setting time (min)[3]	163	10.7	90 days	360.03	10.23
Final setting time (min)[3]	370	14.3	365 days	405.63	9.33

Mix Symbol (SC280C258F112S75)	Cylindrical Strength (ksc)	Cement (kg/m³)	Fly Ash (kg/m³)	Water (kg/m³)	Sand (kg/m³)	Recycled Aggregate 3/4" (kg/m³)	Recommended Quantity Admixture (Liters/m³)	Fiber (%)
	180	215	54	155	800	1,085	1.40	–

Properties	Average	Standard deviation	Compressive strength (ksc)	Average	Standard deviation
Air content (%)[1]	1.2	0.3	7 days	160.36	33.59

(Continued)

TABLE 5.8 (Continued)
Typical Mix Proportions and Related Properties of RAC

Mix Symbol (SC280C258F112S75)	Cylindrical Strength (ksc)	Cement (kg/m³)	Fly Ash (kg/m³)	Water (kg/m³)	Sand (kg/m³)	Recycled Aggregate 3/4" (kg/m³)	Recommended Quantity Admixture (Liters/m³)	Fiber (%)
Slump (mm)[2]	63.0		17.2	28 days			365.14	14.84
Initial setting time (min)[3]	135		12.7	90 days			390.57	9.56
Final setting time (min)[3]	276		33.5	365 days			491.98	4.17

- Tap water with a pH 7.0 conforming to ASTM C1602 (ASTM C1602/C1602M–12 Standard specification for mixing water used in the production of hydraulic cement concrete).
- River sand with gradation conforming to the ASTM C33 (ASTM C33/C33M–1601 standard specification for concrete aggregates).
- Crushed lime recycled aggregate rock with gradation conforming to the ASTM C33 (ASTM C33/C33M–1601 standard specification for RAC aggregates).
- The chemical admixture used superplasticizer that conforms to the ASTM C494 (ASTM C494/C494M–16 Standard specification for chemical admixtures for concrete); that is, the superplasticizer had a recommended dosage rate of cementitious materials (per 100 of a kilogram of cementitious materials).

5.3 MONITORING AND REPAIRING OF RCA STRUCTURES

At this point in time, the construction of technological infrastructure is almost entirely reliant on very inexpensive construction chemical mixtures. The investigation into determining the compressive strength of concrete depending on temperature was first motivated by the core objective of concrete structure contractors, which was to increase the predictability, quality, and efficiency of the building projects they were working on. Concrete is the material employed in building new bridges more often than any other material. In addition, the number of structures created out of prestressed concrete is increasing faster than the number of buildings made out of reinforced concrete. In OPC, the ratio was 25.3% to 60.4% in 2007, but by 2016, it had increased from 41% to 44.2%. The ratio was 25.3% to 80.4% in the year 2020. The amount of time necessary for the concrete to achieve the desired compressive strength is the most important component that must be taken into consideration while creating structures out of prestressed concrete. In addition to building bridges, it can now lay the foundation for the post-tension tendons. This opens up a lot of new possibilities. Due to the fact that working with concrete may take place in various climates, it is essential to have proper information on the strength of cast-in-place concrete. A maturity technique that is in line with the ASTM C1074 standard (2020) and its subsequent development into a numerical program that may be used to determine the size of a structure or portion of a structure as well as the strength of concrete that contains RCA over time.

The algorithm used in the maturity approach may be an essential aspect of the design process for the structural components made of concrete, both during the mixing and constructing stages of the process. If one has a complete understanding of the maturity-strength connection, which may be learned in laboratory settings, then it is conceivable to develop the strength of concrete before the final pouring of the foundation concrete for a structure. Before being brought to the job site, the concrete mixture is put through the concrete monitoring system, which determines the mixture's quality. Additionally, as a result of this, it is feasible to alter the activities that had been previously planned in accordance with the characteristics of the new mix. In addition, the strategy that has been suggested restricts the number of compression tests that may be executed in order to determine the condition of the concrete in the structure. These tests are intended to measure the strength of the concrete. By providing 2D maps of the extradosed bridge's sectional concrete strength that can be seen virtually online as it is being constructed, trust in the building process is improved throughout the decision-making stages. This helps to ensure that the bridge is built safely (Yu et al., 2021). After the completion of the project, an economic profit is generated from the advantages connected with the various timetable changes. The strategy of increased maturity applies to the development of any complex engineering structure that is comprised of a wide variety of components. This is conceivable because the algorithm and the numerical answers for predicting the development of temperature and strength in concrete may be used concurrently. This makes it possible for this to happen. Up to this point, efforts have been undertaken to improve the mechanical characteristics of concrete that can be predicted in the future. In particular, the tensile strength and Young's modulus of concrete have been the focus of these efforts.

In RC constructions, CO_2 is a common cause of deterioration and failure. The reinforcing bars' susceptibility to corrosion is a significant issue. It diminishes the cross-sectional area of the bar, which, among other things, causes concrete to crack and causes the cover of the bar to peel off. Additionally, it influences the bond between the bar and the concrete. Because corrosion is reinforcing itself on many structures, they are breaking down; it is imperative that accurate techniques for assessing the performance of these buildings be developed immediately. The significant effect corrosion has on the structural strength of RC beams has been investigated in several different research studies. Per experimental findings, corrosion can alter the failure mode of a structure (for example, from shear to flexure or anchorage); it also reduces the structure's ability to carry much weight. Corrosion has been shown to make a structure less able to hold much weight. In the past, ribbed reinforcement bars had been the norm in building, since at least the mid-1950s in the United States and Canada, as well as in Europe, at the time of their publication. These bars had been used in construction since the mid-1960s (Martinez-Arguelles et al., 2019).

6 Sustainable RCA for Sustainable Concrete Construction

6.1 SUSTAINABLE RCA IN CONCRETE

In 2015, the United Nations (UN) settled the sustainable development goals (SDGs) for this time frame, 2015–2030. The SDGs consist of 17 and 169 targets that address the social, economic, and surroundings, which are also important parts of growth, respectively (UN, 2015). The SDG Goal 9, which is centered relating to business, infrastructure, and innovation, calls for infrastructure that is sturdy and flexible in emerging and least developed countries (LDC), as well as the retrofitting of existing infrastructure and the building of new infrastructure in these countries. Thus, the building industry must expand exponentially to keep up with the need for new infrastructure. As a result, cement consumption would increase significantly, resulting in increased emissions of greenhouse gases (GHGs) in the future. In order to mitigate the negative concerns of OPC manufacturing, the use of more future research motivation focus on SCM as a partial substitute for cement. Additionally, embodied energy (EE) is essential when evaluating the long-term sustainability of construction material. In the words of the Cement & Concrete Institute (2011), basically, a product's "embodied energy" is the total amount of energy it consumes during its lifetime for the mining of raw materials, transportation, manufacturing, assembling, disassembling and removing the product system from its base. Ecologically friendly materials are those that have low EE. SCMs are an ideal material for sustainable concrete applications since, as shown in Table 6.1 (Samad and Shah, 2017) that have a much lower CO_2 content than OPC. Thus, concrete has a rather substantial ecological imprint. The OPC part accounts for about 5%–7% of all manufactured CO_2 emissions. The burning of coal accounts for about 40% of this, while the calcination of limestone powder accounts for 60%–65% (Khatib, 2016). A 7% increase in the CO_2 emissions per ton of concrete mix production compared to 2019, but lower than the 2010 baseline, was seen in 2021. Figure 6.1 shows the 3-year CO_2 emissions from the manufacturing of concrete mixes (Samad and Shah, 2017).

Concrete recycling is an essential step toward environmentally friendly construction techniques. Simultaneously, it helps to decrease the quantity of construction and demolition waste generated, hence contributing to the mitigation of natural resource depletion; however, it reduces strength and resistance to degradation mechanisms. In order to remedy these undesirable behaviors, it is necessary to remove the adhering mortar paste that has been applied to RCAs, which adversely affects the RAC characteristics. It is possible to increase the mechanical

DOI: 10.1201/9781003257097-6

TABLE 6.1

Embodied CO$_2$ of OPC, GGBS, and FA (Samad and Shah, 2017)

Materials	Embodied CO$_2$ (kg/ton)
OPC	913
GGBS	67
FA	4
Limestone powder	75

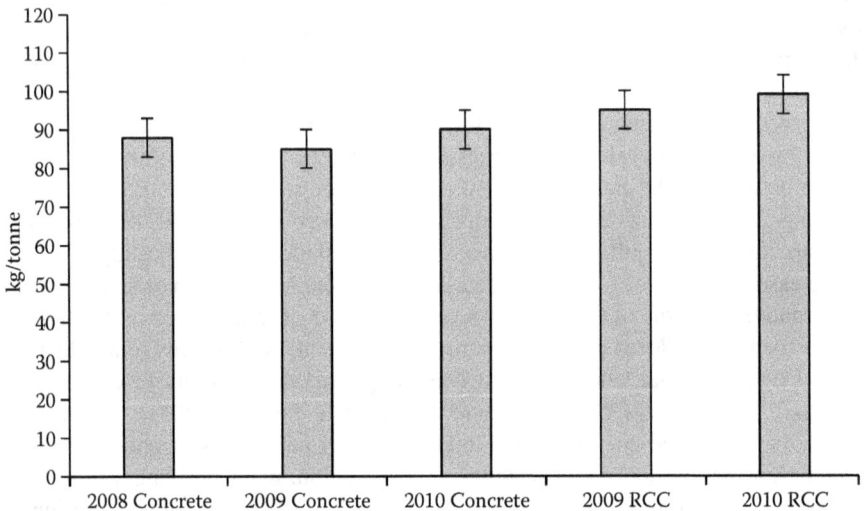

FIGURE 6.1   CO$_2$ emissions for the production of concrete mix (Samad and Shah, 2017).

qualities of concrete by adding extra admixtures such as FA, SF, and fibers into the mix when used in conjunction with a suitable concrete mix design. As a result, if the suggested procedures are implemented, RCA can be a useful instrument for preserving the environment while still offering, in engineering applications, a suitable degree of structural performance.

The three "Rs" (reduce, reuse, recycle) form the foundation of the key hierarchy of sustainability: reduction comes first, then repurpose, and lastly, recycling. Recycling is the most efficient approach to cut down on the number of primary sources discharged into the environment when all other options for reusing the materials used in building have been exhausted. The 1940s saw some early experiments with RCA concrete being carried out. The fact that concrete is the material that is used the second most frequently, after water, underscores the critical significance of using RAC as an aggregate for freshly mixed concrete. As a result of the exceptional qualities that it possesses, concrete is anticipated to continue to be one of the most widely used materials in the construction industry. This prediction suggests that its application around the globe will continue to grow, if not accelerate, in the near future. Simultaneously, the number of demolished concrete structures will increase, as will the demand for reusing and recycling RAC.

To provide knowledge would help to increase the use of RAC in a manner that is less detrimental to the natural environment. Using RCA in fresh concrete mixes can reduce waste disposal while decreasing the requirement for virgin aggregate, which are essential environmental considerations in responsible, sustainable management. A significant contribution to the achievement of Goal 12 (SDG12) of 2030, as a direct outcome of this initiative, the United Nations Agenda for Sustainable Development also makes progress toward the goal of ensuring sustainable consumption and production patterns. At the same time, concentrating on economic growth is based on the effective usage of resources and the least amount of damage to the environment possible concerning improving people's well-being. Changes in resource consumption, as well as advancements in manufacturing technologies, can all contribute to this goal. Sustainable production and consumption policies are required to raise living conditions without risking future generations' ability to meet resource demands. It is the goal of these programs to decouple economic development from environmental damage.

It is a huge problem, not just locally or regionally, but globally because mineral resources are being depleted. Developmental countries have nearly doubled their material footprint per capita over the last 8 years, signaling an important rise in their material standard of living. An increase in the demand for non-metallic minerals, mainly due to the expansion of infrastructure and buildings in these areas, accounts for most of this increase. All this includes how concrete structures, which are becoming more common as the population grows, have an impact on the earth. So, concrete reuse and recycling are important parts of this procedure and are an excellent barrier to implementation.

According to the International Energy Agency, OPC production accounts for around 5%–7% of the world's total CO_2 emissions. Approximately 30% of this comes from coal combustion, and the remaining 60% comes from limestone calcination (Khatib, 2016). The necessity of reducing natural resources and energy consumption, as well as reducing emissions of greenhouse gases, while simultaneously enhancing the satisfaction of demands, can be used for almost any element of life or business such as 20 billion tons of aggregates, 1.5 billion tons of cement, and 800 million tons of water used each year by the concrete industry. In the future, more cement and concrete will be produced. To reduce natural resource consumption, it may be necessary to use RCA in making concrete.

In some cases, the EN-206:2013 (Collivignarelli et al., 2020) standard allows RCA to be used instead of some of the NCA (it does not apply to fractions 0–4 mm that cannot be substituted). Because of this, it seems likely that the popularity of RCAs will continue to grow over time. Besides the most straightforward use, which is to replace NCA with RCA, there are a few other ways that RCA could be used that are worth looking into. The authors of this article wanted to improve the gradation of NCA using RCA. However, the final product did not meet the criteria because it had a lot of 2–4 mm fraction (about 70%). It has been looked into how this kind of technology affects the consistency of a concrete mix and other things about how hardened concrete looks. Using two fractions of RCA (4–8 mm and 8–16 mm), it was possible to change the aggregate composition of NCA to fit the desired gradation curve. The Gradation Index (GI) was made so that it is easy and quick to compare an experimental gradation curve to the expected one.

RCA has become a lot more popular and used in recent years because of the reduction of natural resources and the demolition of structures because they are no longer profitable or useful. Another problem is insufficient landfills; natural habitats are being destroyed by landfilling and quarrying, and energy is being used excessively. According to current trends, if the current trend lasts for 7 years, the amount of waste concrete made each year will reach 90 million tons by 2021 (198 million kips). The same things happen in other countries, as well. Environmental requirements for reducing emissions are also problems that need to be solved immediately. Many countries have already tried RAC and concrete made from recycled materials, but for the time being, the majority of aggregate recycling ends up in landfills instead of being used to make new concrete (McNeil and Kang, 2013).

The construction industry's persistent desire for natural resources has threatened the natural ecosystem's ability to function properly. In the building industry, it is common knowledge that there is a high need for raw materials and electricity. As a result, it is one of Europe's most resource-intensive industries. According to estimates, it consumes more than 50% of all extracted resources, 50% of all energy, and 30% of all water. The construction industry has been scrutinized for its environmentally friendly building methods under investment. Concrete must face the bulk of the blame for the disaster because it is such an excellent building material. Concrete has a high embedded carbon footprint (for example, CO_2), involves a massive quantity of natural resources, and generates a substantial amount of waste after demolition, all of which are significant difficulties. The growing worldwide construction industry necessitates the quest for sustainable resources that may be used to make concrete instead of natural resources. Consequently, the quantity of concrete aggregates and other materials recycled from demolished buildings and other construction detritus has significantly increased. In this regard, alternative options include RCAs and other mineral components recovered from construction and demolition activities.

The use of RCA aggregate, which used to be called RCA, is growing every year. This is because more people are looking for more environmentally friendly materials to buy and build with. There is still much uncertainty about how well RCAs will work in structural applications. On the other hand, RCA producers are constantly working to improve CCA's quality and performance to be used in more high-value applications.

There are no clear answers from scientific research on whether more NCAs can be used to make concrete that is strong enough to last for years and years. Before coarse CCA may be utilized as a possible alternative in high-value applications, additional research must be conducted to determine how it affects the resistance of structural concrete to chloride ion invasion. Due to concrete is static one of the world's most essential and widely used building materials. About 850 kg of CO_2 is released for every ton of clinker. This means that the process of making clinker has a large carbon footprint. This is mainly because making cement takes many resources in terms of energy and raw materials. The International Energy Agency says that improving energy efficiency and using alternative resources like biofuels or recycled raw materials are the best ways to reduce the environmental impact of concrete products (IEA). The investigations found that the building industry is the most cost-effective way to reduce greenhouse gases. According to this point of view, aggregates are vital because they make up about 80% to 85% of a typical concrete mix and cause wasteful use of locally available natural resources. Sand and gravel, the main materials used to make aggregates and traded in massive amounts, cause river deltas and coastlines to erode. Substituting and using alternative materials is a potential approach for preventing or decreasing damage to the river and marine ecosystems.

Following the circular economy principle, incorporating alternative materials into concrete helps conserve natural resources while lowering waste disposal. This is accomplished by combining municipal and industrial wastes into concrete. In most cases, LCA is used to figure out how much better the environment drive is if recycled materials are used in building parts like walls and roofs, for example. Doing this kind of study is called "cradle-to-grave" because the concrete industry will face challenges in the future related to the utilization of increasingly alternative and sustainable raw materials derived from appropriate waste streams. Fine and coarse aggregates can be created from recycled resources such as scrap tires, glass, and foundry sands; however, this process is limited by technological obstacles and national limits. Typical production fuels such as coal and oil can be substituted with biomass or other types of waste. By ensuring that a workable waste management solution is in place, contributing significantly to the circular economy is possible by using alternative materials instead of fossil fuels and raw materials. Because it is the building material used the most and is also the most recyclable, concrete is one of the most important components in implementing circular economy strategies in the construction industry.

To ensure that waste materials can be utilized to build concrete, it is critical to have sustainable practices and increasing demand for it. Alternative materials may be a practical choice for producing effects in light of the rising demand for natural resources and the resulting rise in raw material costs. Furthermore, their efforts to make the world a better place have special consequences on the performance of concrete. Consider aggregates while considering other materials. Fine and coarse aggregates comprise four types of trash: garbage generated during construction and demolition, waste left over after waste management treatment, by-products from the cement industry, and waste that does not fit into other recycling processes. When considering different materials, examine both the mechanical capabilities and the durability of the material, which are both impacted by porosity and water absorption. The finished specimens must fulfill technical resistance standards.

An excellent technique to reduce CO_2 emissions from concrete production is to use wastes and by-products as part of the mix. However, the amount of impact they have is directly related to the characteristics of the final product, which must be designed to fulfill the prerequisites of a wide variety of typical uses. There are so many different mechanical properties and environmental compatibilities that need to be looked into more thoroughly, especially those that deal with the release of heavy metals. Because density and workability directly affect how things are made, more research needs to be done on the physical properties.

Various C&DW materials might be substituted for aggregate or used in different ways to manufacture concrete. People often discard broken or shattered glass, bricks, and RAC debris. In addition, they discard broken or splintered wood, broken or splintered plastic, broken or fractured tile, cardboard, paper, steel scrap, and steel scrap. A significant amount of study has also examined novel and inventive methods to reuse waste materials such as building and demolition debris, municipal solid waste, industrial garbage, and other waste. Thomas et al. (2013) studied the structural performance of concrete, including steel slag and copper tailings as aggregate. Siddique et al. (2008) and Meng et al. (2018) investigated the use of recovered waste plastic in concrete blocks. Wartman et al. (2004) evaluated the efficiency of broken glass as an engineering material since it can be acquired reasonably cheaply. They discovered that it was on par with the majority of NCAs. It was also said that broken glass might be substituted for NCA in building, as stated in research by Olofinnade et al. (2017). Disfani et al. (2017) stated that recycled glass could be utilized in road construction. In a 2013 research, Arulrajah et al. investigated the viability of using construction and demolition waste as pavement sub-bases.

According to Thomas et al. (2016), there is a good chance that waste tire rubber can be used as fine particles in cement concrete. Until then, an increasing proportion of these waste materials will be used as aggregate in various building projects, such as the paving of roads and the embankment of road cuts. In 2011, Puppala et al. evaluated the efficiency of recovered asphalt pavement (RAP) materials as aggregate for pavement building. Hoyos et al. examined the performance of RAP materials as aggregate for pavement building in 2011. The maintenance of sand mining and dredging activities has caused riverbank erosion and the degradation of aquatic and fish habitats. According to Tavakoli et al. (2014), the ongoing depletion of resources needed to produce concrete may have led to a shortage of raw materials and an increase in the price of those commodities in recent years. These resources may have become more expensive due to their scarcity. However, continually searching for new, less expensive materials is essential. Most construction waste is disposed of in landfills or repurposed as reclamation material, which may be detrimental to the surrounding ecosystem.

Waste materials are being encouraged to make cleaner concrete with a lower global warming potential than traditional concrete. This will support the aggregates and cement industries use less natural resources, which will also help the environment. There is a lot of C&DW in 40 significant countries worldwide, and C&DW is the name given to this garbage. Conventional backfilling and landfilling procedures were found to be inefficient compared to C&DW in concrete manufacturing and another waste disposal, which are still used today.

Natural resources such as limestone, clay, and shale are being depleted at an alarming rate as cement manufacturing soars. This industry produces a lot of heat and carbon dioxide. Approximately 7% of human-generated CO_2 emissions are attributed to the cement industry. The use of pozzolanic materials such as FA, granulated blast furnace slag, and other materials in concrete is being examined as a way to reduce the amount of cement that is produced. Using less cement may be an advantage in this situation. Concrete strength and durability can be improved over time by adding pozzolans like FA and SF. Using RCA in concrete has been demonstrated to weaken the product and increase the production cost. Using RCAs increases the cost of making concrete, making it more expensive to produce. When mixed with additional cementitious materials, RCAs can be used up to 50 wt% in concrete, according to Akhtar et al., reducing their negative impact on the concrete's strength and durability.

Many countries are worried about how to use the waste from coal-fired electricity production to manufacture ecologically friendly concrete. Several countries are worried about how to use the waste from coal-fired power production to manufacture concrete. A waste product from coal-fired electricity production may be better for the environment than natural fine aggregates for creating by-product-based concrete.

In a life cycle assessment (LCA), fossil resources used by living creatures can be measured by abiotic depletion of fossil resources (ADP), which is a measure of the number of fossil resources that have been consumed (abiotic depletion of fossil resources). Three types of environmental impact can be measured: global warming potential (GWP), acidification potential (AP), and the ADP. All evaluations based on these parameters can use environmental data, current knowledge, and standardization. The impact categories were identified in such a way that the process was as transparent as possible.

6.2 LIFE CYCLE ASSESSMENT (LCA) OF RAC

The sustainability of the built environment (SBE) follows three fundamental tenets: the use of fewer resources, life cycle costing (LCC), and human-friendly design (HFD). Reduce, reuse, and recycle (or the 3Rs) are three concepts that are widely used in resource conservation, particularly among those working in the building and making of things industries. In building materials, concrete reigns supreme since it is so versatile that architects and designers have a vast range of options to work with. Cement content in concrete mix designs has been shown to affect the amount of embodied CO_2 (ECO_2) produced by the material, in accordance with The Cement and Concrete Institute (2011). Concrete has an environmental impact of approximately 100 kg CO_2 per ton. For durable concrete structures, consideration is given to the embodied CO_2 content of concrete. According to the environmental protection agency (EPA), OPC is essential in manufacturing concrete. The manufacture of OPC results in a significant amount of CO_2 emission. OPC is a part of it and can be replaced by SCMs such as ground GGBS, FA, RHA, SF, and others to minimize the amount of CO_2 embodied in concrete. A lot of the time, SCMs are used as a replacement for clinker or cement. This method makes cheaper concrete, has less of an impact on the environment, has more long-term strength, and is more durable. Regarding SCMs, SF and FA are the two most common ones. SF is a by-product of making silicon and ferrosilicon alloys. It comprises more than 90% SiO_2 and is found in spheres around 100 times smaller than surface cement particles. The material has a large surface area of 10 to 20 times that of typical pozzolanic materials. This provides it with much pozzolanic activity and allows it to move around in the pozzolan. Over the last few decades, SCMs have been widely employed in concrete manufacturing, which is expected to continue. It has been primarily motivated by the desire to offset the cost of concrete production while simultaneously reducing the expense of building. However, the sustainability dimension of concrete is a more relevant issue in using SCM, anticipating that CO_2 emissions will go down a lot if using SCM, reducing the environmental effect of concrete manufacturing. Using a circular economy (CE) approach, buildings' energy consumption and greenhouse gas emissions are predicted to decrease over time, recycling, material recovery, and less material use, among other things (Mata Kihila et al., 2022). In addition, other phases of life cycle, EoL, are less highlighted than the use phase of buildings, which is understandable given the significant energy use of buildings. Each person consumes three tons of concrete every year; on average, concrete has become the most commonly utilized structural material in construction (Gagg, 2014). Given that the reuse of concrete materials is still in its infancy, it is necessary to remove the concrete framework of buildings during the EoL phase (Salama, 2017).

Combining LCA and LCC, eco-efficiency assessment is a business tool for analyzing the environmental effects, benefits, and cost implications of different recycling systems. However, although the idea of eco-efficiency is neither novel nor challenging to understand, a more precise definition is necessary to assess the co-benefits of technological advances. Regarding high-quality recycling concrete, a series of technical breakthroughs offers an ideal case study for examining how technology innovation may affect changes in C&DW management efficiency. Consider yourself a believer that high-quality concrete recycling is an environmental and economic win-win scenario. In addition, a sharing economy may exist in the construction industry. In addition, this case study on concrete recycling provided an eco-efficiency technique for LCA/LCC evaluations.

EoL concrete constitutes most of the composition of C&DW and has a significant possibility of reusing or recycling. In the European Union, concrete that has reached the end of its useful life is typically downcycled as a base course or even thrown away in dumps. It is important to move away from concrete treatment and disposal methods that are not as popular and towards ones that use resources more efficiently. Europe has been a leader in creating new technologies that make it possible to recycle concrete that has reached the end of its useful life and turn it into high-quality secondary raw material that can be used to make new concrete products. This is the end of the cycle of concrete. Eco-efficiency analysis is an excellent way to make decisions about managing resources in a way that is good for both the environment and the economy. An eco-efficiency analysis technique compares the wet process of the business as usual (BAU) approach to the environmental and economic performance of technical improvements like advanced dry recovery (ADR) and heating air classification system (HAS) for EoL concrete recovery. This technique is used to evaluate these technical improvements' environmental and economic performance. It is suggested that a LCA/LCC framework procedure be used to measure how eco-friendly something is. Similar to the "environmental impact assessment" step that is recommended for LCA to describe the project's effect on the environment, an "economic impact assessment" step is proposed for LCC to explain the standard cost structures, types of costs expressed, and cost stressors.

Recycling EoL concrete waste, which accounts for most C&DW, has advanced significantly over the previous half-century, with numerous technologies devised. Most of these methods are aimed to demonstrate that concrete waste may be transformed into a product that resembles the original concrete rather than replacing it. Many RCAs are classified using RCA, recycled fine aggregates (RFA), and a cement-rich fine fraction (CRFF) as a basis for further analysis and comparison. Using selective fragmentation and microwave-assisted beneficiation, EoL concrete can be reused in the construction industry. These are just two examples of the newest tools in the toolbox. There is a difference in how much energy mortar and aggregates can store. To put it another way, this results in higher thermal stress at the mortar-aggregate contact, which causes the adhering mortar and aggregate surface to separate more quickly. A lot of the cement is turned into sludge when this method is used, so it has to be thrown away or used for something else. Using this method, clean coarse aggregates and a cement-rich powder can be utilized to stabilize soil or to blend with blast furnace slag cement, depending on the application. The strength and durability of concrete constructed using aggregates produced using this technology and concrete used as a reference were nearly identical. RAC can be pumped, and the cast is produced using this technology. Waste-free cleaning of RCAs and cement powder is claimed to be possible using a high-temperature furnace and grinding process. For approximately 1 hour, the concrete rubble is exposed to a temperature of 650°C using this method. As a result of the extended residence time, the procedure may become more expensive, and its viability may also be called into question. That is not all: It also makes high-quality RCA that can make concrete 10% stronger without affecting other essential factors. According to the manufacturer, this material has excellent water absorption, water permeability, and frost resistance. The hydrated cement powder can also be crushed with this method, which is one of the many methods available. In the mining industry, this is referred to as "smart crushing." Recycling gravel and sand claims that they are as good as natural gravel and sand. The continuous milling technique results in silica-rich powder rather than calcium-rich powder. It does not matter how much we want to use recycled materials. Most of these technologies cannot be used on a large scale because they cost too much to process. C2CA technologies are also suitable for recycling EoL concrete because they are a good balance between quality and cost in terms of quality and cost referred to as advanced dry recovery ADR and HAS. By way of illustration, the ADR process allows for the separation of wet RCA waste into coarse RCAs (4–12 mm) and fine RCAs (0–4 mm), both of which can be reused in the production of freshly RCAs. There are clean RCAs in the coarse fraction, meaning they can be used immediately in the concrete mix without having to go through any more steps. When the fine fraction (0–4 mm) is used in the HAS, heat treatment is used to make it more durable. This makes it either recycled sand or concrete ultrafine/hydrated cement-rich ultrafine.

The LCA method, based on ISO 14040 (Medine et al., 2020), is used to examine the material and energy flows, as well as any potential environmental implications, during the duration of a concrete's lifespan. NCAs such as gravel and sand in concrete can have a negative impact on climate change, eutrophication, acidification, particulate matter (PM), smog potential, and ozone depletion if they are substituted with synthetic materials. For instance, when rubber aggregates replace natural resources, the environmental effect of concrete is minimized. As a result of the low cement content and the low substitution rate, as well as the lack of data on energy consumption or savings over the use and end-of-life phases, the deterioration has been kept constant. The process-based life cycle assessment method is the most common way to look at the environmental impact of building materials. LCA is noble at providing information about materials, but it has a problem because it does not have a reasonable system boundary. While the system boundary completeness of input-output analysis, which is based on interindustry monetary flow data, is good, there are several drawbacks. For example, it has issues with price volatility, aggregation, and lack of material-specific data (Dixit, 2017).

In monetary values such as dollars, purified input-output tables may capture the direct and indirect relationships between sectors. This makes it easy to locate instantly applicable economic data. In the conventional monetary input-output table (MIOT) model, physical quantity exchanges are not necessarily proportionate to their monetary worth. This indicates that environmental impact apportionment is not always accurate. This indicates that outputs per physical unit are more valuable monetarily; this phenomenon is known as "economy of scale" (Bullard and Herendeen, 1975). Based on this approach, defining the word "functional unit" in the life cycle of structural concrete is difficult. This unit must include all of the concrete's basic materials and emissions across all assessed concrete mixtures. Minimal amounts of concrete are used in this experiment (typically 1 m^3 of concrete mix). There should be no difference between the structural concretes you evaluate in terms of freshness and hardness, with compressive strength being the defining characteristic of hard concrete.

6.3   ECONOMIC AND COST ANALYSIS OF RAC

RAC provides an environmental and natural resource conservation benefit by minimizing the usage of non-renewable materials such as NCAs in construction projects. Construction waste can be recycled with this type of application. Various features of RAC have been studied in several studies, which can be divided into two types: coarse RCA and fine RCA.

C&DW is almost universally acknowledged as a substantial contributor to the amount of waste produced in modern society. A good case in point is Europe, where the majority of the existing stock of buildings and infrastructure was built during World War II. As a result, restoration and deconstruction of this stock are presently the most significant operations carried out by the building and construction business. According to projections made by Eurostat, the 27 member states that compose the European Union (EU) produce 970 million tons of C&DW per year (Construction and demolition waste, 2018). It has been decided that the

Construction and demolition waste (2018) will be Europe's priority stream since there is a great deal of trash, and the materials have a great deal of potential for reuse and recycling. Because of this, the European Commission issued the Trash Framework Directive in 2008, which demands that member states take the appropriate procedures to collect at least 70% of their waste (by weight) of C&DW by 2020; the remaining waste is recycled. According to the European Commission, current C&DW recycling rates significantly differ from one European nation to the next, ranging from less than 5% in Montenegro to more than 90% in many European countries like Belgium, Portugal, and the Netherlands, among others. It is anticipated that the great majority of C&DW will be recycled, for example, in the construction of road foundations or even disposed of in landfills in some European countries. For example, just 10.3% of C&DW was recycled in the Spanish construction industry in 2003. The other 64.1% of the waste was dumped illegally in waste sites, pits, and watercourses since there were no restrictions to prevent it. According to Hincapié et al., Switzerland recycled 51% of its rubbish in 2012, disposed of 26% of its garbage in landfills, burnt 8% (combustible materials such as wood), and reused 15% of its garbage on-site (2015). According to Gálvez-Martos et al. (2018), up to 85% of European municipal solid trash is made up of stony debris. This includes concrete that has reached the end of its usable life. An alternative market for RCAs, which are produced as a byproduct of EoL concrete, already exists in Europe, where EoL concrete is reused as material for road bases (Anastasiou et al., 2015). According to the opinions of several specialists, the exploitation of recycled concrete aggregates in the construction of roads is anticipated to contribute significantly toward the goal of recycling 70% of C&DW set by the European Union (EU) (Bio Intelligence Service, 2011).

There are financial advantages to recycling concrete, however, as the price of RAC for construction will be less high than that of natural sand for use in concrete production. RAC has many advantages, including ecologically sound by preserving existing natural resources, decreasing building waste as an example of an environmentally friendly practice, and economical: by reducing the price of the concrete used in construction. Based on the footprint analysis of RAC production, treating RCA in a mobile plant rather than a fixed plant is preferable, not only in terms of the product's carbon footprint but also in terms of its energy and water footprints. RAC produced in a mobile production has a one-third reduced climate footprint than concrete made in a permanent facility. This is due to shorter transit distances, decreased energy usage, and the use of diesel fuel rather than electricity. Similarly, the energy footprint of the product is lowered to a close range. Because of the wet treatment technique used in the permanent plant, the stationary plant's water consumption might be up to ten times that of the mobile unit. More treatment stages in a stationary plant require more energy and water than additional treatment steps in a mobile plant, resulting in greater resource consumption and global warming for RAC making. As a result, selective deconstruction and pollutant removal can reduce the amount of energy and water required for waste treatment processes. However, if it is to be processed in a mobile plant, the destroyed concrete must be of excellent quality in terms of consistency and low levels of contaminants (Müller, 1998).

Increased population growth, urbanization, and improved economic conditions in developing countries contributed to the construction industry's rapid growth. As a result, existing structures must be demolished to construct new ones. Each year, over 850 million tons of building waste are generated globally, according to data, accounting for nearly one-third of total waste generated globally. An example of a waste item resulting from the destruction of historical constructions is aged concrete, which can be recycled. RCA is a term used to describe recycled material that contains at least 95% aggregate. Reducing the demand for natural stone and its environmental and social impacts can be achieved by replacing some or all of it with RCA in the concrete mix. Side construction elements such as wood chips, ceramics, bricks, or reinforcing pieces may blame RCA inferior quality rather than NCAs. Because of this, the use of RCA in producing concrete is restricted. Analyzing the qualities of concrete infused with RCA is crucial, as it, like NCAs, has the potential to be exploited in developing new concretes with improved performance characteristics. Rapid urbanization has also led to a rise in the number of people living in cities for construction jobs. Construction supplies are in high demand right now, and this is true across the country. The demolition of ancient structures generates a large amount of C&DW, which poses a threat to the environment. C&DW necessitate a more significant amount of room in disposal sites. It accounts for the lion's share of the waste created throughout the country (Shaikh, 2016). Because NCAs and landfill space are in short supply, the only way to get rid of waste concrete is to recycle it, which will help save both. It is an extensively utilized material in other nations and is mainly suggested for nonstructural applications. In order to employ waste concrete from various sources in the manufacturing of needed grade concrete, it must first be crushed, screened, and used as coarse aggregates, either with or without NCAs, before they have been cleaned and dried C&DW is made when concrete structures get old and need to be demolished. This leads to the formation of C&DW. Waste that is not used again could be used to make concrete. Although RCA features are quite adaptable, it is recommended that their properties be thoroughly examined before they are used in any applications. Before utilizing this concrete in large quantities, it is important to look very carefully at how concrete made with RCA and NCAs behaves so that it can be strong and last for a long time.

In the meantime, the destruction of old concrete structures, natural disasters including earthquakes, avalanches, and tornadoes, as well as human causes such as war and bombing. A lot of C&DW is made every year when buildings fall or are torn down. China's building and tearing down projects made more than 5 billion tons of construction and demolition waste in 2020. C&DW has become a problem worldwide, particularly when dumping it into landfills. Many communities are confronted with the problem of improper construction and demolition trash disposal. It has a responsibility, along with other elements, to destroy landscapes and other harms to human safety and the lives of living things. For example, a landslide occurred near the end of 2020 at the Hongao construction waste landfill in the Chinese city of Shenzhen. It demolished buildings and killed workers who lived in the surrounding area. In addition to the tragedies that can occur as a result of incorrect demolition waste disposal, there is a significant detrimental influence on the long-term viability of our environment. Because of its significant effect on the environment, there is a great need for environmental sustainability measures in our current infrastructure projects, and proper construction and demolition waste management is critical to accomplishing these goals. Old concrete debris must be crushed, processed, and reused in new concrete constructions before RCAs are reused. This recycling helps the construction industry become more sustainable by solving the problem of how to deal with C&DW and limiting the depletion of NCAs by giving people access to alternative aggregates. Many countries are relaxing their infrastructure rules to encourage the use of RCA in construction projects. However, because of the lack of a clear code and standard, many issues of the design and implementation of RAC structures must be considered and handled carefully. This review paper discusses how to deal with the problems that come with using RCA, like the old mortar paste that is still on the concrete and the high porosity, permeability, and water absorption of RAC. It also talks about how RCA affects fresh concrete mix's rheological properties and the mechanical properties and long-term service of RCA concrete.

The potential influence of alternative materials will vary depending on the scale of the construction sector. There has been a slight rise in concrete production in Italy. It went from 35.7 million cubic meters in 2016 to 47.3 million cubic meters in 2020. This trend has been confirmed by the Italian Association of Building Contractors, which says that concrete production in Italy is going up a little (ANCE). Since 2015, aggregate production in Europe has climbed by approximately 2.5%, with Russia being the leading producer, with 618 million tons produced in 2020.

As people become more aware of climate change, people who work with transportation infrastructure are looking for new ways to save resources while also cutting down on energy use and emissions from transportation infrastructure. Environmentally friendly technology and materials should be used more often in constructing and maintaining highway infrastructures, especially when building new roads. When NCA and mixing temperature are cut back, the environmental benefits are outweighed by the poorer efficiency and need for more optimal asphalt content in warm mix asphalt (WMA) with RCA, which are both terrible things. Compared to hot mix asphalt (HMA), only one mixture has environmental benefits in almost all of the effect categories that were looked at WMA.

When developing a brand-new RAC product, the objective is to strike a balance between technical viability, manufacturing costs, and end product quality. More study is needed to understand the economic effect of a policy properly. It is feasible to tackle the issue of unpredictable or uniform pricing by employing physical units rather than money. On this screen, we can only view the physical input-output table (PIOT) in physical units. It displays the movement of resources and trash from one industry to the next, as well as the impacts on the local area, the system, and the environment (Hoekstra and van den Bergh, 2006). A PIOT can accomplish many things effectively, but it is not particularly good at capturing money movements in the economy, which is especially true for service-based companies. Furthermore, the data required to generate a PIOT is often limited, inconsistent, and time-consuming. As a result, just a few low-resolution aggregated tables are created (Altimiras-Martin, 2014).

A strategy can utilize monetary and physical units, or the system can employ both process and input-output (IO) data (Suh, 2004). In the first scenario, gaps in data may be filled with more relevant data, but this does not work as effectively in the second (e.g., unit mass for raw materials and unit dollar for services). Because they do not have to convert prices while estimating a product's cost, approximation approaches are less prone to mistakes. There are various benefits to the comprehensive LCA framework used by Suh et al. (2004). It combines the precision and breadth of input-output analysis (IOA) with process-based material flow analysis (MFA). This component provides the missing supply networks upstream to complete the MFA. Incorporating actual units into the model minimizes the risk associated with price uncertainty. In this manner, it is possible to study a product's life cycle from beginning to finish, including how it is used, discarded, and recycled. Mayer and Flachmann (2011) investigated how energy and emissions move through an industrial system using mixed-unit input-output table (MUIOT) models. For energy and emissions, they employed extended input-output (IO) models, integrated heavy metals sector monitoring, ecological input-output analysis (IOA), and a multi-layered mixed-unit input-output table (MUIOT) (Majeau-Bettez et al., 2016). The use of mixed-unit hybrid life cycle assessment (MU-hLCA) to predict how recycled building materials may be utilized throughout the Australian economy demonstrates the benefits of MU-hLCA. Using this methodology, it will be simpler to determine how much carbon is incorporated in construction materials and commodities manufactured from recycled resources. It combines physical and monetary units from industrial systems related to commodities and products, among other things.

Using the MU-hLCA approach, which employs data from the life cycle inventory (LCI), IO, and material flow, the potential of recycled construction materials may be used across the whole Australian economy. The notion mixes monetary units from economic sectors, such as IO, with physical units from industrial systems, such as structures and recovered materials (Medine et al., 2020). As an illustration of the adaptability of MU-hLCA, this methodology uses recycled building materials and by-products. Physical units may be used to describe the use, disposal, and recycling of a product. MU-hLCA is accountable for presenting all of these phases. This research may use these goods without negatively impacting the input-output table (IoT) industry as a whole, which is a plus. Geopolymer concrete (GPC) and OPC concretes may be improved by substituting 100 wt% RCA for NCA. If exact waste data is available, trash information modeling may be utilized to calculate the real-world movement of rubbish and recyclable items. This method, also known as MU-hLCA, employs non-waste, non-recycled materials. Due to the complete detail of the processing system used for analysis, MU-hLCA may analyze the adaptability of certain goods or processes (such as individual chemicals).

The circular economy to promote 3Rs can benefit significantly from the use of MU-hLCA, which models how much pollution is emitted when recycled materials are used in place of virgin materials across the entire economy (Stahel and Clift, 2016). The circular economy has the potential to have a substantial influence on the economy and the environment over time. Consequently, employment, welfare, and GDP may increase (Agarwal et al., 2016). Recycling, reusing, and rejuvenating local recycling activities are Australia's most excellent options (Stahel, 2016) for avoiding the adverse effects of this new law and reducing its reliance on foreign waste management service providers.

According to specialists in the construction sector, concrete is now the most popular product on the market. Cost and strength are important factors when choosing a concrete mixture, so long as a certain degree of durability is met. Taxes are not the primary determinant of the cost of hiring and transporting employees and the types of materials used to produce concrete. A given price may not be the primary consideration when categorizing concrete. The price of concrete is a significant factor, but little research in the scientific literature examines the relationship between the price and the strength of concrete prepared using FA and RCA. Numerous research have examined how combining FA and RCA in concrete may make it stronger and more durable, but comparatively few have examined how this combination would impact the price of concrete. This research considers the combined impacts of FA and RCA (of type F) while examining the strength and durability of concrete. When examining the strength of concrete, it also considers the economic effect of FA and RCA. Extremely high transportation costs may affect the price of FA at the point of sale for the customer. FA is created as waste, and since it is scarce in the majority of areas of nations, its price may be influenced by the cost of transportation. FA examined two transportation scenarios for its economic research: the worst-case scenario, in which transportation costs were high, and the most probable situation.

6.4 POTENTIAL ENVIRONMENTAL HAZARDS OF RAC

Because RCA is a possible source of environmental pollution, the physical and chemical properties of RCA leachate have been extensively explored (Butera et al., 2014). According to the findings of RCAs heavy metal flushing investigations, heavy metals in RCA leachate originate primarily from hydrated cement paste. The leaching patterns of typical construction binders (e.g., OPC, FA, and/or slag) are used to define heavy metal leaching (Galván et al., 2014). The pH of the leachate and the degree of carbonation in the concrete material significantly impact the leaching mechanisms. In order to appreciate the potential environmental consequences of RCA usage, it is vital to comprehend the chemical and physical mechanisms underlying the substance's carbonation during intake. Current field experiments of RCA used as a road subbase indicated that the pH of the leachate initially increases but then returns to neutral within a year of construction completion (Engelsen et al., 2017). Because the most predominant method of disposal, which is to deposit waste in landfills, is rapidly becoming a significant issue especially in terms of humans and the environment. As a result, in developed countries, laws have been put into effect to limit waste generation, whether in the form of restrictions or particular levies levied against those who create waste areas.

The concrete industry's base is produced of RCAs, as opposed to NCAs, which is an alternate solution to this unsustainable management and aids in reducing the consumption of NCA in the process. A significant saving in the environmental effect of concretes can be achieved by substituting in the mix design of cement-based concrete mixes that include slag; a substantial proportion of RCAs is substituted for NCAs. Conversely, the mechanical qualities of concretes containing slag and a significant percentage of recycled particles have received less attention. The early mechanical characteristics of slag-containing RCA concrete are projected to be lower than those of RAC due to the delayed pozzolanic reaction of GGBS. Small quantities of SF can compensate for this shortage or add other ultrafine SCMs. Recycling waste materials and using them in concrete manufacturing can be a realistic option for saving natural resources while safeguarding the environment.

Despite the unfavorable aspects of RAC's decreased carbonation resistance, one perspective is worthy of consideration. In general, concrete can absorb CO_2 from the atmosphere. High-porosity RAC with deeper carbonation absorbs more CO_2 than conventional concrete, providing higher environmental benefits. Masonry aggregate, substantial amounts of waste masonry, such as red bricks, calcium silicate bricks, ceramic blocks, aerated concrete, plaster, and other similar materials, and mixed RCA, a mixture of masonry aggregate (RMA) and RCA, have not yet found widespread application. According to Czech legislation, RCA containing less than 90% of waste concrete cannot be used as an aggregate due to its harmful influence on mechanical qualities and durability. RMA is characterized by its porosity, water absorption, density, and resistance, such as wear, abrasion, and freeze (thaw) resistance. Numerous components of RMA increase its porosity and water absorption. These materials include red bricks, ceramic blocks, aerated concrete, and mortar-bound concrete particles. Therefore, although recycled MAC is highly porous and water-absorbent, it has somewhat different effects than RAC but has comparable effects on mechanical properties and durability. The water-absorbing capability of RMA should be added to the concrete mixture either during or after hardening. This will increase the quality of both fresh and cured concrete. The RMA should be soaked in water for 24 hours before mixing.

RCA concrete's mechanical properties and durability deteriorate as the quantity of RMA-replaced aggregate increases. In coarse RMA with a coarseness of 15%, the highest replacement rate may be obtained without compromising the mechanical characteristics of RMAC, which are equivalent to those of NAC. When the replacement rate of 100 wt% RCA, the compressive strength of RAC is unaffected by replacing natural sand with fine RMA. This is likely owing to the high concentration of small particles and the presence of silica and alumina in broken bricks, which may contribute to the pozzolanic activity. In addition, when the aggregate replacement ratio is the same, the carbonation depth of concrete generated with RMA is greater than that of concrete made using RCA. In addition, the carbonation depth of concrete with mixed RCA, a combination of RMA and RCA, increases as the proportion of RMA in the mixture rises.

There is a significant issue with infrastructure, and fixing it will cost around $44 billion. Seventy-nine percent of the usable life of public infrastructure has already been used (Ijaz et al., 2007). The country's infrastructure is reaching the end of its useful life. As part of the Canada Economic Action Plan (CEAP) (2011), the federal government would finance $12 billion in infrastructure projects (Mirza, 2007). Numerous obsolete and inefficient buildings will be eliminated to create a place for future development. During C&DW, a great deal of waste will be created.

The bulk of this garbage is composed of cement. Due to the absence of appropriate alternatives, RCA might be a feasible option for meeting the rising demand for NCAs. It saves money on locating, transporting, and conserving the environment since it uses less virgin aggregate; it reduces the amount of rubbish in landfills. A lack of suitable land has severely hampered the expansion of communities. Recent landfilling increases are anticipated to continue considerably (Poon and Chan, 2007). To restrict the use of virgin materials, several fees have been imposed on the disposal and incineration of C&DW waste (Poon, 2007). By reusing and recycling it, C&DW can be placed to noble use.

The amount of concrete waste generated by new construction and demolition of old structures has risen substantially over time, and it is now a primary environmental concern on a global scale. One long-term answer to the expanding waste disposal problem and the depletion of NCA supplies is reusing and recycling rejected concrete. The development of RCA, or coarse aggregate, is taking place for many applications. Several studies have been undertaken to discover the essential characteristics of RCA and the optimum approach to mixing it. Scientists have discovered that RCA contains intrinsic unpredictability and variability due to the variances in the raw ingredients and technical factors used in concrete. Certain researchers have demonstrated the strength of RAC to be comparable to that of NAC. Sami et al. found that RAC effectiveness is less than that of NAC. This was a finding that came as a surprise to the researchers. The lower elastic modulus of RAC can be attributed to the presence of RCA. The bond strength between RAC and plain rebar diminishes as the amount of RCA replacement in the mix increases. On the other hand, there does not appear to be a direct correlation between the bond strength between RAC and deformed rebar and the percent of RCA replacement in the mix.

To look at the environmental and practical properties of RAC manufactured using recycled cement and NCAs to find new ways to improve the long-term sustainability of the finished product. Environmentally acceptable and economically effective means of reducing urban surface waterproofing, such as pervious concrete, provide a promising avenue for progress in this area. Although achieving the right balance between the material is complex, the major challenge in creating this material is its mechanical and hydraulic qualities.

Worldwide, in terms of energy, resource use, and emissions, the construction industry is one of the worst (UNEP-2, 2007). Over the previous two centuries, the need for more environmentally friendly housing in cities has been made clear globally (SDSN, 2013). Formal material and construction providers have generated incentives to improve their products and processes, while environmental problems have gained significant attention in the construction industry (Frischknecht et al., 2015). Even though the built environment of low-income groups and the so-called semi-formal or informal building sector has often been overlooked, it has much to do with the environment and how well it performs. It is essential because it is a fast-growing part of the construction industry in countries that are still developing. Asia Pacific: The fact that the total number of people living in informal settlements in the region is still growing despite the fact that the number of people living in informal settlements in the region is decreasing demonstrates that poverty and inequality are prevalent in the Burban century when more than half of the world's population lives in cities. Globally, nearly 40% of urban dwellers in the Asia Pacific area lack suitable housing, according to the United Nations World Urbanization Prospects Survey (UN-Habitat, 2011). When we talk about structural inadequacy, we are referring to issues such as poor land tenure, people having limited access to essential services, and unhealthy live-in conditions.

In addition, the usage of fine and coarse materials in the manufacturing process of concrete is responsible for a significant amount of acidity and the eutrophication induced by the creation of concrete. Because it requires travel and the use of energy, this phase is the one that has the most substantial effect on the surrounding environment. The construction sector is responsible for about 18% of worldwide greenhouse gas emissions, 40% of the depletion of natural resources, and 25% of the creation of rubbish (Yu et al., 2017). As a consequence, environmentally friendly choices and processes need to be developed in order to decrease the influence that the industry has on the surrounding environment. Building materials and waste reduction strategies, such as reusing C&DW, could assist the construction sector in contributing to the economy's more sustainable growth. However, only a comprehensive quantitative study can verify whether or not the potential advantages of low-carbon and recycled building materials are achieved when all of the activities that occur during their life cycle are considered. The IoT has a severe flaw in that it does not consider the end-of-life phase, which includes recycling and disposal of materials (Nakamura and Nansai, 2016). A worthy technique is necessary to estimate alternative building materials' carbon footprint and activities. Additionally, many studies have been conducted on the environmental effect of RAC using the conventional LCA technique. When RAC is pitted against NAC in comparative research, the findings have varied from negligible to substantial reductions in environmental impacts. This disparity may be partially explained by the wide variety of approaches utilized in an LCA framework. In general, it may be ascribed to the variety of fractional unit (FU) alternatives and LCI modeling approaches (attributional or consequential). RCA has a more unusual deflection behavior (i.e., lesser serviceability) in its environmental performance, regardless of the modeling method used. The FU must encompass all critical functional parameters of the concrete construction, such as strength and service life (durability). Because of the growing FU capacity, the environmental effect of the RAC is increasing; consequently, the RAC's FU needs to be increased in order to create a similar deflection behavior between the structural components of the NAC and the RAC.

Due to the fact that sand contributes so little to the entire environmental assessment of concrete, many factors might influence the environmental evaluation of concrete. When alternatives for cement and sand are utilized, many environmental effects might be anticipated. Cement and sand have distinct environmental implications in the life cycle evaluation of byproduct-based concretes. Pushkar utilized this instance to explore the impacts of LCAs on by-product-based concretes, including bottom ash (BA) at concentrations up to 100 wt%, utilizing alternate design techniques (fixed slump or fixed water/cement ratios). The LCA of the w/c ratios of concretes with a set slump range was more destructive to the environment than the LCA of those with a fixed slump range. In addition to the different by-product modeling approaches (consequential or waste), the transportation distance for by-product delivery to the concrete batch plant (short or long) and the approach to by-product-based concrete design (with fixed slump ranges or fixed w/c) may affect the responsiveness of the LCAs of by-product-based concretes. These variables altered the LCAs of concretes made from by-products. Using w/c produced negative environmental implications when the by-product-based concrete design approach was used. However, it was shown that fixed slump ranges favorably influence the environment (Alhazmi et al., 2020).

6.5 FURTHER TRENDS IN RESEARCH AND PRACTICE ON RAC

Green accreditation has been a target for many construction projects for at least a decade and maybe longer. Several projects have used a unique concrete blend with environmental benefits to achieving this goal. This concrete was produced utilizing RAC as an alternative to NCA. The resulting material is concrete that is stronger and lasts longer. RCA can be found and graded in destroyed concrete, as illustrated in Figure 6.2. This ground-breaking material has been successfully implemented in construction projects in Asia, Europe, and North America. At least ten buildings that use sustainable concrete and RCA are listed in Table 6.2.

FIGURE 6.2 RCA.

TABLE 6.2

Sustainable RAC for Buildings (Gales et al., 2016)

Building	Country	Coarse RCA (%)	Date	Floors	Usage
Wessex water	UK	40	2001	2	Commercial
60 Leicester	Australia	60	2004	4	Commercial
Council house 2	Australia	NR	2006	10	Residential
Workplace 6	Australia	20	2008	4	Commercial
Enterprise park	USA	NR	2008	3	Industrial
Samwoh research	Singapore	100	2010	3	Industrial
Middlehaven	UK	50	2011	10	Residential
J-Cube	Singapore	50	2012	6	Commercial
Athletes village	UK	20 to 50	2012	10	Residential
Okanagan	Canada	NR	2013	1	Residential

New RCA construction materials should be used carefully to protect against accidents. Possible measures for this defense include a test of how flammable a new substance is. Investigating a material's mechanical properties in a fire may lead to the creation of new materials, increased trust in already-existing materials, and streamlined applications in the construction sector. New building materials should show structural responses in a fire equivalent to those of older related materials in countries where fire rules are based on objective and performance criteria. One might conduct a full-scale fire test (new and conventional) to evaluate the relative fire safety of two construction materials. Predicting mechanical property changes is standard practice for fire-resistant concretes like RCA since mechanical qualities fluctuate significantly when aggregates are modified. However, it may be more practical to study the mechanical properties of these groundbreaking sustainable concrete materials on a smaller scale before committing to costly full-scale fire testing.

When looking forward, it is important to keep in mind that recycling, which can be defined as the practice of returning used products and materials to the material cycle, is not a recent development. Instead, most of the structures and constructions from antiquity and the middle centuries that have remained can be shown to have used older materials. Once mass-scale production of construction materials became possible because of the achievements of the industrial revolution, the necessity to recycle existing building materials gradually diminished. When exactly "modern" recycling processes for construction materials were originally put into place is unclear. We are now seeing a transition from bare demolition to deconstruction and disassembly, from the disposal of building components to their future reuse. It is demolished if a building is no longer required, does not meet its users' needs, or does not conform to the required technical specifications and standards. At this point, it is safe to affirm that the process of demolition, dismantling, and recycling has become an integral and well-recognized part of the building industry. Some individuals are still concerned about using recycled materials in construction.

As long as the manufacturing process adheres by all applicable requirements and legislation, it is acceptable to employ reprocessed concrete waste in the production of fresh concrete using a cycle for the various shapes and sizes of coarse aggregate. Even if, for safety reasons, just these uses are considered, since they do not substantially impair the durability, the whole quantity of coarse RCAs from concrete trash might be utilized to manufacture fresh concrete. Although around 20 million tons of recycled materials are available, the use of NCAs in concrete production is almost ten times more. This shows that it might be viable to create a closed cycle for the coarse sections of the RAC debris. Fine sizes resulting from concrete debris recycling have not been included in this cycle. However, if crushed to a consistency comparable to cement, these fines might be used as a component in concrete production. According to the results of one's research, using this technology would result in cement savings and improvements in the performance characteristics of the concrete. All of the problems mentioned above only apply to pure-grade concrete. In contrast, concrete fragments resulting from the removal of buildings generally has extra secondary components that must be removed prior to the rubble's use as an RCA—included in these secondary components are glass, metal, and plastic. Similar technologies and procedures would need to be developed to recycle concrete. Other recycling subindustries, such as the plastics and glass sectors, rely on sensor-based categorization systems. These processes detect and categorize the individual particles of a bulk solid.

To better understand recycled-aggregate concretes, it is vital to consider that RCAs are composite materials consisting of cement paste and NCAs when developing more comprehensive quality requirements. Water is required for most reactions that diminish durability and can penetrate concrete farther than other substances. If these RCAs are used in producing fresh concrete, the cement paste ratio of the secondary concrete will increase. Because RCAs contain less water than virgin aggregates, this is the case. Consequently, its mechanical characteristics exhibit both consistent and more dramatic variations. Because it increases porosity, the ratio of better cement paste plays a significant role in determining the material's durability. Another significant, influential parameter is the chemical condition of the old cement paste and the original aggregates. For instance, concrete that is very old or RCAs that have been held for a significant amount of time may have undergone complete carbonation, which is why they do not contain any hydration products anymore. Determining the reactivity of the initial aggregates in an alkali-silica reaction is just as difficult as it sounds.

One aspect that will significantly affect recycling in the future is the introduction of cutting-edge construction materials. Concrete is a commodity construction material that can be recycled easily when made conventionally. However, this high degree of recyclable quality may be compromised by future advancements. As a result, alternatives are needed that harness concrete's potential as a raw material rather than relying on the material qualities of concrete alone. Relevant examples include using novel mix designs and fusing concrete with other materials to create composites. Thus, quality standards need to be developed more comprehensively due to our continued lack of understanding about the characteristics of RCAs and concretes created with them. Statistics play an even more crucial role than natural-aggregate concrete in the building industry.

Bibliography

Abed, M. and de Brito, J. 2020. Evaluation of high-performance self-compacting concrete using alternative materials and exposed to elevated temperatures by non-destructive testing. *J. Build. Eng.* 32, 101720.

Abid, S., Nahhab, A., Al-aayedi, H. and Nuhair, A. 2018. Expansion and strength properties of concrete containing contaminated recycled concrete aggregate. *Case Studies Construct. Mater.* 9, e00201.

Agarwal, R., Chandrasekaran, S. and Sridhar, M. 2016. Imagining construction's digital future. *McKinsey & Company*, 24.

Ahmad, S., Zubair, A. and Maslehuddin, M. 2015. Effect of key mixture parameters on flow and mechanical properties of reactive powder concrete. *Constr. Build. Mater.*, 99, 73–81.

Ahmed, S. F. U. (2013). Properties of concrete containing construction and demolition wastes and fly ash. *Journal of Materials in Civil Engineering.* 25(12), 1864–1870.

Akbarnezhad, A., Ong, K. C. G., Zhang, M. H., Tam, C. T. and Foo, T. W. J. 2011. Microwave-assisted beneficiation of recycled concrete aggregates. *Construct. Build. Mater.* 25(8), 3469–3479.

Akhtaruzzaman, A. A. and Hasnat, A. 1983. Properties of concrete using crushed brick as aggregate. *Concrete Int.* 5(2), 58–63.

Alhazmi, H., Shah, S. and Mahmood, A. 2020. Sustainable development of innovative green construction materials: A study for economical eco-friendly RCA based geopolymer concrete. *Materials* 13(21), 4881.

Alhozaimy, A., Soroushian, P. and Mirza, F. 1996. Effects of curing conditions and age on chloride permeability of fly ash mortar. *Materials J.* 93(1), 87–95.

Al-Jubory, N. H. 2013, Mechanical properties of reactive powder concrete (RPC) with mineral admixture. *Al-Rafidain Engin.* 21(5), 92–101.

Altimiras-Martin, A. 2014. Analysing the structure of the economy using physical input–output tables. *Economic Syst. Res.* 26(4), 463–485.

Amiri, M., Hatami, F. and Golafshani, E. 2021. Evaluating the synergic effect of waste rubber powder and recycled concrete aggregate on mechanical properties and durability of concrete. *Case Studies Construc. Mater.* 15, e00639.

Anastasiou, E., Liapis, A. and Papayianni, I. 2015. Comparative life cycle assessment of concrete road pavements using industrial by-products as alternative materials. *Resources, Conservation Recycling*, 101, 1–8.

Anike, E., Saidani, M., Ganjian, E., Tyrer, M. and Olubanwo, A. 2020. Evaluation of conventional and equivalent mortar volume mix design methods for recycled aggregate concrete. *Mater. Struct.* 53(1), 1–15.

Annual Report 2010 | UN-Habitat. (n.d.). Retrieved January 2, 2023, from https://unhabitat.org/annual-report-2010

Arezoumandi, M., Drury, J., Volz, J. and Khayat, K. 2015. Effect of recycled concrete aggregate replacement level on shear strength of reinforced concrete beams. *ACI Mater. J.* 112(4), 559–567.

Arjomandi, K., Estekanchi, H. and Vafai, A. 2009. Correlation between structural performance levels and damage indexes in steel frames subjected to earthquakes. *Iranian J. Sci. Technol., Trans. Civil Eng.* 16(2), 147–155.

Assaad, J. and Harb, J. 2017. Formwork pressure of self-consolidating concrete containing recycled coarse aggregates. *ACI Mater. J.* 114(3), 491–500.

Astray, D., Ogawa, W. and Shitote, S. 2018. Compressive and flexural strength of recycled reactive powder concrete containing finely dispersed local wastes. *Open J. Civ. Eng.* 08(01), 12–26.

Aydin, S., Yazici, H., Yardimci, M. Y., and Yiğiter, H. 2010. Effect of aggregate type on mechanical properties of reactive powder concrete. *ACI Mater. J.* 107(5), 441–449.

Ayub, T., Khan, S. U. and Memon, F. A. 2014. Mechanical characteristics of hardened concrete with different mineral admixtures: a review. *Scientific World J.* 2014.

Babalola, O., Awoyera, P., Tran, M., Le, D., Olalusi, O., Viloria, A. and Ovallos-Gazabon, D. 2020. Mechanical and durability properties of recycled aggregate concrete with ternary binder system and optimized mix proportion. *J. Mater. Res. Tech.* 9(3), 6521–6532.

Bali, I., Kushartomo, W. and Jonathan. 2016. Effect of in-situ curing on compressive strength of reactive powder concrete. *MATEC Web Conf.* 67, 03013.

Belaidi, A. S. E., Azzouz, L., Kadri, E. and Kenai, S. 2012. Effect of natural pozzolana and marble powder on the properties of self-compacting concrete. *Constr. Build. Mater.* 31, 251–257.

Bektas, F., Wang, K. and Ceylan, H. 2009. Effects of crushed clay brick aggregate on mortar durability. *Constr. Build. Mater.* 23(5), 1909–1914.

Belarouf, S., Samaouali, A., Gueraoui, K. and Rahier, H. 2020. Mechanical properties of concrete with recycled concrete aggregates. *Int. Rev. Civ. Eng.* 11(6), 268.

Bhasya, V. and Bharatkumar, B. 2018. Mechanical and durability properties of concrete produced with treated recycled concrete aggregate. *ACI Mater. J.* 115(2).

Bio Intelligence Service contract on management of construction and demolition waste – SR1, final report task 2. European commission (DG ENV) ENV.G.4/FRA/2008/0112(2011).

Blomfors, M. G., Berrocal, C., Lundgren, K. and Zandi, K. 2021. Incorporation of pre-existing cracks in finite element analyses of reinforced concrete beams without transverse reinforcement. *Eng. Struct.* 229, 111601.

Boyd, M. 2014. A method for prioritizing interventions following root cause analysis (RCA): Lessons from philosophy. *J. Eval. Clin. Pract.* 21(3), 461–469.

Boyd, M. 2015. Lessons from philosophy are a method for prioritizing interventions following root cause analysis (RCA). *J. Eval. Clin. Pract.* 21, 461–469.

Brandt, A. M. 2008. Fibre reinforced cement-based (FRC) composites after over 40 years of development in building and civil engineering. *Compos. Struct.* 86(1-3), 3–9.

Bullard, C. W. and Herendeen, R. A. 1975. The energy cost of goods and services. *Energ. Policy*, 3(4), 268–278.

Butera, S., Christensen, T. H. and Astrup, T. F. 2014. Composition and leaching of construction and demolition waste: inorganic elements and organic compounds. *J. Hazard. Mater.* 276, 302–311.

Candy, S. 2010. The futures of everyday life: Politics and the design of experiential scenarios University of Hawaii (2010). https://www.researchgate.net/publication/305280378_The_Futures_of_Everyday_Life_Politics_and_the_Design_of_Experiential_Scenarios

Cement & Concrete Institute. 2011. Sustainable concrete: ISBN 978-0-9584779-4-9 Cement & Concrete Institute, Midrand, South Africa. P 37. http://www.sustainableconcrete.org.uk/

Cement Association of Canada. 2004. Concrete thinking for a sustainable future. http://www.cement.ca/cement.nsf/e/6ABDCDE126A87A6C85256D2E005CC53B?OpenDocument (Nov. 15, 2005).

Chakradhara Rao, M., Bhattacharyya, S. and Barai, S. 2010. Influence of field recycled coarse aggregate on properties of concrete. *Mater. Struct.* 44(1), 205–220.

Chalee, W., Cheewaket, T. and Jaturapitakkul, C. 2021. Utilization of recycled aggregate concrete for marine site based on 7-year field monitoring. *Int. J. Concr. Struct. Mater.* 15(1), 1–11.

Cheewaket, T., Jaturapitakkul, C. and Chalee, W. 2014. Concrete durability presented by acceptable chloride level and chloride diffusion coefficient in concrete: 10-year results in marine site. *Mater. Struct.* 47, 1501–1511.

Chen, Z., Xu, J., Chen, Y. and Jing, C. 2016. A case study on utilization of 50-year-old concrete in recycled aggregate. *J. Residuals Sci. Tech.* 13(4), S147–S152.

Cheyrezy, M., Maret, V. and Frouin, L. 1995. Microstructural analysis of RPC (reactive powder concrete). *Cement Concrete Res.* 25(7), 1491–1500.

Chidiac, S. E. and Habibbeigi, F. 2005. Modeling the rheological behavior of fresh concrete: An elasto-viscoplastic finite element approach. *Comput. Concrete*, 2(2), 97–110.

Chidiac, S. and Mahmoodzadeh, F. 2009. Plastic viscosity of fresh concrete – A critical review of predictions methods. *Cement Concrete Comp.* 31(8), 535–544.

Chidiac, S. and Mahmoodzadeh, F. 2013. Constitutive flow models for characterizing the rheology of fresh mortar and concrete. *Can. J. Civ. Eng.* 40(5), 475–482.

Choi, H., Choi, H., Lim, M., Inoue, M., Kitagaki, R. and Noguchi, T. 2016. Evaluation on the mechanical performance of low-quality recycled aggregate through interface enhancement between cement matrix and coarse aggregate by surface modification technology. *Int. J. Concrete Struct. Mater.* 10(1), 87–97.

Circular economy action plan. 2023, January 1. Retrieved January 2, 2023, from https://environment.ec.europa.eu/strategy/circular-economy-action-plan_en

Clarke, J., Heinonen, J. and Ottelin, J. 2017. Emissions in a decarbonised economy? Global lessons from a carbon footprint analysis of Iceland. *J. Clean. Prod.* 166, 1175–1186.

Collivignarelli, M., Cillari, G., Ricciardi, P., Miino, M., Torretta, V., Rada, E. and Abbà, A. 2020. The production of sustainable concrete with the use of alternative aggregates: A Review. *Sustainability* 12(19), 7903.

Construction and demolition waste. 2018, September 18. Retrieved January 2, 2023, from https://environment.ec.europa.eu/topics/waste-and-recycling/construction-and-demolition-waste_en

Corinaldesi, V. and Moriconi, G. 2009. Behaviour of cementitious mortars containing different kinds of recycled aggregate. *Constr. Build. Mater.* 23(1), 289–294.

Corinaldesi, V., Orlandi, G. and Moriconi, G. 2002. Self-compacting concrete incorporating recycled aggregate. RK Dhir, PC Hewlett and LJ Csetenyi (Eds.), 4, 455–464.

Cwirzen, A., Penttala, V. and Vornanen, C. 2008. Reactive powder-based concretes: Mechanical properties, durability and hybrid use with OPC. *Cem. Concr. Res.* 38(10), 1217–1226.

Demiss, B., Ogawa, W. and Shitote, S. 2018. Mechanical and microstructural properties of recycled reactive powder concrete containing waste glass powder and fly ash at standard curing. *Cogent Eng.* 5(1), 1464877.

Ding, X., Hao, J., Chen, Z., Qi, J. and Marco, M. 2020. New mix design method for recycled concrete using mixed source concrete coarse aggregate. *Waste Biomass Valorization,* 11(10), 5431–5443.

Disfani, M. M., Tsang, H. H., Arulrajah, A. and Yaghoubi, E. 2017. Shear and compression characteristics of recycled glass-tire mixtures. *J. Mater. Civil Eng.,* 29(6). ISSN 0899-1561.

Dixit, M. K. 2017. Embodied energy analysis of building materials: An improved IO-based hybrid method using sectoral disaggregation. *Energy,* 124, 46–58.

Dodds, W., Christodoulou, C., Goodier, C., Austin, S. and Dunne, D. 2017. Durability performance of sustainable structural concrete: Effect of coarse RCA aggregate on rapid chloride migration and accelerated corrosion. Construct. *Build. Mater.* 155, 511–521.

Dodds, W., Goodier, C., Christodoulou, C., Austin, S. and Dunne, D. 2017. Durability performance of sustainable structural concrete: Effect of coarse RCA aggregate on microstructure and water ingress. Construct. *Build. Mater.* 145, 183–195.

Dumlao-Tan, M. I. and Halog, A. 2017. Moving towards a circular economy in solid waste management: Concepts and practices. In *Advances in solid and hazardous waste management* (pp. 29–48). Springer, Cham.

Ebrahim, Z. and Abdel-jawad, Y. 2020. A modified semi-probabilistic approach for assessing the residual service life of reinforced concrete structures subjected to carbonation. *Procedia Manuf.* 44, 148–155.

Ebrahim Abu El-Maaty Behiry, A. 2013. Utilization of cement-treated recycled concrete aggregates as base or subbase layer in Egypt. *Ain Shams Eng. J.* 4(4), 661–673.

Engelsen, C. J., van der Sloot, H. A. and Petkovic, G. 2017. Long-term leaching from recycled concrete aggregates applied as sub-base material in road construction. *Sci. Total Environ.* 587, 94–101.

Erdem, S., Dawson, A. R. and Thom, N. H. 2011. Microstructure-linked strength properties and impact response of conventional and recycled concrete reinforced with steel and synthetic macro fibres. *Constr. Build. Mater.* 25(10), 4025–4036.

Etxeberria, M., Marí, A. R., & Vázquez, E. (2007). Recycled aggregate concrete as structural material. *Mater. Struct.* 40, 529–541.

EU construction and demolition waste protocol. (n.d.). Retrieved January 2, 2023, from https://single-market-economy.ec.europa.eu/news/eu-construction-and-demolition-waste-protocol-2018-09-18_en

Eurostat. 2018. Eurostat Circular economic indicators-Recovery rate of construction and demolition waste [WWW Document]URL https://ec.europa.eu/eurostat/tgm/table.do?tab=table&init=1&language=en&pcode=cei_wm040&plugin=1 (2018).

Fan, C. C., Huang, R., Hwang, H. and Chao, S. J. 2016. Properties of concrete incorporating fine recycled aggregates from crushed concrete wastes. *Constr. Build. Mater.* 112, 708–715.

Fang, S., Hong, H. and Zhang, P. 2018. Mechanical property tests and strength formulas of basalt fiber reinforced recycled aggregate concrete. *Materials* 11(10), 1851.

Fathifazl, G., Razaqpur, A. G., Isgor, O. B., Abbas, A., Fournier, B. and Foo, S. 2011. Shear capacity evaluation of steel reinforced recycled concrete (RRC) beams. *Eng. Struct.* 33(3), 1025–1033.

Fernandez, R., Martirena, F. and Scrivener, K. L. 2011. The origin of the pozzolanic activity of calcined clay minerals: A comparison between kaolinite, illite and montmorillonite. *Cement Concrete Res.* 41(1), 113–122.

Ferraris, C. F. and deLarrard, F. 1998. *Testing and modeling of fresh concrete rheology.* Maryland: Building and Fire Research Laboratory National Institute of Standards and Technology Gaithersburg.

Ferraris, C. F. and Brower, L. E. 2003. Comparison of concrete rheometers, 25(8), 41–47.

Ferraris, C. F., deLarrard, F. and Martys, N. 2001. Fresh concrete rheology-recent developments. *Materials science of concrete VI, Sidney Mindess and Jan Skalny (Eds.).* The American Ceramic Society. Westerville, OH, 215–241.

Friend, R. M., Anwar, N. H., Dixit, A., Hutanuwatr, K., Jayaraman, T., McGregor, J. A., ... and Roberts, D. 2016. Re-imagining inclusive urban futures for transformation. *Current Opinion Environ. Sustainability,* 20, 67–72.

Frischknecht, R., Wyss, F., Knöpfel, S. B. and Stolz, P. 2015. Life cycle assessment in the building sector: Analytical tools, environmental information and labels. *Int. J. Life Cycle Assess.* 20(4), 421–425.

Gagg, C. R. 2014. Cement and concrete as an engineering material: An historic appraisal and case study analysis. *Engineering Failure Anal.* 40, 114–140.

Gales, J., Parker, T., Cree, D. and Green, M. 2016. Fire performance of sustainable recycled concrete aggregates: Mechanical properties at elevated temperatures and current research needs. *Fire Tech.* 52(3), 817–845.

Gálvez-Martos, J. L., Styles, D., Schoenberger, H. and Zeschmar-Lahl, B. 2018. Construction and demolition waste best management practice in Europe. *Resources, Conservation Recycling*, 136, 166–178.

Galvín, A. P., Agrela, F., Ayuso, J., Beltrán, M. G. and Barbudo, A. 2014. Leaching assessment of concrete made of recycled coarse aggregate: Physical and environmental characterisation of aggregates and hardened concrete. *Waste Manag.* 34(9), 1693–1704.

Gao, D., Zhang, L., Nokken, M. and Zhao, J. 2019. Mixture proportion design method of steel fiber reinforced recycled coarse aggregate concrete. *Materials* 12(3), 375.

GarduñoG arcía, C. and Gaziulusoy, D. 2021. Designing future experiences of the everyday: Pointers for methodical expansion of sustainability transitions research. *Futures*, 127, 102702.

GarduñoG arcía, C. and Gaziulusoy, İ. 2021. Designing future experiences of every day: Pointers for methodical expansion of sustainability transitions research. *Futures*, 127, 102702.

Gebremariam, A., Vahidi, A., Di Maio, F., Moreno-Juez, J., Vegas-Ramiro, I., Łagosz, A., Mróz, R. and Rem, P. 2021. Comprehensive study on the most sustainable concrete design made of RAC, glass and mineral wool from C&D wastes. *Construct. Build. Mater.* 273, 121697.

Geels, F. W. 2010. Ontologies, socio-technical transitions (to sustainability), and the multi-level perspective. *Research Policy*, 39(4), 495–510.

Gicala, M. and Halicka, A. 2019. The influence of the demolition process on the environmental impact of reinforced concrete structures based on recycled aggregate. *Czasopismo Techniczne.* 116(9), 81–96.

Gil-Martín, L., González-López, M., Grindlay, A., Segura-Naya, A., Aschheim, M. and Hernández-Montes, E. 2012. Toward the production of future heritage structures: Considering durability in building performance and sustainability – A philosophical and historical overview. *Int. J. Sustain. Built Environ.* 1(2), 269–273.

Gil-Martín, L. M., González-López, M. J., Grindlay, A., Segura-Naya, A., Aschheim, M. A. and Hernández-Montes, E. 2012. Toward the production of future heritage structures: Considering durability in building performance and sustainability – A philosophical and historical overview. *Int. J. Sustainable Built Environ.* 1(2), 269–273.

Gino, D., Castaldo, P., Bertagnoli, G., Giordano, L. and Mancini, G. 2019. Partial factor methods for existing structures according to fib Bulletin 80: Assessment of an existing prestressed concrete bridge. *Struct. Concr.* 21(1), 15–31.

Gómez-Soberón, J. M. 2002. Porosity of recycled concrete with substitution of recycled concrete aggregate. *Cement Concrete Res.* 32(8), 1301–1311.

González, J. S., Gayarre, F. L., Pérez, C. L. C., Ros, P. S. and López, M. A. S. 2017. Influence of recycled brick aggregates on properties of structural concrete for manufacturing precast prestressed beams. *Constr. Build. Mater.* 149, 507–514.

Grabiec, A. M., Klama, J., Zawal, D. and Krupa, D. 2012. Modification of recycled concrete aggregate by calcium carbonate biodeposition. *Constr. Build. Mater.* 34, 145–150.

Griffiths, P. and Cayzer, S. 2016. Design of indicators for measuring product performance in the circular economy. In *International Conference on Sustainable Design and Manufacturing* (pp. 307–321). Springer, Cham.

Gu, C., Ye, G. and Sun, W. 2015. Ultrahigh performance concrete-properties, applications and perspectives. *Sci. China Technol. Sci.* 58(4), 587–599.

Gunasekara, C., Seneviratne, C., Law, D. W. and Setunge, S. 2020. Feasibility of developing sustainable concrete using environmentally friendly coarse aggregate. *Appl. Sci.* 10(15), 5207.

Hadjieva-Zaharieva, R., Dimitrova, E., & Buyle-Bodin, F. (2003). Building waste management in Bulgaria: challenges and opportunities. *Waste Manag.* 23(8), 749–761.

Hahladakis, J., Purnell, P. and Aljabri, H. 2020. A mini-review and a case study will assess the role and use of recycled aggregates in the sustainable management of construction and demolition waste. *Waste Manag. Res.: J. Sustainable Circular Econ.* 38(4), 460–471.

Hamad, B. and Dawi, A. 2017. Sustainable normal and high strength recycled aggregate concretes using crushed tested cylinders as coarse aggregates. *Case Studies Construct. Mater.* 7, 228–239.

Hammerl, M. and Kromoser, B. 2021. The influence of pretensioning on the load-bearing behaviour of concrete beams reinforced with carbon fibre reinforced polymers. *Compos. Struct.*, 273, 114265.

Hansen, T. C. 1986. Recycled aggregates and recycled aggregate concrete second state-of-the-art report developments 1945–1985. *Materials and Structures*, 19(3), 201–246.

Hincapié, I., Caballero-Guzman, A., Hiltbrunner, D. and Nowack, B. 2015. Use of engineered nanomaterials in the construction industry with specific emphasis on paints and their flows in construction and demolition waste in Switzerland. *Waste Manag.* 43, 398–406.

Ho, N. Y., Lee, Y. P. K. and Tan, J. K. 2008. Beneficial Use of Recycled Concrete Aggregate for Road Construction in Singapore 6th International Conference on Road and Airfield Pavement Technology, Sapporo, Japan, p. 8.

Ho, N., Lee, Y., Lim, W., Zayed, T., Chew, K., Low, G., and Ting, S. 2013. Efficient utilization of recycled concrete aggregate in structural concrete. *J. Mater. Civ. Eng.* 25(3), 318–327.

Hoekstra, R. and van den Bergh, J. C. J. M. 2006. Constructing physical input-output tables forenvironmental modeling and accounting: Framework and illustrations. *Ecol. Econ.* 59(3), 375–393.

Hossain, A. B., Islam, S. and Copeland, K. D. 2007. Influence of ultrafine fly ash on the shrinkage and cracking tendency of concrete and the implications for bridge decks. In Transportation Research Board 86th Annual Meeting, Washington DC, United States.

Hou, Y., Ji, X. and Su, X. 2017. Mechanical properties and strength criteria of cement-stabilised recycled concrete aggregate. *Int. J. Pave. Eng.* 20(3), 339–348.

Hou, Y., Ji, X. and Su, X. 2020. Mechanical properties of rubberized concrete containing recycled concrete aggregate. *ACI Mater. J.* 117(3).

Hu, H., Shinde, S., Adrian, S., Chua, Z. L., Saxena, P. and Liang, Z. 2016. Data-oriented programming: On the expressiveness of non-control data attacks. In *2016 IEEE Symposium on Security and Privacy (SP)* (pp. 969–986). IEEE.

Iacovidou, E., Millward-Hopkins, J., Busch, J., Purnell, P., Velis, C. A., Hahladakis, J. N., ... and Brown, A. 2017. A pathway to circular economy: Developing a conceptual framework for complex value assessment of resources recovered from waste. *J. Clean. Prod.* 168, 1279–1288.

Ijaz, F., Mirza, A., Nisar, H., Rasool, N., Javed, H. A., Sarwar, S., ... and Durani, Z. 2007. Impact of weighted average cost of capital and value of firmon firm's investment decision; No. 1 - 2. *GCU Economic Journal*, 1–16.

Infrastructure Canada - Canada's Economic Action Plan. 2011, October 26. Retrieved January 2, 2023, from https://www.infrastructure.gc.ca/prog/eap-pae-eng.html#:~:text=Canada's%20Economic%20Action%20Plan%20(%20EAP,economy%20around%20and%20create%20jobs

Inman, M., Thorhallsson, E. R. and Azrague, K. 2017. A mechanical and environmental assessment and comparison of basalt fibre reinforced polymer (BFRP) rebar and steel rebar in concrete beams. *Energy Procedia*, 111, 31–40.

Jagan, S., Neelakantan, T., Reddy, L. and Gokul Kannan, R. 2020. Characterization study on recycled coarse aggregate for its utilization in concrete - A review. *J. Phys.: Conf. Ser.* 1706(1), 012120.

Jamadin, A., Ibrahim, Z., Jumaat, M. and Hosen, M. 2020. Serviceability assessment of fatigued reinforced concrete structures using a dynamic response technique. *J. Mater. Res. Tech.* 9(3), 4450–4458.

Jones, P. H. 2014. Systemic design principles for complex social systems. In Metcalf, G. (Eds.), *Social systems and design. Translational systems sciences*, vol 1. Springer, Tokyo.

Kang, T., Kim, W., Kwak, Y. and Hong, S. 2014. Flexural testing of reinforced concrete beams with recycled concrete aggregates. *ACI Struct. J.* 111(3), 607–616.

Kapoor, K., Singh, S. and Singh, B. 2021. Improving the durability properties of self-consolidating concrete made with recycled concrete aggregates using blended cements. *Int. J. Civ. Eng.* 19(7), 759–775.

Kayali, O. and Ahmed, M. S. 2013. Assessment of high volume replacement fly ash concrete–Concept of performance index. *Constr. Build. Mater.* 39, 71–76.

Khan, I., Castel, A. and Gilbert, R. I. 2017. Effects of fly ash on early-age properties and cracking of concrete. *ACI Mater. J.* 114(4).

Khatib, J. (Ed.). 2016. *Sustainability of construction materials.* Woodhead Publishing Limited.

Kim, S. B., Song, J. H., Huh, H. and Lim, J. H. 2007. Dynamic tensile test and specimen design of auto-body steel sheet at the intermediate strain rate. *WIT Trans. Eng. Sci.* 57.

Knaack, A. and Kurama, Y. 2018. Modeling time-dependent deformations: Application for reinforced concrete beams with recycled concrete aggregates. *ACI Struct. J.* 115(1), 175–190.

Kodur, V. 2014. Properties of concrete at elevated temperatures. *ISRN Civil Eng.* 1–15.

Kou, S. C., Poon, C. S. and Wan, H. W. 2012. Properties of concrete prepared with low-grade recycled aggregates. *Constr. Build. Mater.* 36, 881–889.

Krishna, D., Priyadarsini, R. and Narayanan, S. 2019. Effect of elevated temperatures on the mechanical properties of concrete. *Procedia Struct. Integrity*, 14, 384–394.

Kromoser, B., Preinstorfer, P. and Kollegger, J. 2019. Building lightweight structures with carbon-fiber-reinforced polymer-reinforced ultra-high-performance concrete: Research approach, construction materials, and conceptual design of three building components. *Struct. Concrete*, 20(2), 730–744.

Kromoser, B., Pachner, T., Tang, C., Kollegger, J. and Pottmann, H. 2018. Form finding of shell bridges using the pneumatic forming of hardened concrete construction principle. *Adv. Civil Eng.* 2018.

Kubissa, W., Jaskulski, R., Koper, A. and Szpetulski, J. 2015. Properties of concretes with natural aggregate improved by RCA addition. *Procedia Eng.* 108, 30–38.

Kurda, R., Silvestre, J. D. and de Brito, J. 2018. Life cycle assessment of concrete made with high volume of recycled concrete aggregates and fly ash. *Resources, Conservation Recycling*, 139, 407–417.

Kurda, R., Silvestre, J. and de Brito, J. 2018. Toxicity and environmental and economic performance of FA and RCAs use in concrete: A review. *Heliyon*, 4(4), e00611.

Kurda, R., de Brito, J. and Silvestre, J. 2018. Combined economic and mechanical performance optimization of RCA concrete with high volume of FA. *Appl. Sci.* 8(7), 1189.

Kurda, R., Silvestre, J.D. and de Brito, J. 2018. Toxicity and environmental and economic performance of fly ash and recycled concrete aggregates use in concrete: A review. *Heliyon*, 4(4), e00611, 1–45.

Kuryłowicz-Cudowska, A., Wilde, K. and Chróścielewski, J. 2020. Prediction of cast-in-place concrete strength of the extradosed bridge deck based on temperature monitoring and numerical simulations. Construct. *Build. Mater.* 254, 119224.

Landa-Sánchez, A., Bosch, J., Baltazar-Zamora, M., Croche, R., Landa-Ruiz, L., Santiago-Hurtado, G., Moreno-Landeros, V., Olguín-Coca, J., López-Léon, L., Bastidas, J., Mendoza-Rangel, J., Ress, J. and Bastidas, D. 2020. Corrosion behavior of steel-reinforced green concrete containing recycled coarse aggregate additions in sulfate media. *Materials*, 13(19), 4345.

Lei, B., Li, W., Tang, Z., Li, Z. and Tam, V. 2020. Effects of environmental actions, recycled aggregate quality, and modification treatments on durability performance of recycled concrete. *J. Mater. Res. Tech.* 9(6), 13375–13389.

Lei, H., Li, L., Yang, W., Bian, Y. and Li, C. Q. 2021. An analytical review on application of life cycle assessment in circular economy for built environment. *J. Build. Eng.* 44, 103374.

Levy, S. M. and Helene, P. 2004. Durability of recycled aggregates concrete: A safe way to sustainable development. *Cement Concrete Res.* 34(11), 1975–1980.

Li, J., Xiao, H. and Gong, J. 2008. Granular effect of fly ash repairs damage of recycled coarse aggregate. *J. Shanghai Jiaotong Univ. (Sci.)*, 13(2), 177–180.

Limbachiya, M. C., Leelawat T. and Dhir R. K. 2000. Use of recycled concrete aggregate in high-strength concrete. *Mater. Struct.* 33, 574–880.

Liu, F., Yu, Y., Li, L. and Zeng, L. 2017. Experimental study on reuse of recycled concrete aggregates for load-bearing components of building structures. *J. Mater. Cy. Waste Manag.* 20(2), 995–1005.

Lu, G., Wanga, K. and Rudolphi, T. J. 2008. Modeling rheological behaviour of highly flowable mortar using concepts of particle and fluid mechanics. *Cement Concrete Composites*, 30(1), 1–12.

Mahmood, W., Khan, A. and Ayub, T. 2021. Mechanical and durability properties of concrete containing recycled concrete aggregates. *Iranian J. Sci. Tech. Transac. Civ. Eng.* 46.

Mahmoodzadeh, F. and Chidiac, S. 2013. Rheological models for predicting plastic viscosity and yield stress of fresh concrete. *Cement Concrete Res.* 49, 1–9.

Majeau-Bettez, G., Wood, R. and Strømman, A. H. 2016. On the financial balance of input–output constructs: Revisiting an axiomatic evaluation. *Economic Syst. Res.* 28(3), 333–343.

Mankar, A., Bayane, I., Sørensen, J. and Brühwiler, E. 2019. Probabilistic reliability framework for assessment of concrete fatigue of existing RC bridge deck slabs using data from monitoring. *Eng. Struct.* 201, 109788.

Marie, I. and Mujalli, R. 2019. Effect of design properties of parent concrete on the morphological properties of recycled concrete aggregates. *Eng. Sci. Tech. Int. J.* 22(1), 334–345.

Martinez-Arguelles, G., Coll, M., Pumarejo, L., Cotte, E., Rondon, H., Pacheco, C., Martinez, J. and Espinoza, R. 2019. Characterization of recycled concrete aggregate as potential replacement of natural aggregate in asphalt pavement. IOP Conference Series: Materials Science and Engineering, 471, 102045.

Martínez-Lage, I., Vázquez-Burgo, P. and Velay-Lizancos, M. 2020. Sustainability evaluation of concretes with mixed recycled aggregate based on holistic approach: Technical, economic and environmental analysis. *Waste Manag.*, 104, 9–19.

Marzouk, M. and Azab, S. 2014. Environmental and economic impact assessment of construction and demolition waste disposal using system dynamics. *Resources, Conservation Recycling*, 82, 41–49.

Mata Kihila, J., Wanemark, J., Cheng, S., Harris, S., Sandkvist, F., ... and Yaramenka, K. 2022. Non-technological and behavioral options for decarbonizing buildings – A review of global topics, trends, gaps, and potentials. *Sustainable Prod. Consumption*, 29, 529–545.

Mayer, H. and Flachmann, C. 2011. Extended input-output model for energy and greenhouse gases. *Final Report, Federal Statistical Office of Germany*.

McNeil, K. and Kang, T. H. K. 2013. Recycled concrete aggregates: A review. *Int. J. Concrete Struct. Mater.* 7(1), 61–69.

Medine, M., Trouzine, H., De Aguiar, J. B. and Djadouni, H. 2020. Life cycle assessment of concrete incorporating scrap tire rubber: Comparative study. *Revue Nature Tech.* 12(2), 1–11.

Mehta, K. and Burrows, R. W. 2001. Building durable structures in the 21st century. *Concrete Int.* 23(2001), 57–63.

Meng, Y., Ling, T. C. and Mo, K. H. 2018. Recycling of wastes for value-added applications in concrete blocks: An overview. *Resources, Conservation Recycling*, 138, 298–312.

Mikhailenko, P., Kakar, M. R., Piao, Z., Bueno, M. and Poulikakos, L. 2020. Incorporation of recycled concrete aggregate (RCA) fractions in semidense asphalt (SDA) pavements: Volumetrics, durability and mechanical properties. *Construct. Build. Mater.* 264, 120166.

Milicevic, I., Bjegovic, D. and Stirmer, N. 2015. Optimisation of concrete mixtures made with crushed clay bricks and roof tiles. *Mag. Concrete Res.* 67(3), 109–120.

Moosberg-Bustnes, H., Lagerblad, B. and Forssberg, E. 2004. The function of fillers in concrete. *Mater. Struct.* 37(2), 74–81.

Moreno-Juez, J., Vegas, I. J., Gebremariam, A. T. and Garcia-Cortes, F. D. 2020. Treatment of end-of-life concrete in an innovative heating-air classification system for circular cement-based products. *J. Clean. Prod.* 263, 121515.

Mostert, C., Sameer, H., Glanz, D. and Bringezu, S. 2021. Climate and resource footprint assessment and visualization of recycled concrete for circular economy. *Res. Conserv. Recycle.* 174, 105767.

Müller, C. 1998. Requirements on concrete for future recycling R.K. Dhir, N.A. Henderson, M.C. Limbachiya (Eds.), *Proceedings of the International Symposium on Sustainable Construction: Use of Recycled Concrete Aggregate*, Thomas Telford, London, UK, 445–457.

Musch, A. K. and von Streit, A. 2020. (Un)intended effects of participation in sustainability science: A criteria-guided comparative case study. *Environ. Sci. Policy*, 104, 55–66.

Naik, T. R. and Moriconi G. 2005. Environmental friendly durable concrete made with recycled materials for sustainable concrete construction. Proceedings of International Symposium on Sustainable Development of Cement, Concrete and Concrete Structures, 5–7 October 2005, Toronto, 485–505.

Nakamura, S. and Nansai, K. 2016. Input–output and hybrid LCA. In *Special types of life cycle assessment*, 219–291. Springer, Dordrecht.

Naouaoui, K., Bouyahyaoui, A. and Cherradi, T. 2019. Experimental characterization of recycled aggregate concrete. *MATEC Web of Conferences*, 303, 05004.

Natarajan, B., Kanavas, Z., Sanger, M., Rudolph, J., Chen, J., Edil, T. and Ginder-Vogel, M. 2019. Characterization of recycled concrete aggregate after eight years of field deployment. *J. Mater. Civ. Eng.* 31(6), 04019070.

Nawaz, M., Qureshi, L., Ali, B. and Raza, A. 2020. Mechanical, durability and economic performance of concrete incorporating FA and RCAs. *SN Appl. Sci.* 2(2), 1–8.

Nicoara, A., Stoica, A., Vrabec, M., Šmuc Rogan, N., Sturm, S., Ow-Yang, C., Gulgun, M., Bundur, Z., Ciuca, I. and Vasile, B. 2020. End-of-life materials used as supplementary cementitious materials in the concrete industry. *Materials*, 13(8), 1954.

Nixon, P. J. 1978. Recycled concrete as an aggregate for concrete—A review. *Matériaux Et Constructions*, 11(5), 371–378.

Noumowe, A. and Debicki, G., 2002. Effect of elevated temperature from 200 to 600 oC on the permeability of high performance concrete. *Proceedings of the 6th International Symposium on Utilization of High Strength/Performance Concrete*, Vol. 1, Leipzig, Germany.

Ntaryamira, T., Quansah, A. and Zhang, Y. 2017. Assessment of recycled concrete aggregate (RCA) usage in concrete. *Int. J. Res. Eng. Tech.* 06(12), 72–78.

Nuaklong, P., Jongvivatsakul, P., Pothisiri, T., Sata, V. and Chindaprasirt, P. 2020. Influence of rice husk ash on mechanical properties and fire resistance of recycled aggregate high-calcium fly ash geopolymer concrete. *J. Clean. Prod.* 252, 119797.

Olofinnade, O. and Ogara, J. 2021. Workability, strength, and microstructure of high strength sustainable concrete incorporating recycled clay brick aggregate and calcined clay. *Cleaner Eng. Tech.* 3, 100123.

Olofinnade, O. M., Ede, A. N. and Ndambuki, J. M. 2017. Sustainable green environment through utilization of waste soda-lime glass for production of concrete. *J. Mater. Environ. Sci.* 8, 1139–1152.

Omary, S., Ghorbel, E., Wardeh, G. and Nguyen, M. 2017. Mix design and recycled aggregates effects on the concrete's properties. *Int. J. Civ. Eng.* 16(8), 973–992.

Paula Junior, A., Jacinto, C., Oliveira, T., Polisseni, A., Brum, F., Teixeira, E. and Mateus, R. 2021. Characterisation and life cycle assessment of pervious concrete with RCAs. *Crystals*, 11(2), 209.

Pavlu, T., Fortova, K., Divis, J. and Hajek, P. 2019. The utilization of recycled masonry aggregate and recycled eps for concrete blocks for mortarless masonry. *Materials*, 12(12), 1923.

Pavlů, K. and Hájek, P. 2019. Environmental assessment of two use cycles of recycled aggregate concrete. *Sustainability*, 11(21), 6185.

Pavlů, T., Kočí, V. and Hájek, P. 2019. Environmental assessment of two use cycles of recycled aggregate concrete. *Sustainability*, 11(21), 6185.

Pavlů, T., Fořtová, K., Řepka, J., Mariaková, D. and Pazderka, J. 2020. Improvement of the durability of recycled masonry aggregate concrete. *Materials*, 13(23), 5486.

Pavlů, T., Khanapur, N. V., Fořtová, K., Mariaková, D., Tripathi, B., Chandra, T. and Hájek, P. 2022. Design of performance-based concrete using sand reclaimed from construction and demolition waste–comparative study of Czechia and India. *Materials*, 15(22), 7873.

Pellizzer, G. P. and Leonel, E. D. 2020. Probabilistic corrosion time initiation modelling in reinforced concrete structures using the BEM. *Revista IBRACON De Estruturas E Materiais*, 13(4).

Peng, C. 2016. Calculation of a building's life cycle carbon emissions based on Ecotect and building information modeling. *J. Clean. Prod.* 112, 453–465.

Poon, C. S. and Chan, D. 2007. A review on the use of recycled aggregate in concrete in Hong Kong. *In International Conference on Sustainable Construction Materials and Technologies. Coventry*, 144–155.

Pourkhorshidi, S., Sangiorgi, C., Torreggiani, D. and Tassinari, P. 2020. Using recycled aggregates from construction and demolition waste in unbound layers of pavements. *Sustainability*, 12, 938.

Pourtahmasb, M. and Karim, M. 2014. Utilization of recycled concrete aggregates in stone mastic asphalt mixtures. *Adv. Mater. Sci. Eng.* 2014, 1–9.

Pushkar, S. 2019. The effect of different concrete designs on the life-cycle assessment of the environmental impacts of concretes containing furnace bottom-ash instead of sand. *Sustainability*, 11(15), 4083.

Pushkar, S. 2019. The effect of additional byproducts on the environmental impact of the production stage of concretes containing bottom ash instead of sand. *Sustainability*, 11(18), 5037.

Qin, Y., Chen, J., Li, Z. and Zhang, Y. 2019. The mechanical properties of recycled coarse aggregate concrete with lithium slag. *Adv. Mater. Sci. Eng.* 2019, 1–12.

Rahal, K. N. 2007. Mechanical properties of concrete with recycled coarse aggregate. *Build. Environ.* 42, 407–415.

Ranta, V., Aarikka-Stenroos, L. and Mäkinen, S. J. 2018. Creating value in the circular economy: A structured multiple-case analysis of business models. *J. Clean. Prod.* 201, 988–1000.

Rao, M. C., Bhattacharyya, S. K. and Barai, S. V. 2019. *Systematic approach of characterisation and behaviour of recycled aggregate concrete* (pp. 39–63). Singapore: Springer.

Rattanachu, P., Toolkasikorn, P., Tangchirapat, W., Chindaprasirt, P. and Jaturapitakkul, C. 2020. Performance of recycled aggregate concrete with rice husk ash as cement binder. *Cement Concrete Composites*, 108, 103533.

Reichenbach, S., Preinstorfer, P., Hammerl, M. and Kromoser, B. 2021. A review on embedded fibre-reinforced polymer reinforcement in structural concrete in Europe. *Constr. Build. Mater.* 307, 124946.

Reid, J. M., Hassan, K. E. G. and Al-Kuwari, M. B. S. 2016. Improving the management of construction waste in Qatar. In *Proceedings of the Institution of Civil Engineers-Waste and Resource Management* (Vol. 169, No. 1, pp. 21–29). Thomas Telford Ltd.

Reiner, M., Durham, S. A. and Rens, K. L. 2010. Development and analysis of high-performance green concrete in the urban infrastructure. *Int. J. Sustainable Eng.* 3(3), 198–210.

Reto, F., Gianluca, S. and Andrea, F. 2021. Comparison of reliability- and design-based code calibrations. *Structural Safety*, 88, 102005.

Rigamonti, L., Grosso, M. and Biganzoli, L. 2012. Environmental assessment of refuse-derived fuel co-combustion in a coal-fired power plant. *J. Ind. Ecol.* 16(5), 748–760.

Rohden, A., Kirchheim, A. and Molin, D. 2020. Strength optimization of reactive powder concrete. *Rev. IBRACON Estrus. Mater.* 13(5), e13507.

Rong, C., Ma, J., Shi, Q. and Wang, Q. 2021. The simple mix design method and confined behavior analysis for recycled aggregate concrete. *Materials*, 14(13), 3533.

Rutz, D., Mergner, R., Janssen, R., Hoffstede, U., Hahn, H., Kulisic, B., ... and Zapora, D. 2013. A challenge in the waste sector: The use of organic urban waste for biomethane production. In *Proceedings of the 21th European Biomass Conference and Exhibition*, Copenhagen, Denmark.

Safiuddin, M. D., Salam, M. A. and Jumaat, M. Z. 2011, Effects of recycled concrete aggregate on the fresh properties of self-consolidating concrete. *Arch. Civ. Mech. Eng.* 11(4), 1023–1041.

Sakthieswaran, N. and Renisha, M. 2020. Mutual effect of coal bottom ash and recycled fines on reactive powder concrete. *Rom. J. Mater.* 50(3), 395–402.

Salahuddin, H., Qureshi, L., Nawaz, A., Abid, M., Alyousef, R., Alabduljabbar, H., Aslam, F., Khan, S. and Tufail, R. 2020. Elevated temperature performance of reactive powder concrete containing recycled fine aggregates. *Materials*, 13(17), 3748.

Salama, W. 2017. Design of concrete buildings for disassembly: An explorative review. *Int. J. Sustainable Built Environ.* 6(2), 617–635.

Salzer, C., Wallbaum, H., Ostermeyer, Y. and Kono, J. 2017. Environmental performance of social housing in emerging economies: Life cycle assessment of conventional and alternative construction methods in the Philippines. *Int. J. Life Cycle Assess.* 22(11), 1785–1801.

Samad, S. and Shah, A. 2017. Role of binary cement including supplementary cementitious material (SCM), in production of environmentally sustainable concrete: A critical review. *Int. J. Sustain. Built Environ.* 6(2), 663–674.

Samir, A. 2000. Effect of compressive strength and tensile reinforcement ratio on flexural behavior of high-strength concrete beams. *Eng. Sturct.* 22(5), 413–423.

Sandoval, G., Galobardes, I., Teixeira, R. and Toralles, B. 2017. Comparison between the falling head and the constant head permeability tests to assess the permeability coefficient of sustainable pervious concretes. *Case Studies Construct. Mater.* 7, 317–328.

Schladitz, F., Frenzel, M., Ehlig, D. and Curbach, M. 2012. Bending load capacity of reinforced concrete slabs strengthened with textile reinforced concrete. *Eng. Struct.* 40, 317–326.

Scrivener, K., Martirena, F., Bishnoi, S. and Maity, S. 2018. Calcined clay limestone cements (LC3). *Cement Concrete Res.* 114, 49–56.

SDSN. 2013. An Action Agenda for Sustainable Development: Report for the UN Secretary-General. *Prepared by the Leadership Council of the Sustainable Development Solutions Network (SDSN)*. New York: United Nations.

Shahbazpanahi, S., Tajara, M., Faraj, R. and Mosavi, A. 2021. Studying the C–H crystals and mechanical properties of sustainable concrete containing recycled coarse aggregate with used nano-silica. *Crystals*, 11(2), 122.

Shahria Alam, M., Slater, E. and Muntasir Billah, A. 2013. Green concrete made with RCA and FRP scrap aggregate: Fresh and hardened properties. *J. Mater. Civ. Eng.* 25(12), 1783–1794.

Shaikh, F. 2016. Mechanical and durability properties of fly ash geopolymer concrete containing recycled coarse aggregates. *Int. J. Sustain. Built Environ.* 5(2), 277–287.

Shaikh, F. 2016. Effect of ultrafine fly ash on the properties of concretes containing construction and demolition wastes as coarse aggregates. *Struct. Concr.* 17(1), 116–122.

Shaikh, F. 2017. Mechanical properties of recycled aggregate concrete containing ternary blended cementitious materials. *Int. J. Sustain. Built Environ.* 6(2), 536–543.

Shaikh, F. 2017. Mechanical properties of concrete containing recycled coarse aggregate at and after exposure to elevated temperatures. *Struct. Concr.* 19(2), 400–410.

Shaikh, F., Chavda, V., Minhaj, N. and Arel, H. 2017. Effect of mixing methods of nano-silica on properties of recycled aggregate concrete. *Struct. Concr.* 19(2), 387–399.

Shaikh, F. U. A., and Nguyen, H. L. (2013). Properties of concrete containing recycled construction and demolition wastes as coarse aggregates. *Journal of Sustainable Cement-Based Materials.* 2(3–4), 204–217.

Siddika, A., Mamun, M., Alyousef, R. and Mohammadhosseini, H. 2021. State-of-the-art-review on rice husk ash: A supplementary cementitious material in concrete. *J. King Saud Univ. Eng. Sci.* 33(5), 294–307.

Siddique, R., Khatib, J. and Kaur, I. 2008. Use of recycled plastic in concrete: A review. *Waste Manag.* 28(10), 1835–1852.

Sobotka, A. and Sagan, J. 2017. An environmental LCA of selected concrete recycling processes. *Czasopismo Techniczne*, 10, 123–130.

Somna, R., Jaturapitakkul, C. and Amde, A. M. 2012a. Effect of ground fly ash and ground bagasse ash on the durability of recycled aggregate concrete. *Cement Concrete Composites*, 34(7), 848–854.

Somna, R., Jaturapitakkul, C., Rattanachu, P. and Chalee, W. 2012b. Effect of ground bagasse ash on mechanical and durability properties of recycled aggregate concrete. *Mater. Design (1980–2015)*, 36, 597–603.

Spadea, S., Farina, I., Carrafiello, A. and Fraternali, F. 2015. Recycled nylon fibers as cement mortar reinforcement. *Constr. Build. Mater.* 80, 200–209.

Stahel, W. R. and Clift, R. 2016. Stocks and flows in the performance economy. In *Taking stock of industrial ecology* (137–158). Springer, Cham.

Stierschneider, E., Tamparopoulos, A., McBride, K. and Bergmeister, K. 2021. Influencing factors on creep displacement assessment of bonded fasteners in concrete. *Eng. Struct.* 241, 112448.

Stoiber, N., Hammerl, M. and Kromoser, B. 2021. Cradle-to-gate life cycle assessment of CFRP reinforcement for concrete structures: Calculation basis and exemplary application. *J. Clean. Product.* 280, 124300.

Struble, L. and Tebaldi, G. (Eds.). 2017. Materials for sustainable infrastructure. *Proceedings of the 1st GeoMEast International Congress and Exhibition, Egypt 2017 on Sustainable Civil Infrastructures*. Springer.

Suh, S. 2004. A note on the calculus for physical input–output analysis and its application to land appropriation of international trade activities. *Ecol. Econ.* 48(1), 9–17.

Sun, H., Gao, Y., Zheng, X., Chen, Y., Jiang, Z. and Zhang, Z. 2019. Meso-scale simulation of concrete uniaxial behavior based on numerical modeling of CT images. *Materials* 12(20), 3403.

Supit, S. W., Shaikh, F. U. and Sarker, P. K. 2014. Effect of ultrafine fly ash on mechanical properties of high volume fly ash mortar. *Constr. Build. Mater.* 51, 278–286.

Tabsh, S. W., & Abdelfatah, A. S. (2009). Influence of recycled concrete aggregates on strength properties of concrete. *Constr. Build. Mater.* 23(2), 1163–1167.

Tahar, Z., Benabed, B., Kadri, E., Ngo, T. and Bouvet, A. 2020. Rheology and strength of concrete made with recycled concrete aggregates as replacement of natural aggregates. *Epitoanyag J. Silicate Based Compos. Mater.* 72(2), 48–58.

Tahmoorian, F. and Samali, B. 2018. Laboratory investigations on the utilization of RCA in asphalt mixtures. *Int. J. Pavement Res. Tech.* 11(6), 627–638.

Talwar, M. and Arora, S. 2017. Analysis of flexural fatigue of concrete containing coarse recycled concrete aggregates (RCA). *Int. J. Adv. Res. Comput.* 8(4), 153–156.

Tam, V. W. and Tam, C. M. 2006. A review on the viable technology for construction waste recycling. *Resources, Conservation Recycling*, 47(3), 209–221.

Tam, V. W., Tam, L. and Le, K. N. 2010. Cross-cultural comparison of concrete recycling decision-making and implementation in construction industry. *Waste Manag.* 30(2), 291–297.

Tamanna, N. and Tuladhar, R. 2020. Sustainable use of recycled glass powder as cement replacement in concrete. *Open Waste Manag. J.* 13(1).

Tamanna, K., Tiznobaik, N., Banthia, N. and Alam, M. S. 2020. Mechanical properties of rubberized concrete containing recycled concrete aggregate. *ACI Mater. J.* 117(3), 169–180.

Tan, T., Al-Khalaqi, A. and Al-Khulaifi, N. 2014. Qatar national vision 2030. *Sustainable Develop.: Appraisal Gulf Region*, 19(1), 65–81.

Tang, X. G., Xie, Y. J. and Long, G. C. 2016. Experimental study on performance of imitative RPC for sulphate leaching. *J. Asian Ceramic Societies*, 4(1), 143–148.

Tavakoli, D., Heidari, A. and Hayati Pilehrood, S. 2014. Properties of concrete made with waste claybrick as sand incorporating nano SiO_2. *Indian J. Sci. Technol.* 7(12), 1899–1905.

Tayeh, B., Saffar, D. and Alyousef, R. 2020. The utilization of recycled aggregate in high performance concrete: A review. *J. Mater. Res. Tech.* 9(4), 8469–8481.

Teh, S., Wiedmann, T. and Moore, S. 2018. Mixed-unit hybrid life cycle assessment applied to the recycling of construction materials. *J. Econ. Struct.* 7(13), 1–25.

Thomas, B. S., Damare, A. and Gupta, R. 2013. Strength and durability characteristics of copper tailing concrete. *Constr. Build. Mater.*, 48, 894–900.

Thomas, B. S., Gupta, R. C. and Panicker, V. J. 2016. Recycling of waste tire rubber as aggregate in concrete: Durability-related performance. *J. Clean. Prod.* 112, 504–513.

Tošic, N., Marinkovic, S. and Kurama, Y. 2021. Improved serviceability and environmental performance of one-way slabs through the use of layered natural and RCA concrete. *Sustainability*, 12, 10278.

Tüfekçi, M. and Çakır, Ö. 2017. An investigation on mechanical and physical properties of recycled coarse aggregate (RCA) concrete with GGBFS. *Int. J. Civ. Eng.* 15(4), 549–563.

Umanath, U. and Muthukkumaran, K. 2021. Assessment of innovative dented sheet liner on the improvement of hydraulic properties of pervious concrete pile. *Eng. Sci. Tech. Int. J.* 26, 1–11.

UNEP-2. 2007. Buildings and Climate Change—Status, Challenges and Opportunities. *Sustainable Construction and Building Initiative (SBCI) of United Nations Environment Programme (UNEP) France.* Available at: http://www.unep.fr/pc/sbc/documents/Buildings_and_climate_change.pdf, accessed on 2 January 2023.

United Nations. 2015. Transforming our world: the 2030 Agenda for sustainable development. Available at: https://www.un.org/ga/search/view_doc.asp?symbol=A/RES/70/1=E

Vega-Araujo, D., Martinez-Arguelles, G. and Santos, J. 2020. Comparative life cycle assessment of warm mix asphalt with RCAs: A Colombian case study. *Procedia CIRP.* 90, 285–290.

Velenturf, A. P. and Purnell, P. 2017. Resource recovery from waste: Restoring the balance between resource scarcity and waste overload. *Sustainability*, 9(9), 1603.

Wang, Q., Liang, J., He, C. and Li, W. 2021. Axial compressive behavior of steel fiber-reinforced recycled coarse aggregate concrete-filled short circular steel columns. *Adv. Mater. Sci. Eng.* 2021, 1–9.

Wardeh, G., Ghorbel, E. and Gomart, H. 2014. Mix design and properties of recycled aggregate concretes: Applicability of Eurocode 2. *Int. J. Concr. Struct. Mater.* 9(1), 1–20.

Wartman, J., Grubb, D. G. and Nasim, A. S. M. 2004. Select engineering characteristics of crushed glass. *J. Mater. Civil Eng.* 16(6), 526–539.

Wei, W., Shao, Z., Zhang, Y., Qiao, R. and Gao, J. 2019. Fundamentals and applications of microwave energy in rock and concrete processing–A review. *Appl. Thermal Eng.* 157, 113751.

Wu, L., Kou, X. and Jiang, M. 2015. Probabilistic corrosion initiation time assessment of existing concrete structures under marine environment. *Arabian J. Sci. Eng.* 40(11), 3099–3105.

Xiao, J. and Ding, T. 2013. Research on recycled concrete and its utilization in building structures in China. Front. *Struct. Civ. Eng.* 7(3), 215–226.

Xiao, J., Li, J. and Zhang, C. 2005. Mechanical properties of recycled aggregate concrete under uniaxial loading. *Cement Concrete Res.* 35(6), 1187–1194.

Xiao, J., Zhang, K. and Akbarnezhad, A. 2018. Variability of stress-strain relationship for recycled aggregate concrete under uniaxial compression loading. *J. Clean. Prod.* 181, 753–771.

Xiao, J., Wang, C., Ding, T. and Akbarnezhad, A. 2018. A recycled aggregate concrete high-rise building: Structural performance and embodied carbon footprint. *J. Clean. Prod.* 199, 868–881.

Xu, J., Wang, Y., Ren, R., Wu, Z. and Ozbakkaloglu, T. 2020. Performance evaluation of recycled aggregate concrete-filled steel tubes under different loading conditions: Database analysis and modelling. *J. Build. Eng.* 30, 101308.

Xu, F., Wang, S., Li, T., Liu, B., Li, B. and Zhou, Y. 2020. The mechanical properties of tailing recycled aggregate concrete and its resistance to the coupled deterioration of sulfate attack and wetting–drying cycles. In *Structures* (Vol. 27, pp. 2208–2216). Elsevier.

Yang, S. and Lim, Y. 2018. Mechanical strength and drying shrinkage properties of RCA concretes produced from old railway concrete sleepers using by a modified EMV method. *Constr. Build. Mater.*, 185, 499–507.

Yang, Y., Chen, B., Su, Y., Chen, Q., Li, Z., Guo, W. and Wang, H. 2020. Concrete mix design for completely recycled fine aggregate by modified packing density method. *Materials*, 13(16), 3535.

Ye, T., Cao, W., Zhang, Y., and Yang, Z. 2018. Flexural behavior of corroded reinforced recycled aggregate concrete beams. *Adv. Mater. Sci. Eng.* 2018, 1–14.

Yeheyis, M., Hewage, K., Alam, M. S., Eskicioglu, C. and Sadiq, R. 2013. An overview of construction and demolition waste management in Canada: A lifecycle analysis approach to sustainability. *Clean Technol. Environ. Policy*, 15(1), 81–91.

Yu, M., Wiedmann, T., Crawford, R. and Tait, C. 2017. The carbon footprint of Australia's construction sector. *Procedia Eng.* 180, 211–220.

Yu, F., Yin, L., Fang, Y., and Jiang, J. 2019. Mechanical behavior of recycled coarse aggregates self-compacting concrete-filled steel tubular columns under eccentric compression. *Struct. Concr.* 20(6), 2000–2014.

Yu, X., Robuschi, S., Fernandez, I. and Lundgren, K. 2021. Numerical assessment of bond-slip relationships for naturally corroded plain reinforcement bars in concrete beams. *Eng. Struct.* 239, 112309.

Zameeruddin, M. and Sangle, K. 2021. Damage assessment of reinforced concrete moment resisting frames using performance-based seismic evaluation procedure. *J. King Saud Univ. Eng. Sci.* 33(4), 227–239.

Zhang, C. 2014. Life cycle assessment (LCA) of fibre reinforced polymer (FRP) composites in civil applications. In *Eco-efficient construction and building materials* (pp. 565–591). Woodhead Publishing.

Zhang, M. H. and Islam, J. 2012. Use of nano-silica to reduce setting time and increase early strength of concretes with high volumes of fly ash or slag. *Constr. Build. Mater.*, 29, 573–580.

Zhang, P. and London, K. 2013. Towards an internationalized sustainable industrial competitiveness model. *Compet. Rev.* 23(2), 95–113.

Zhang, P., Yang, Y., Wang, J., Hu, S., Jiao, M. and Ling, Y. 2020. Mechanical properties and durability of polypropylene and steel fiber-reinforced recycled aggregates concrete (FRRAC): A review. *Sustainability*, 12(22), 9509.

Zhang, C., Hu, M., Dong, L., Xiang, P., Zhang, Q., Wu, J., … Shi, S. 2018. Co-benefits of urban concrete recycling on the mitigation of greenhouse gas emissions and land use change: A case in Chongqing metropolis, China. *J. Clean. Prod.* 201, 481–498.

Zhang, C., Hu, M., Dong, L., Gebremariam, A., Miranda-Xicotencatl, B., Di Maio, F. and Tukker, A. 2019. Eco-efficiency assessment of technological innovations in high-grade concrete recycling. *Resour. Conservat. Recycle.* 149, 649–663.

Zhang, C., Hu, M., Yang, X., Miranda-Xicotencatl, B., Sprecher, B., Di Maio, F., ... and Tukker, A. 2020. Upgrading construction and demolition waste management from downcycling to recycling in the Netherlands. *J. Clean. Prod.* 266, 121718.

Zimbili, O., Salim, W. and Ndambuki, M. 2014. A review on the usage of ceramic wastes in concrete production. *Int. J. Civil, Environmen., Struct., Construct. Architect. Eng.* 8(1), 91–95.

Index

Note: Page number in **bold** and *italics* indicates tables and figures respectively.

For Product Safety Concerns and Information please contact our EU
representative GPSR@taylorandfrancis.com
Taylor & Francis Verlag GmbH, Kaufingerstraße 24, 80331 München, Germany